Communications
in Computer and Information Science 2361

Series Editors

Gang Li ⓘ, *School of Information Technology, Deakin University, Burwood, VIC,
Australia*
Joaquim Filipe ⓘ, *Polytechnic Institute of Setúbal, Setúbal, Portugal*
Zhiwei Xu, *Chinese Academy of Sciences, Beijing, China*

Rationale

The CCIS series is devoted to the publication of proceedings of computer science conferences. Its aim is to efficiently disseminate original research results in informatics in printed and electronic form. While the focus is on publication of peer-reviewed full papers presenting mature work, inclusion of reviewed short papers reporting on work in progress is welcome, too. Besides globally relevant meetings with internationally representative program committees guaranteeing a strict peer-reviewing and paper selection process, conferences run by societies or of high regional or national relevance are also considered for publication.

Topics

The topical scope of CCIS spans the entire spectrum of informatics ranging from foundational topics in the theory of computing to information and communications science and technology and a broad variety of interdisciplinary application fields.

Information for Volume Editors and Authors

Publication in CCIS is free of charge. No royalties are paid, however, we offer registered conference participants temporary free access to the online version of the conference proceedings on SpringerLink (http://link.springer.com) by means of an http referrer from the conference website and/or a number of complimentary printed copies, as specified in the official acceptance email of the event.

CCIS proceedings can be published in time for distribution at conferences or as post-proceedings, and delivered in the form of printed books and/or electronically as USBs and/or e-content licenses for accessing proceedings at SpringerLink. Furthermore, CCIS proceedings are included in the CCIS electronic book series hosted in the SpringerLink digital library at http://link.springer.com/bookseries/7899. Conferences publishing in CCIS are allowed to use Online Conference Service (OCS) for managing the whole proceedings lifecycle (from submission and reviewing to preparing for publication) free of charge.

Publication process

The language of publication is exclusively English. Authors publishing in CCIS have to sign the Springer CCIS copyright transfer form, however, they are free to use their material published in CCIS for substantially changed, more elaborate subsequent publications elsewhere. For the preparation of the camera-ready papers/files, authors have to strictly adhere to the Springer CCIS Authors' Instructions and are strongly encouraged to use the CCIS LaTeX style files or templates.

Abstracting/Indexing

CCIS is abstracted/indexed in DBLP, Google Scholar, EI-Compendex, Mathematical Reviews, SCImago, Scopus. CCIS volumes are also submitted for the inclusion in ISI Proceedings.

How to start

To start the evaluation of your proposal for inclusion in the CCIS series, please send an e-mail to ccis@springer.com.

Prasanna Devi Sivakumar · Raj Ramachandran ·
Chitra Pasupathi · Prabha Balakrishnan
Editors

Computing Technologies for Sustainable Development

First International Research Conference, IRCCTSD 2024
Chennai, India, May 9–10, 2024
Proceedings, Part II

Editors
Prasanna Devi Sivakumar
SRM Institute of Science and Technology
Chennai, Tamil Nadu, India

Raj Ramachandran
Cardiff Metropolitan University
Cardiff, UK

Chitra Pasupathi
SRM Institute of Science and Technology
Chennai, Tamil Nadu, India

Prabha Balakrishnan
SRM Institute of Science and Technology
Chennai, Tamil Nadu, India

ISSN 1865-0929 ISSN 1865-0937 (electronic)
Communications in Computer and Information Science
ISBN 978-3-031-82382-4 ISBN 978-3-031-82383-1 (eBook)
https://doi.org/10.1007/978-3-031-82383-1

© The Editor(s) (if applicable) and The Author(s), under exclusive license to Springer Nature Switzerland AG 2025

This work is subject to copyright. All rights are solely and exclusively licensed by the Publisher, whether the whole or part of the material is concerned, specifically the rights of translation, reprinting, reuse of illustrations, recitation, broadcasting, reproduction on microfilms or in any other physical way, and transmission or information storage and retrieval, electronic adaptation, computer software, or by similar or dissimilar methodology now known or hereafter developed.
The use of general descriptive names, registered names, trademarks, service marks, etc. in this publication does not imply, even in the absence of a specific statement, that such names are exempt from the relevant protective laws and regulations and therefore free for general use.
The publisher, the authors and the editors are safe to assume that the advice and information in this book are believed to be true and accurate at the date of publication. Neither the publisher nor the authors or the editors give a warranty, expressed or implied, with respect to the material contained herein or for any errors or omissions that may have been made. The publisher remains neutral with regard to jurisdictional claims in published maps and institutional affiliations.

This Springer imprint is published by the registered company Springer Nature Switzerland AG
The registered company address is: Gewerbestrasse 11, 6330 Cham, Switzerland

If disposing of this product, please recycle the paper.

Preface

SRM Institute of Science and Technology (Vadapalani Campus), India, a prominent unit of the prestigious SRM Group of Institutions located in Chennai, India, is known for its commitment to fostering academic excellence and innovation. SRMIST (Vadapalani Campus) is distinguished by the emphasis placed on experiential learning, industry partnerships, and research initiatives. With specialized research centers and active industry collaborations, the campus provides an ideal environment for innovative research and learning.

SRMIST (Vadapalani Campus) takes immense pride in presenting the proceedings of the International Research Conference on Computing Technologies for Sustainable Development (IRCCTSD 2024). This distinguished event was jointly organized by SRMIST (Vadapalani Campus) and Cardiff Metropolitan University, UK, on May 9th and 10th, 2024. This Research Conference brought together a wide range of global scholars, industry professionals, and innovative researchers. The event served as an energetic platform for impactful discussions, groundbreaking ideas, and forward-thinking innovations in computing technologies focused on sustainable development. We are confident that the ideas shared here will make a lasting contribution to both academic research and industrial practices, shaping the future of sustainable technology solutions.

The conference received an outstanding 264 research articles. After a rigorous peer-review process, 82 articles were selected for presentation and publication. The conference featured a diverse range of tracks, including AI Solutions, Machine Learning Strategies, Deep Learning Techniques, Industrial Innovation, Image Processing Applications and IoT Innovations related to Computing Technologies domains, namely Healthcare, Security, Agriculture and Climate Mitigation. Each track provided a vibrant and inclusive space for academicians and professionals to exchange ideas, address challenges and contribute to the conversation on leveraging computing technologies to solve today's most pressing sustainability challenges.

Eminent industry expert P Ilango, HCL Technologies, India delivered the keynote address in the inauguration ceremony emphasizing the role of technological innovation in sustainable development; Raj Ramachandran, Cardiff Metropolitan University, UK; Deepan Raj, HCL Technologies, India; and Mohanraj Vengadachalam, Standard Chartered Bank, India delivered the session keynote addresses.

The research outcomes shared during this conference foster continuous innovation and collaboration in the field of computing technology for sustainable development.

November 2024

Prasanna Devi Sivakumar
Raj Ramachandran
Chitra Pasupathi
Prabha Balakrishnan

Acknowledgement

The SRM Institute of Science and Technology (Vadapalani Campus), India, being a premier Institution, focuses on diverse and interdisciplinary research activities emphasizing both applied and fundamental research, aligning with National and Global priorities for technological advancement and self-reliance.

The apex management of SRMIST hones the academicians and students to actively engage in pioneering research projects to address real-world challenges, particularly in the fields of Computing Technologies, Engineering, and Sustainable Development. The commitment to high-quality research output is reflected in the numerous research articles published by the faculty and students, contributing to both national and international academic communities.

SRMIST (Vadapalani campus) recognizes the exceptional contributions of all individuals and institutions who helped in making the International Research Conference on Computing Technologies for Sustainable Development (IRCCTSD 2024) such a remarkable success. This conference showcased the productive partnership between SRMIST, India and Cardiff Metropolitan University, UK. The collaboration between these renowned institutions fostered an environment of innovation and academic excellence, advancing the frontiers of sustainable technology development.

We express our sincere gratitude to the management of SRMIST for their continuous support and encouragement in advancing academic research and knowledge dissemination. The resources, facilities, and collaborative environment of the institution were instrumental in the successful completion of this work. We extend our heartfelt thanks to the administration of SRMIST and the faculty, research scholars, and staff of the CSE department for their guidance and support throughout the execution of this conference and publication of the research proceedings.

Our deepest thanks also go to Cardiff Metropolitan University, UK, for their incredible support and partnership throughout the conduct of the conference. The guidance and insights provided by the distinguished faculty from Cardiff Metropolitan University played a crucial role in shaping the conference's trajectory. Collaborating with such a prestigious institution has been a truly rewarding experience, and we are thankful for the dedication of the team to promoting academic growth and innovation.

We sincerely thank Springer for their outstanding support in publishing this proceedings. It is a privilege to collaborate with such a prestigious publishing platform, renowned for its dedication to high-quality publications.

We sincerely appreciate the invaluable contributions of all the authors for their research efforts and the reviewers whose meticulous evaluations were pivotal in finalizing the selection of articles. We also wish to acknowledge the tireless work and commitment of the organizing and technical program committees, whose efforts were crucial to the conference's success.

In conclusion, we extend our deepest gratitude to all individuals and institutions whose contributions made this conference a success. The collaboration and support provided by the authors, reviewers, organizing committee, and sponsors were pivotal in advancing key subtopics such as AI solutions for sustainability, IoT applications for environmental monitoring, green computing practices, and smart city innovations. These proceedings reflect the ongoing exploration of cutting-edge Computing Technologies for Sustainable Development, and we are confident that the insights shared will inspire further innovation and drive meaningful progress in this essential field.

Organization

Chief Patrons

Ramasamy Paarivendhar	SRM Institute of Science and Technology, India
Ravi Pachamuthu	SRM Institute of Science and Technology, India
Pachamuthu Sathyanarayanan	SRM Institute of Science and Technology, India
Harini Ravi	SRM Group, India

Patrons

Chellamuthu Muthamizhchelvan	SRM Institute of Science and Technology, India
Suruttaiyaudiyar Ponnusamy	SRM Institute of Science and Technology, India
Kandhasami Gunasekaran	SRM Institute of Science and Technology, India
Santhanagopalan Ramachandran	SRM Institute of Science and Technology, India
Chandrathil Velappan Jayakumar	SRMIST (Vadapalani Campus), India
Chidambaram Gomathy	SRMIST (Vadapalani Campus), India
Sadagoparaman Karthikeyan	SRMIST (Vadapalani Campus), India
Krishnamoorthy Ramachandran	SRMIST (Vadapalani Campus), India

Program Committee

Jedsada Tipmontian	International Academy of Aviation Industry - King Mongkut's Institute of Technology Ladkrabang, Thailand
Ram Kumar Jayaseelan	AMD, USA
Anand Lakshmanan	Ericsson, India
Dhananjay Kumar	MIT Campus, Anna University, India
Yamini Jagadeesan	Facebook, USA
Nickolas Savarimuthu	NIT Trichy, India
Anandhakumar Palanisamy	MIT Campus, Anna University, India
Baskaran Ramachandran	Anna University, India
Masilamani Vedhanayagam	IIITDM, India
Roy Antony Arnold	Infosys Ltd, India
Jothi Periyasamy	DeepSphere AI, USA
Ruso Tamilarasan	Cognizant, India

Renuka Devi Saravanan	VIT Chennai, India
Mirnalinee Thanganadar Thangathai	SSN College of Engineering, India

Steering Committee and International Reviewers

Catherine Tryfona	Cardiff Metropolitan University, UK
Karl Jones	Cardiff Metropolitan University, UK
Fiona Carroll	Cardiff Metropolitan University, UK
Khoa Phung	University of the West of England, UK
Chandru Sandrasekaran	International College of Business and Technology, Sri Lanka
Pakkianathan Prabu Premkumar	International College of Business and Technology, Sri Lanka

External Reviewers

Sandra Johnson
P. Latchoumy
Poonkodi Mariappan
Boopathi Raja Govindasamy
Subhashini Palaniswamy
Jyostna Devi Bodapati
Gokul Chandrasekaran
Geetha Chellaian
Karthikeyan Vedanandam
Inbamalar Tharcis Mariapushpam
Boomija Malaisamy Duraipandian
Balasundaram Ananthakrishnan
Sandhya M. Kumar
Immanuvel Arokia James
Dass Purushothaman
Chidambarathanu Krishnan
Surender Shanmugam
Rajan Subramaniam
Subburaj Varadharajaperumal
Meena Rajeswaran
Rohith Bhat
Victo Sudha George
Anuradha Muthukrishnan
Kavin Kumar Kandasamy
Madhavi Thiruvengadam
Prameeladevi Chillakuru
Priya Vijay

Organizing Committee

Golda Dilip
Rajasekar Velswamy
Bharathi Navaneetha Krishnan
Paavai Anand Gopalan
Neelam Sanjeev Kumar
Arun Nehru Jawaharlal Nehru
Durgadevi Palani
Sridhar Srinivasan
Niveditha Satiyamoorthy
Karthikayani Kaliyaperumal
Sangeetha Subramaniam Karuppaiyah Bharathi

Akila Krishnamoorthy
Sridevi Sridhar
Maheswari Sendur Pandy
Jessy Sujana Godwin
Vidhusavarshini Suresh Kumar
Anusha Thamaraichelvan
Gayathri Ramanakumar

Muthurasu Nallappan
Manohar Shanmugavel
Punitha Dhandapani
Indumathy Mayuranathan
Deepa Ravichandran
Rajavel Manickam
Jayanthi Palraj

Contents – Part II

Video and Image Processing for Security Analysis

Intrusion Detection from Surveillance Video Using Frame Based Image Processing Techniques 3
    M. S. Mohamed Yaseen, P. Radeesh, P. Chitra, and Kishore Ramesh

Blockchain Based Research Data Repository 18
    M. Madhurani, S. Saraswathi, S. Harini, M. Janaki, and R. Buvana

DeepFake Image, Video and Audio Detection 29
    A. H. Mahima, M. Monica, S. Neha, and Deepti Balaji Raykar

Targeted Image Manipulation 42
    S. Sanjay Sen, Sameer Dushyant Patel, A. Bedict, and Golda Dilip

Efficient Deep Fake Image Detection Using Dense CNN Architecture 53
    P. Chitra, N. Venkateswarlu, V. Prem Kumar, and O. Abhishek Rithik

Innovation in Vehicle Tracking: Harnessing YOLOv8 and Deep Learning Tools for Automatic Number Plate Detection 66
    Raj Purohith Arjun, R. Akshitha, Navneet Ranjan, and K. Karthikayani

Network Anomaly Detection Mitigation of DDoS Attack Using Machine Learning 80
    R. Harish, Shashank Suresh, Saguturu Kishan Sai, and R. Deepa

Leveraging the Trio of 5G, MIMO, and IoT for Proactive Vehicular Crash Prevention in VANET 93
    S. Bharathi, S. Sathiyapriya, K. Nivethika, S. Sivachitralakshmi, and P. Durgadevi

Envision, Enhanced, Envisage (EEE) IoT Device Prediction Using Neural Network 104
    S. Menaka, Saswat Biswal, Pulibandla Sri Surya Teja, and Mothukuri Koushik

Image Tampering Detection Using Deep Learning 117
    S. S. Nagamuthu Krishnan, Saran Chowdam, Sandeep Badarla, and C. S. Nithin Tejesh

Innovations for Smart Cities

Artificial Intelligence-Powered Advanced Driver Assistance Systems in Vehicles 131
   S. Kathiresh and M. Poonkodi

Drive Safe: AI & IoT Powered Driver Alertness for Enhanced Passenger Safety 147
   S. Deepti, A. Anitha Pai, and S. Anandhi

Driver Safety Advancements: Drowsiness Detection with YOLO V8 162
   C. G. Balaji, V. R. Sai Krishnaa, S. Shyam Sundar, and S. Rajesh Kumar

Elevator Management System with SRTF Scheduling 176
   J. Arunnehru, P. Jayakrishnaa, and P. Vetrivel

Methods for Mitigating Leakage Power in VLSI Design 185
   Ramya Belde, K. Niranjan Reddy, and E. John Alex

Next-Gen Home Automation: Sensor-Based Connectivity and Appliance Control 194
   E. Emerson Nithiyaraj, S. Srinivass, K. Mukilan, M. Muthukumar, R. Nanmaran, and S. Srimathi

Navigating the Complexities of Municipal Waste Management: Enhancing Cost Prediction for Sustainable Urban Solutions 204
   M. S. Minu, Tushar Samal, Khushi Bisani, and Aishwarya Tewari

Sustainable Practices in E-Commerce: Challenges and Trends

Analyzing Bank Customer Behavior: Segmentation and Prediction Using Big Data Analytics 223
   D. Doreen Hephzibah Miriam and C. R. Rene Robin

Strategic BPNN Forecasting: Integrating Indicators, Bonds, Gold, and Indices for Enhanced Stock Trend Analysis 251
   Sachin M, Vishwas S. Shastry, Niharika J, and Trupti Hegde

Enhancing Product Review Understanding: Text Summarization with BART Large XSum 263
   S. Gopika, Mayuri Mahimaa Balaji, M. Vishwanath, and K. Karthikayani

Forecasting Sustainability: A Study of Demand Prediction in Circular Economics 275
   C. Harshavardhini, K. Sarayu, C. R. Roshan, and S. K. B. Sangeetha

Gold Price Forecasting Using Machine Learning Techniques 286
   Binu John and K. Nidhina

Credit Card Fraud Detection Using XG Boost 298
   Aryan Vinod Shankar, Megha Ramamurthy, and Golda Dilip

A Remunerative Self-checkout System Designed for Small Scale
Supermarkets 311
   M. Shrinidhi, K. Yeshvanthini, S. Yogitha, and J. Jasmine Hephzipah

SMART FIT- Elevating Fashion with Image Processing 321
   Pavana Nayak, R. U. Pratham, and Sunny Gupta

E-commerce Product Sentiment Assessment and Aspect Analysis 330
   M. Praveen, R. R. Vijay, R. S. Aaditya Shreeram, and S. Manohar

Author Index 343

Video and Image Processing for Security Analysis

Intrusion Detection from Surveillance Video Using Frame Based Image Processing Techniques

M. S. Mohamed Yaseen, P. Radeesh, P. Chitra[✉], and Kishore Ramesh

Department of Computer Science and Engineering (Emerging Technologies),
SRM Institute of Science and Technology, Vadapalani, Chennai, Tamil Nadu, India
{mm5978,rp2209,chitrap1,kr1524}@srmist.edu.in

Abstract. In India, the number of surveillance cameras has increased rapidly, making it challenging for law enforcement to go through the vast amount of recorded footage. We suggest a technique for quickly reviewing large amounts of video feed in order to address this issue. By utilizing sophisticated algorithms such as facial recognition, lightweight YuNet model face detection, and motion detection through background subtraction, our system accelerates the analysis of videos from various camera systems. Our model is a great tool for any agency looking to improve its operational efficiency when handling large amounts of monitoring data because it can quickly identify important events and produce concise summaries with timestamps. This piece of research contributes towards development within this field since such new methods are combined into one single solution thereby advancing surveillance technology while at the same time providing concrete benefits for everyday challenges faced by law enforcement agencies.

Keywords: Video Surveillance · YuNet · Background Subtraction · Intrusion Detection · Summarization

1 Introduction

Due to the exponential growth in the quantity of surveillance cameras throughout India, law enforcement agencies are confronted with a significant obstacle in manually examining and analyzing the vast volume of footage produced daily. The task of examining the footage has become increasingly difficult due to the enormous number of cameras, with over 1.54 million cameras installed across the country and over 280,000 cameras in the city of Chennai alone. This has resulted in well over 1.2 million hours of footage which needs scrutiny. The importance of handling this issue stems from the extensive deployment of surveillance cameras in residential and commercial areas in recent years, requiring efficient approaches to maximize the use of this data. Moreover, in the context of investigations or instances of intrusion, the capacity to promptly examine important footage and identify significant events is essential for prompt intervention by the authorities. Furthermore, Manual review processes are susceptible to human error, fatigue, and inefficiencies, which can impact the overall dependability of surveillance footage.

To tackle these challenges, our research proposes a solution that utilizes complex algorithms to enhance the analysis of surveillance footage more effectively. Our system aims to enhance public safety and security by combining important tasks such as motion detection, face recognition, and event summarization using timestamps in a single solution. This process decreases the workload on law enforcement agencies, improves effectiveness, and facilitates better decision-making.

1.1 Novelty and Original Contributions

Our research introduces an approach to intrusion detection in surveillance videos by integrating three key components which work together: motion detection using background subtraction, face detection with the lightweight YuNet model, and a facial recognition module. The synergy of these techniques enables efficient identification of crucial events and individuals within video streams, streamlining the analysis process and reducing the overall workload required. Additionally, our summaries in the form of timestamps facilitate easy access to relevant footage, enhancing public safety and security operations. A significant advantage of our proposed system is its ability to analyze every frame in the video, a task that is not feasible to accomplish manually without such an automated system, thereby addressing a critical challenge in handling large volumes of surveillance data.

Our method differs from other approaches because it analyzes surveillance footage frame-by-frame, addressing the drawbacks of conventional systems that typically sample frames at random, missing important events. The intrusion detection process is improved overall because the precise timestamps and frames guarantee prompt access to possible incidents.

2 Literature Review

Jin Su Kim et al. (2021) [1] developed an intelligent video surveillance system using embedded modules to address limitations of conventional CCTV systems, such as the inability to identify 95% of incidents after 22 min of single-person monitoring, and challenges like high power consumption and system costs. The authors had utilized embedded modules for intruder, fire, loitering and fall detection by leveraging information learning, color/motion data, and human body motion information, respectively.

Wei Wu et al. (2023) [2] highlighted several key contributions aimed at enhancing the efficiency-accuracy trade-off in face detection models. The authors analyzed face detection models and proposed a strategy to reduce model size of such systems. They had introduced a lightweight feature extraction backbone along with a simplified pyramid feature fusion neck in YuNet, aiming to achieve the best trade-off between accuracy and speed. The model only had 75,856 parameters, which is significantly less than other small-size detectors, making it highly efficient.

Ibrahim Delibaşoğlu (2023) [3] integrated spatio-temporal features, tracking mechanisms, and frame differencing techniques which adapted the background models to changing scenes, validated detected regions, reduced false positives, and improved performance. Also, addressing challenges like homography errors efficiently for moving object detection using background subtraction.

Yizhong Yang et al. (2023) [4] proposed a Multi-Scale Inputs and Labels (MSIL) model with an encoder-decoder network and channel attention for enhanced background subtraction performance. It included a Multi-Scale Fusion Encoding (MSFE) module designed to effectively utilize multi-scale inputs, fusing high-level and low-level feature details crucial for capturing global and local scene context to accurately identify moving objects - essential for applications like video surveillance and autonomous driving.

Anirudha Shetty et al. (2021) [5] focused on optimizing resource usage and achieving high efficiency, making it relevant for security systems, surveillance, and attendance management applications. The emphasis was on practical application and potential impact highlighting its contribution to advancing facial recognition technology.

Shivalila Hangaragi et al. (2023) [6] presented an innovative approach that combined face mesh and deep neural networks for face detection and recognition. The authors leveraged face mesh, a 3D facial geometry model, to extract structured facial features which were then fed into a deep neural network. This allowed the extraction of detailed features crucial for accurate face recognition in applications like security and surveillance.

Junjiao Tian et al. (2023) [7] addressed challenges of zero-shot segmentation for images with unknown categories using stable diffusion, emphasizing the importance of semantic segmentation, and addressing challenges of zero-shot segmentation for images with unknown categories. The proposed method, DiffSeg, used pre-trained stable diffusion models to segment images in an unsupervised and zero-shot manner, setting new performance benchmarks.

Khawla Alhanaee et al. (2021) [8] presented a facial recognition attendance system using deep learning CNNs with transfer learning from three pre-trained models, focusing on face identification for biometric authentication in attendance management and access control applications. The models showcased high prediction accuracy and reasonable training time, which is suitable for real-world deployment.

Yunzuo Zhang et al. (2023) [9] introduced a key frame extraction method for video sequences that leveraged the Quaternion Fourier transform (QFT), an extension of the traditional Fourier transforms to handle complex color image data, and multiple features. The QFT aimed to capture spatial and temporal video frame characteristics which are crucial in identifying key frames representing significant events or changes.

Jie Xu (2021) [10] focused on automated object detection and face recognition, aiming to integrate the above system into commercial video surveillance systems. The deep learning approach addressed this by leveraging deep learning methods for both tasks. Deep learning was chosen for its ease of deployment, scalability, and ability to process data in its raw form.

Yaniv Taigman et al. (2014) [21] developed the "DeepFace" system, significantly advancing face verification with a neural network and 3D alignment strategy. The system achieved 97.35% accuracy on the LFW benchmark, utilized over 120 million parameters, and was trained on 4 million images from 4,000 identities. This approach set new standards and has been adapted in this project.

Changxing Ding and Dacheng Tao (2015) [23] significantly enhanced face verification by incorporating a sophisticated ensemble of convolutional neural networks (CNNs) optimized for multimodal data, achieved over 99% recognition accuracy on the LFW benchmark. This approach leveraged diverse sets of facial features extracted through various modalities, which were then combined using a deep learning framework to handle the complex variations typically found in social media images.

3 Model

3.1 Model Description

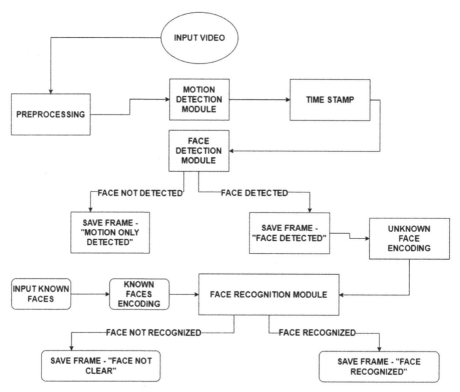

Fig. 1. Architecture Diagram for the proposed model

This model, improved by YuNet [2], is a cutting-edge surveillance system intended to fortify security protocols and facilitate proactive monitoring in a variety of settings. It functions as a security measure against unwanted access and questionable activities by utilizing techniques like motion detection, face detection, and face recognition. By using background subtraction for motion detection, the system minimizes false positives and quickly detects possible intrusions. Quick face recognition is made possible by YuNet, which improves the system's surveillance capabilities (Fig. 1).

Furthermore, the integration of deep learning-based face recognition, coupled with YuNet's enhanced facial feature extraction and analysis, empowers our project to accurately identify known individuals, reinforcing security protocols. With its adaptable architecture, scalable design, and tailored functionalities, this solution emerges as an invaluable asset for enhancing security in residential areas, ushering in a new era of comprehensive surveillance capabilities.

3.2 Motion Detection

Motion detection using background subtraction is used to detect moving objects in a video stream. The basic idea is to subtract the current frame of the video from a reference background frame, which is usually the first frame of the video or a combination of multiple frames. The resulting difference image is thresholded to produce a binary image, where pixels representing moving objects are white and the background pixels are black. This binary image is then processed to extract contours, and the contours are analyzed to determine if there is motion in the video [16]. If the area of a contour is above a certain threshold, it is considered to be motion. This technique is widely used for motion detection in surveillance systems and other video processing applications. However, it may be affected by changes in lighting conditions, camera noise, and other factors, which may result in false detections or missed detections.

3.3 Face Detection

In this system, face detection is executed utilizing YuNet, a lightweight deep learning model tailored specifically for real-time face detection tasks, as opposed to Haar Cascades. YuNet employs convolutional neural networks (CNNs) to accurately detect faces within video streams or images, providing bounding box coordinates around the detected faces [14, 15]. Unlike Haar Cascades, which rely on handcrafted features, YuNet learns hierarchical features directly from data, allowing for more robust and accurate face detection. The process begins by feeding the frame after motion detection into the YuNet model, which then identifies faces present in the frame. If faces are detected, the function returns True along with the motion flag and the current frame with faces marked by rectangles around them as seen in Fig. 3. Conversely, if no faces are detected, the function returns False along with the motion flag and the unaltered current frame. This integration of YuNet into this system enhances face detection capabilities, offering improved accuracy and reliability in identifying faces within video streams (Fig. 2).

Fig. 2. Face Detection using the YuNet model

3.4 Face Recognition

The facial recognition module "facial_recognition" uses deep learning algorithms to compare a target face to a database of known faces, and it can also be used to identify specific individuals in real-time. Face_Recognition_func function that takes in 4 parameters, the destinations for detected and not detected faces, and the directories for known and unknown faces. The function starts by loading all the known faces and names, where the faces are stored in subdirectories of KNOWN_FACES_DIR with the subdirectory names being the labels (names) of the faces. It then processes the unknown faces in the UNKNOWN_FACES_DIR by finding the locations of the faces in the images and then encoding those faces.

The function then applies the face_recognition.compare_faces method to perform a face comparison between the unknown and known faces. The function locates the name with the most favorable results from the comparison, uses the OpenCV library's cv2.rectangle and cv2.putText functions to draw a rectangle around the face, and then uses those functions to write the name on the image. The function then, based on whether or not a face was detected, moves the image to the designated destination folder. It accomplishes this by using the change_location method.

4 Methodology

First, motion is detected in the video streams. This is an important first step because it guarantees that only relevant segments are subjected to further analysis, effectively filtering out static background and focusing on potential security concerns. When motion is detected, the YuNet model is used by the system to determine whether any faces are visible within the segments. This stage is crucial because it involves using the frames that were collected during facial detection to classify people using a pre-existing database in the facial recognition process. Two things come out of this process: first, security is improved by instantaneous identification; second, data is arranged for effective retrieval. The motion triggers and faces that are detected are organized into specific directories to facilitate easy access and review in the future. This process helps users to swiftly sift through extensive footage, identifying key events and generating actionable summaries using the timestamps from the categorized frames. It's particularly beneficial for law enforcement and security operations, where rapid data analysis may significantly impact

4.1 Background Subtraction Subsystem

In order to identify moving objects in a video stream, motion detection using background subtraction is employed. Essentially, the process involves deducting the current frame of the video from a reference background frame, which is typically the first frame or a composite of several frames [11, 16].

In order to create a binary image with moving object pixels colored white and background pixels black, the difference image that results is stored at threshold. Next, contours are extracted from this binary image and analyzed to see if the video contains any motion. A contour's area is deemed to be moving if it exceeds a predetermined threshold. This technique is simple, efficient, and widely used for motion detection in surveillance systems and other video processing applications which is accurate and it effectively captures all the important events in a video [12, 13].

Our system utilizes background subtraction for efficient motion detection in surveillance footage [17, 18]. The mathematical basis for this technique is given by:

$$D(i, j) = |Frame_{current}(i, j) - Frame_{background}(i, j)| \tag{1}$$

where the Absolute Difference is calculated between the previous frame and current frame for each pixel which effectively acts as identification of motion, where the previous frame is the background, which is crucial for Background subtraction (Figs. 3, 4 and 5).

Algorithm 1   Motion detection using the Background subtraction function

Input: A video with multiple frames
Output: Motion detected from the input video to be stored in a separate folder along with timestamp with maximum accuracy and fast computational speed

Begin
1. Take frames with dimensions of 1280x720 pixels.
2. Take the background frame which is a grayscale image of the first image of the scene/video (fig 3).
3. Take the current frame which is a grayscale image of the current frame of the scene (fig 4).
4. Calculate the Absolute Difference between the background Frame and the current Frame. The resulting image is named Image Z.
5. For each pixel (i,j) :
$Z(i,j) = |\ backgroundFrame(i,j) - currentFrame(i,j)\ |$

The image that is obtained is the one where all the pixels that don't change (i.e., pixels that have same values) are zeroed out, and all the pixels that change (i.e., regions of motion) will be highlighted.
6. For each Pixel (x) in Z:
If Pixel Value > Pixel Threshold
        pixel(x)=Black
Else
        pixel(x)=white
7. Use findcontours() in OpenCV and check area threshold to decide if one should classify a contour as motion or not, based on the size/area of the contour
If Contour Area > Threshold Area
        Motion=True
Else
        Motion=False
8. Filter out and neglect irrelevant areas like movement of tree leaves in wind etc. using appropriate threshold values.
9. Highlight the resulting image in the unaltered Current Frame to indicate areas of motion and replace the background image with the current image if motion was not detected for some time (fig 5).
End

4.2 Face Detection Using YuNet Algorithm

YuNet face detection works by using a single-stage convolutional neural net-work (CNN) [2, 23] to predict the offsets of pre-defined anchor boxes to the nearest face. Upon detecting motion, our next step involves identifying human faces within the affected areas using the YuNet model, a robust convolutional neural network designed for efficient face detection.

Fig. 3. Motion Detection - Background Frame

Fig. 4. Motion Detection – Current Frame

Fig. 5. Motion highlighted in the unaltered current frame

The process is as follows:

Anchor Box Initialization.: They are predefined bounding boxes that consists of various scales and aspect ratios that cover different areas of the input image.

Feature Extraction.: YuNet processes the image to extract features relevant for face detection process.

Offset Prediction.: For each anchor box, YuNet predicts offsets to the nearest face, refining the position of the box to closely encompass the face.

Face Detection.: Anchor boxes with high confidence scores are selected, and their refined positions are output as the detected faces.

Algorithm 2 Face detection using YuNet script

Input: A video with multiple frames
Output: Faces detected from the input video to be stored in a separate folder along with timestamp with maximum accuracy and fast computational speed

Begin

1. Initialize video stream.
2. Perform background subtraction on the image to check if there is any motion detected in the image.
3. Perform YuNet face detection if motion is detected from previous step.
i. Anchor box generation: The first step is to generate anchor boxes for the image. This is done by placing pre-defined rectangles at different locations and scales in the image.
ii. Feature extraction: extract features from the image using the YuNet CNN.

iii. Offset prediction: The YuNet CNN then predicts the offset of each anchor box to the nearest face.
iv. Face detection: The final step is to detect faces in the image. This is done by selecting the anchor boxes with the highest predicted offsets.
4. Save image in particular location along with timestamp if face is detected. If only motion is detected then one frame for every 99 frame is saved in a particular location.
End

4.3 Face Recognition Subsystem

OpenCV and dlib are two popular libraries that can be used in Python to accomplish facial recognition [21]. While dlib increases accuracy by identifying facial landmarks, OpenCV offers ready-to-use models for face detection. Developers can create efficient facial recognition systems that meet a variety of needs by combining these tools with machine learning techniques like support vector machines and deep learning [20]. A library for face recognition and related tasks is the "face_recognition" module in Python. It compares a target face to a database of recognized faces using deep learning algorithms.

Algorithm 3   Face recognition using OpenCV Library module
Input: A video with multiple frames
Output: Faces recognized from existing database of "known faces" from the input video to be stored in a separate folder along with timestamp with maximum accuracy and fast computational speed.
Begin
1. Initialize face_recognition module in Python for face recognition and other tasks based on the topic.
2. Compare a target face to the "known faces" database.
3. Perform face comparison between the unknown faces and the known faces using the face_recognition.compare_faces method.
4. If (result = positive)
function finds the name with the most positive results, draws a rectangle around the face and writes the name on the image using the cv2.rectangle and cv2.putText functions from the OpenCV library.
5. Move the image to the specified destination folder depending on whether a face was detected or not using the change_location method.
End

5 Experimental Setup

5.1 Datasets

The dataset, which our team meticulously assembled, is made up of several MP4 video files that are between 12 and 24 h long—a total of one week's worth of footage. With a 640 x 352 pixel frame width and height, the video provides a standard resolution suitable for a range of video processing tasks [19]. With a steady frame rate of 30.00 frames per second, this dataset guarantees accurate temporal analysis and fluid playback. Its carefully chosen content is meant to be relevant and applicable in a variety of contexts, making it an invaluable resource for practitioners, academics, and researchers who want to improve their knowledge and skills in the area of video processing and analysis.

```
The data set used for Input Video:
File type: MP4 File (.mp4)
Frame width: 640
Frame height: 352
Frame rate: 60.00fps

The data set used for known faces video:
File type: MP4 File (.mp4)
Frame rate: 30.00fps
```

5.2 Results

This section presents the results obtained from testing our product on a real-world dataset. We describe the dataset's characteristics, including file type, size, video length, frame dimensions, and frame rate. The results include metrics such as the number of detected faces, the number of recognized faces, and accuracy, measured at various tolerance levels. We analyze these results to evaluate the performance of our product. The accuracy for face recognition is calculated based on Number of faces detected vs Number of recognized faces.

The result is a variety of folders, as shown in Fig. 6 and Fig. 7, where every frame is labeled and separated, as well as an output log, as shown in Fig. 7. Figure 8 also displays multiple faces that were detected in different frames, demonstrating the YuNet model's potential for face detection, even in situations where it is more difficult for other models to detect faces because of their relatively small size. As demonstrated in the output below, tolerance has a direct impact on face detection accuracy because it ignores minute changes that may be essential to the detection process. However, if tolerance is too low, it will also significantly increase the total time needed with little to no benefit to the accuracy of face detection.

Fig. 6. Output Folders based on detections

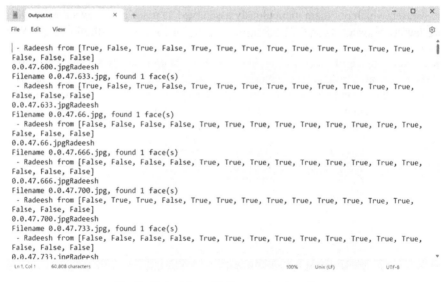

Fig. 7. Output log with frame-wise detections

The Face Detection accuracy in Table 1 and The Face Recognition accuracy in Table 2 is calculated by dividing the "Number of Faces detected" by "total number of faces present".

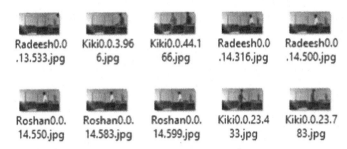

Fig. 8. Detected faces that match with user Database

Table 1 shows the comparison between YuNet and Haar cascades by the metrics of Time taken to compute each frame and the Accuracy which is calculated based on the detection capabilities as mentioned earlier. Haar Cascades is the existing method and the newer model is YuNet and we can notice the performance improvements by comparison in Table 1.

Formula for the motion detection algorithm proposed.

$$Z(i, j) = |\ backgroundFrame(i, j) - currentFrame(, j)\ | \qquad (2)$$

Table 1. Face Detection Comparison.

Video File Number	Total number of faces	Face Detection Algorithm	Number of Faces Detected	Average time take to compute each frame	Accuracy
1	2049	Harr Cascades	1568	0.4784 s	0.765
		YuNet	1843	0.1205 s	0.899
2	1082	Harr Cascades	657	0.3437 s	0.607
		YuNet	879	0.0840 s	0.812

Table 2. Face Recognition Observations.

Tolerance (for face detection)	Number of Detected Faces	Number of Recognized Faces	Accuracy
5	301	71	.816
6	558	328	.730
7	699	469	.688

The best overall accuracy reached for face recognition is 81.6%.

The tolerance variable determines how sensitive the face recognition is. The lower the number, higher the accuracy (Lower false positive, higher false negative) and likewise for the opposite [22]. The accuracy is calculated based on the faces detected divided by the total number of faces present in the video used for observation and is converted to the scale of 100.

6 Conclusion and Future Work

Our approach, in summary, represents a major step forward in the video analysis process, offering a solution that not only improves productivity but also tackles the mounting difficulty of organizing the massive amount of surveillance footage produced every day. Through the use of cutting-edge face and motion detection techniques, our method offers a more efficient way to recognize important events and people in film streams. Faster investigation times and easier access to insightful information from surveillance data are two benefits of this optimized process, which also helps security and surveillance operations make better decisions and allocate resources more wisely. A wider range of users can effectively utilize our method's capabilities because of its user-friendly interface, which also guarantees accessibility and ease of use.

Looking ahead, future research will concentrate on improving adaptability and scalability for various environments and surveillance scenarios. This involves processing video streams in real time so that events and threats can be quickly addressed. The accuracy and resilience of face and motion detection algorithms will be improved to account for changes in lighting, camera angles, and environmental conditions. Advances in AI and machine learning will further automate and optimize the analysis process, allowing for the development of more proactive and intelligent surveillance systems. Ongoing innovation has the potential to transform investigative practices and advance the use of video analysis to protect communities and improve public safety.

References

1. Kim, J.S., Kim, M.G., Pan, S.B.: A study on implementation of real-time intelligent video surveillance system based on embedded module. EURASIP J. Image Video Process. **2021**(1), 35 (2021)
2. Wu, W., Peng, H., Yu, S.: Yunet: a tiny millisecond-level face detector. Mach. Intell. Res. **20**(5), 656–665 (2023)
3. Delibaşoğlu, İ: Moving object detection method with motion regions tracking in background subtraction. SIViP **17**(5), 2415–2423 (2023)
4. Yang, Y., Li, D., Li, X., Zhang, Z., Xie, G.: A multi-scale inputs and labels model for background subtraction. SIViP **17**(8), 4133–4141 (2023)
5. Shetty, A.B., Rebeiro, J.: Facial recognition using Haar cascade and LBP classifiers. Global Transit. Proc. **2**(2), 330–335 (2021)
6. Hangaragi, S., Singh, T., Neelima, N.: Face detection and recognition using face mesh and deep neural network. Procedia Comput. Sci. **218**, 741–749 (2023)
7. Tian, J., Aggarwal, L., Colaco, A., Kira, Z., Gonzalez-Franco, M.: Diffuse, attend, and segment: unsupervised zero-shot segmentation using stable diffusion. arXiv e-prints, arXiv-2308 (2023)

8. Alhanaee, K., Alhammadi, M., Almenhali, N., Shatnawi, M.: Face recognition smart attendance system using deep transfer learning. Procedia Comput. Sci. **192**, 4093–4102 (2021)
9. Zhang, Y., Zhang, J., Liu, R., Zhu, P., Liu, Y.: Key frame extraction based on quaternion Fourier transform with multiple features fusion. Expert Syst. Appl. **216**, 119467 (2023)
10. Xu, J.: A deep learning approach to building an intelligent video surveillance system. Multimedia Tools Appl. **80**(4), 5495–5515 (2021)
11. Kalsotra, R., Arora, S.: Background subtraction for moving object detection: explorations of recent developments and challenges. Vis. Comput. **38**(12), 4151–4178 (2022)
12. Pramanik, A., Sarkar, S., Pal, S.K.: Video surveillance-based fall detection system using object-level feature thresholding and Z− numbers. Knowl.-Based Syst. **280**, 110992 (2023)
13. Mobsite, S., Alaoui, N., Boulmalf, M., Ghogho, M.: Activity classification and fall detection using monocular depth and motion analysis. IEEE Access (2023)
14. Andrejevic, M., Selwyn, N.: Facial Recognition. John Wiley & Sons, Hoboken (2022)
15. Kakadiya, R., Lemos, R., Mangalan, S., Pillai, M., & Nikam, S. (2019, June). Ai based automatic robbery/theft detection using smart surveillance in banks. In 2019 3rd International conference on Electronics, Communication and Aerospace Technology (ICECA) (pp. 201–204). IEEE
16. Lu, N., Wang, J., Wu, Q.H., Yang, L.: An improved motion detection method for real-time surveillance. IAENG Int. J. Comput. Sci. **35**(1) (2008)
17. Haritaoglu, I., Harwood, D., Davis, L.S.: W/sup 4: real-time surveillance of people and their activities. IEEE Trans. Pattern Anal. Mach. Intell. **22**(8), 809–830 (2000)
18. Wang, Y., Wu, Y., Zhang, S., Ogai, H., Hirai, K., Tateno, S.: Smart house system for safety of elderly living alone based on camera and PIR sensor. Artif. Life Rob. **29**(1), 43–54 (2024)
19. Tevissen, Y., Guetari, K., Tassel, M., Kerleroux, E., Petitpont, F.: Inserting faces inside captions: image captioning with attention guided merging (2024)
20. Saini, P., Kumar, K., Kashid, S., Saini, A., Negi, A.: Video summarization using deep learning techniques: a detailed analysis and investigation. Artif. Intell. Rev. **56**(11), 12347–12385 (2023)
21. Taigman, Y., Yang, M., Ranzato, M.A., Wolf, L.: Deepface: closing the gap to human-level performance in face verification. In Proceedings of the IEEE Conference on Computer Vision and Pattern Recognition, pp. 1701–1708 (2014)
22. Deng, J., Guo, J., Xue, N., Zafeiriou, S.: Arcface: additive angular margin loss for deep face recognition. In Proceedings of the IEEE/CVF Conference on Computer Vision and Pattern Recognition, pp. 4690–4699 (2019)
23. Ding, C., Tao, D.: Robust face recognition via multimodal deep face representation. IEEE Trans. Multimedia **17**(11), 2049–2058 (2015)

Blockchain Based Research Data Repository

M. Madhurani[1], S. Saraswathi[2], S. Harini[1,2], M. Janaki[1,2(✉)], and R. Buvana[1,2]

[1] Rajalakshmi Engineering College, Thandalam, Chennai, India
201001034@rajalakshmi.edu.in, janakisaisaran24@gmail.com
[2] Sri Sivasubramaniya Nadar College of Engineering, Kalavakkam, Chennai, India

Abstract. In today's digital era, the management and collaboration of research data pose significant challenges for organizations across various sectors. Research data could be any data that needs to be kept confidential or accessible only to specific members. Traditional systems often fall short in ensuring data integrity, security, and efficient collaboration. To address these challenges, the Blockchain-based Research Data Repository (BRDR) project introduces a transformative solution powered by blockchain technology. The BRDR offers a transparent, secure, and collaborative platform for managing and sharing research data within organizations. By utilizing the decentralized architecture and immutable ledger of blockchain technology, the BRDR reduces the likelihood of tampering and illegal access while guaranteeing data integrity, accountability, and transparency. With features such as multi-tiered authentication, smart contract-driven copyright management, and decentralized access controls, the BRDR empowers organizations to streamline research data management processes and foster collaboration among stakeholders. By providing a reliable and efficient platform, the BRDR aims to revolutionize research data management, enhancing productivity, trust, and innovation in the research ecosystem.

Keywords: Blockchain · Research Data · Data Repository · Collaborative · Data integrity · Immutable · Decentralized · Secure · multi-tiered · authentication

1 Introduction

The Blockchain-based Research Data Repository (BRDR) addresses challenges in research data management [11] by providing a transparent and secure platform [1]. Through multi-tiered authentication, employees gain controlled access to collaborate on research papers and datasets. Traditional systems often lack data integrity and security measures. BRDR utilizes blockchain, ensuring cryptographic security and immutability, reducing data tampering risks. It offers transparency via a decentralized ledger, enabling stakeholders to track data transactions. Utilizing blockchain technology ensures data integrity, traceability, and version control, fostering trust [16] in the collaborative environment. Decentralized architecture protects sensitive data, reducing risks of unauthorized access and tampering. Compliance with data protection regulations maximizes research efforts while upholding ethical standards. The BRDR revolutionizes research data management, promoting efficient collaboration and integrity within organizations.

The project offers versatile applications across industries. In academic institutions, it streamlines collaboration among faculty, researchers, and students, enhancing research productivity.

1.1 Existing System

In today's world, traditional research data management systems face several challenges. These systems struggle to maintain data integrity and security and facilitate effective collaboration between researchers. They often struggle to maintain trust and accountability because they often lack transparency [2]. Existing systems may have vulnerabilities that can be exploited by cyber attackers to gain unauthorized access to sensitive research data. Data security breaches can be caused by inadequate safety measures, faulty access controls, and weak encryption [3]. Researchers and organizations need a more powerful solution, one that addresses these limitations and promotes seamless data management and collaboration. The challenge bridges the gap between existing systems and an ideal platform that can simplify research while complying with data protection regulations.

1.2 Proposed System

In response to the limitations of traditional research data management systems, the proposed Blockchain Research Data Repository (BRDR) aims to transform the way organizations manage research data. The challenge is to ensure data integrity, maintain security, and facilitate effective collaboration between researchers. Current systems lack transparency and maintaining trust is difficult. BRDR introduces innovative features such as multi-level authentication, cryptographically secured files, immutable audit trails, and a decentralized architecture.

Using blockchain technology, BRDR provides a transparent, accountable, and user-friendly platform to manage and share research data. Researchers maintain control over their intellectual property rights while ensuring compliance with data protection regulations. Overall, BRDR enables organizations to improve research and promote innovation while maintaining data security and integrity.

2 System Architecture

System architecture is a comprehensive design framework that defines the structure, components, and interactions of a software system to meet specific requirements. It defines how hardware, software, data, networks and interfaces are organized and integrated. This plan guides decisions about technology, scalability, performance, security and maintainability, ensuring a robust, efficient and adaptable system. It includes models, principles and best practices for system development and improvement.

As Fig. 1, depicts the first step involves user registration. Users provide credentials such as username, password, and their team affiliation within the organization. This information is stored in the database only after the organization is approved. Upon approval, users can log in and upload research data in formats like DOC or PDF [8]. The admin, who holds the highest level of hierarchy, approves organizations by verifying their credentials. This multi-level authentication process ensures data security. The system uses the SHA-256 algorithm for secure file storage.

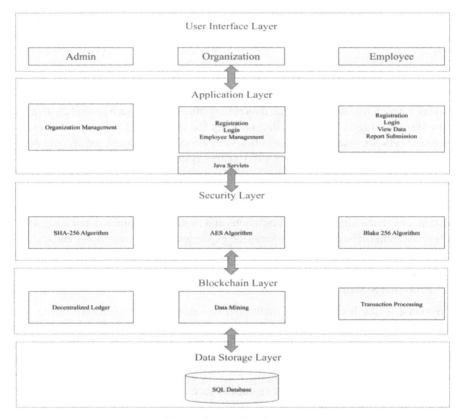

Fig. 1. System Architecture

3 System Design

The BRDR (Blockchain-based Research Data Repository) system design focuses on creating a secure, scalable and user-friendly platform for research data management. Using a multi-tier architecture, the system includes components for user authentication, data storage, access control and blockchain integration [4, 10]. The design emphasizes data integrity and confidentiality through encryption techniques such as SHA-256 hashing and AES encryption. It ensures smooth user interaction through a web-based user interface built using technologies such as Java Servlets and JSP. Integration with and SQL databases improves system performance and reliability. In summary, security, usability and efficiency are priorities in, BRDR's system design to meet the diverse needs of researchers and organizations.

3.1 Eclipse IDE

The Eclipse IDE is used in its comprehensive development environment, which provides features such as code editing, debugging tools, and project management capabilities. Its user-friendly interface streamlines the development process and increases productivity and efficiency. In addition, Eclipse supports various plugins and extensions, providing the flexibility to customize the environment to meet specific project requirements. Its strong ecosystem and community support make it a popular choice among developers. Overall, the Eclipse IDE enables seamless development and maintenance of a research data repository, ensuring a smooth and efficient workflow.

3.2 JSP

The project uses JSP (JavaServer Pages) to facilitate the creation of a dynamic web page. As a server-side technology, JSP allows Java code to be integrated directly into HTML pages, enabling the creation of dynamic and interactive web applications. Using JSP, the Research Data Repository can generate dynamic content based on user input or data retrieved from a database, improving user experience and interaction. In addition, JSP simplifies the development process by enabling reuse of code components and promoting modularity. Overall, JSP is a powerful tool for building dynamic and rich web applications, making it an integral part of the Research Data Repository project.

3.3 Servlets

Servlets are an integral part of the BRDR project because they handle HTTP requests, enabling dynamic content generation and server-side processing. By leveraging servlets, BRDR can securely manage user interactions, data storage and integration with blockchain technology.

Servlets provide scalability, efficiency and portability, and meet BRDR's requirements for a robust and efficient research data repository. Their adherence to the Java Servlet API standard ensures compatibility and easy maintenance in different environments. In short, servlets enable BRDR to provide a smooth and responsive web application tailored to the needs of both researchers and organizations.

3.4 Cryptographic Algorithms

SHA-256, AES, and Blake-256 are important cryptographic algorithms used in the BRDR project to strengthen data security. SHA-256 hashes to ensure data integrity by generating distinct digital signatures for data stored on the blockchain [9]. AES encryption protects sensitive data such as user data and research reports and prevents unauthorized access. Blake-256 complements SHA-256 by improving data integrity and authentication capabilities. Together, these algorithms strengthen BRDR's data security framework and ensure the confidentiality, integrity, and authenticity of research data. Their implementation conforms to industry best practices and provides users with a secure and reliable platform to manage and share research data. SHA-256 algorithm is well-known for having robust security features, such as resistance to preimage attacks and collisions, which make it appropriate for a range of uses in digital signatures and data integrity verification.

3.5 SQL

SQL plays a key role in the BRDR project in relational database management, which facilitates efficient storage, retrieval and processing of data. SQL queries allow BRDR to extract specific subsets of data, such as research reports and user profiles, based on predefined criteria. SQL also ensures data security by implementing access control mechanisms and enforcing privacy policies. Its support for constraints and integrity rules preserves data consistency and accuracy, which is critical to the reliability of a research repository. In addition, SQL databases provide scalability and performance that effectively meet the storage and processing needs of the BRDR platform. Overall, SQL plays a key role in enabling robust database management [14] and data operations in the BRDR project.

4 User Workflow

User Registration and Authentication: Users first register on the BRDR platform by entering required data such as name, email address, and affiliation with an organization.Users must go through a multi-tiered authentication process after registering, which includes organization verification and administrator approval.Users are given special login credentials to access the platform after successful authentication.

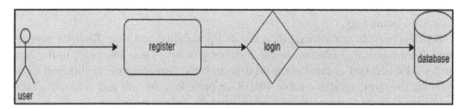

Fig. 2. User Registration

Figure 2 depicts,user after registration will be made by entering details such as Username, password. After this process,user will be able to login. The credentials are stored in database.

Accessing the Platform: Users provide their login credentials to access the BRDR platform following successful authentication. Users see a dashboard display that gives them an overview of their account and available services when they log in.

Figure 3 depicts,when a user is trying to login, the request is sent if that username and password matches with the one that was given during registration and successfully logged in only after the approval from the organization.

Uploading Research Data: Users can upload datasets, papers, and other materials pertaining to research to the BRDR platform. During the upload procedure, the user selects the desired files from their local device.

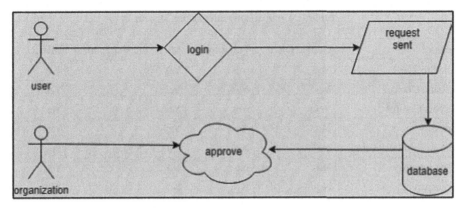

Fig. 3. User Authentication

Collaborating on Research Data: After uploading, users can work together on research data in the BRDR platform with team members and colleagues [5, 6].

Managing Access and Permissions: Users can manage who can view the data by managing access and permissions for uploaded files. Access control capabilities allow users to grant rights depending on project affiliations and roles, protecting the security and confidentiality of user data.

Logging Out and Session Management: Users can safely stop their session and safeguard their account from illegal access by logging out of the BRDR platform after finishing their work.

5 Performance Evaluation

BRDR performance evaluation involves quantifying the efficiency of data transfer processes and evaluating the utilization of system resources such as CPU, memory, and disk I/O. The purpose of this evaluation is to measure the system's ability to effectively manage data transfers while efficiently managing available resources. By analyzing performance metrics such as throughput, latency and resource usage, we gain insight into overall system performance and identify areas for optimization and improvement. Ultimately, performance evaluation ensures that BRDR is meeting its goal of providing a reliable and efficient platform for research data management. The comparison with existing system is shown in Table 1.

5.1 Data Upload Rate

Data upload rate of BRDR refers to the speed with which users can upload research-related files and datasets to the blockchain-based research repository. It indicates the transfer rate, usually measured as the amount of data downloaded per unit of time, such as megabytes per second or files per hour. Higher data rate means faster data transfer capability, allowing users to efficiently import their research results into the BRDR

Table 1. Comparison with existing system

Feature	Google Drive	Dropbox	Microsoft Sharepoint	BRDR
File Storage	Yes	Yes	Yes	Yes
Version Control	Yes	Yes	Yes	Yes
Access Control	Yes	Yes	Yes	Yes
Security	Yes	Yes	Yes	Yes
Blockchain Integration	No	No	No	Yes
Data Integrity	No	No	No	Yes
Transparency	No	No	No	Yes
Decentralization	No	No	No	Yes
Cost	Freemium	Freemium	Paid	Custom

platform. As Fig. 4 depicts, the data upload rate varies throughout the day, indicating changes in the activity of the user or the systems of use,

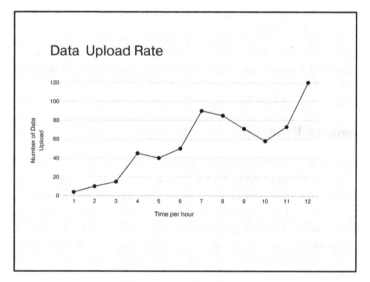

Fig. 4. Data Upload Rate

For example, 20 data uploads were sent in the first hour, which increased to 60 data uploads by the eighth hour, indicating the peak of data transfer. These variations can be further analyzed to optimize system resources and improve overall performance.

5.2 Resource Utilization

Resource utilization of BRDR refers to the extent to which system resources such as CPU, memory, and disk I/O are used during the operation of the blockchain-based research repository. It measures the extent to which these resources are actively used in the system to perform tasks and process data. By monitoring resource utilization, BRDR managers can assess the effectiveness of resource allocation and identify potential bottlenecks or inefficiencies that may affect system performance.

The Fig. 5 gives an idea of how efficiently the BRDR system is using its resources and helps identify potential optimizations or resource constraints. By monitoring resource usage, system administrators can ensure efficient resource allocation and maintain optimal performance. The Fig. 5 gives an idea of how efficiently the BRDR system is using its resources and helps identify potential optimizations or resource constraints. By monitoring resource usage, system administrators can ensure efficient resource allocation and maintain optimal performance.

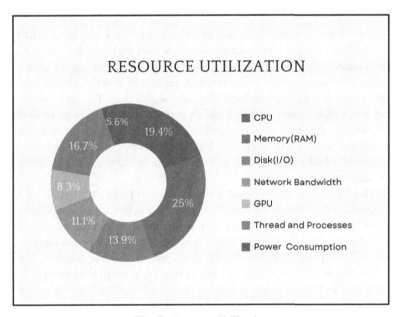

Fig. 5. Resource Utilization

6 Future Enhancements

There are a lot of chances for advancement and development in the field of blockchain technology integration with research data repositories, which is expected to continue growing and evolving. To properly utilize blockchain technology, cross-sector collaboration in scientific research, healthcare [12], and education is essential. It is anticipated that interdisciplinary collaborations will produce cutting-edge methods and best practices, completely changing how research data is managed and shared. By creating interoperability standards for blockchain-based research data repositories, organizations, researchers, and institutions will be able to share data seamlessly, breaking down organizational walls and fostering collaboration.Further work includes putting a real-world database system into operation, optimizing protocol efficiency in terms of message exchange and size, and investigating the use of several algorithms to improve system speed and functionality.

Future Work that can be carried out are setting Up a Database in the Real World where the goal is to enable smooth communication and data exchange between conventional databases and blockchain-based repositories by integrating blockchain technology with real-world database systems. Researchers and organizations will benefit from improved data accessibility, interoperability, and usability as a result of this integration. Next improving the efficiency of Methods can be considered. The emphasis will be on maximizing the effectiveness of blockchain protocols with regard to the quantity and size of messages that are sent. The goal of research will be to create leaner, more effective protocols that will minimize network congestion, use less resources, and improve system performance in general.

Exploring Multiple Algorithm Implementation: Further exploration will be conducted into implementing two or more algorithms within the blockchain-based research data management system [7]. The system can be improved in means of security, fault tolerance, and resilience against different attack vectors by utilizing several algorithms, guaranteeing the integrity and confidentiality of research data. The cross-sector collaboration for innovation is necessary to a full extend. To drive innovation in blockchain-based research data management, cooperation between the academic [13], medical, and scientific research sectors will be crucial. Multidisciplinary collaborations will produce creative ideas and best practices that will transform industry-wide data sharing [17], teamwork, and knowledge exchange.

Establishing Interoperability Standards: Efforts will be made to establish interoperability standards for blockchain-based research data repositories, enabling seamless data sharing [15, 18] and collaboration across institutions, researchers, and organizations. Organizational silos will be broken down and cross-sector interaction will be encouraged by standardization, opening up new opportunities for cooperative research projects. In addition to this Security and Data Privacy are the major concern. The main goals of research and development will be to improve blockchain network security and data privacy [20]. This includes creating consensus-building procedures that prioritize privacy and investigating stronger encryption techniques to shield private research data from malicious hackers and illegal access.

7 Conclusion

In conclusion, the Blockchain-based Research Data Repository (BRDR) project offers a robust solution for managing organizational research data securely and collaboratively. It guarantees smooth access control and user management with its multi-tiered authentication process and interface with current authentication systems. Organizations are empowered by the platform's extensive functionality, which includes file upload, to speed up research activities and promote teamwork among members. Furthermore, the application of blockchain technology guarantees data accountability, openness, and integrity [19] throughout the data management process, and compliance monitoring tools help businesses successfully comply with legal obligations. With capabilities like report submission modules, granular access controls, and client support, the BRDR offers a flexible platform designed to meet the requirements of research companies.The administrative features, such as affiliation management and access permission control, offer administrators the tools needed to oversee and manage the platform effectively, ensuring smooth operations and optimal utilization of research resources. All things taken into account, the BRDR is an innovative approach to research data management that improves collaboration within the organization, security, and compliance.By providing nuanced views within affiliations and enabling efficient client support, the BRDR fosters a conducive environment for research collaboration. Its user-friendly interface makes it easier for users to submit PDF reports to designated application modules, which improves productivity and user experience. Furthermore, the capability of individual grouping facilitates efficient data segmentation within the program, hence enhancing organization and accessibility.

Major benefits of this project are it enhanced data security and integrity. The BRDR uses cryptographic hashing, decentralization, and immutability to guarantee data security and integrity by utilizing blockchain technology. As a result, there is less chance of data breaches, illegal access, and tampering, fostering confidence in the accuracy of the research findings. Next benefit is effective data management and compliance. The BRDR simplifies research data management procedures and makes it simpler for businesses to comply with data protection laws and compliance standards thanks to its extensive functionality and strong features. This guarantees responsible handling of research data, lowers compliance concerns, and lessens administrative cost.

Enhanced Accountability and Transparency is one of the consideration of this work, Version control, immutable audit trails, and transaction traceability are just a few of the ways blockchain technology promotes accountability and transparency. This increases reproducibility of outcomes, improves transparency in research processes, and fortifies the validity of research findings. Next major contribution is cost reduction and resource optimization. Organizations may maximize resources, cut expenses, and boost operational efficiency by using the BRDR to automate data management procedures, streamline workflows, and lessen the need for manual interventions. This maximizes productivity and innovation by allowing researchers to concentrate more on their primary research pursuits.

References

1. Alam, S.: A blockchain-based framework for secure educational credentials. Turk. J. Comput. Math. Educ. (TURCOMAT) **12**(10), 5157–5167 (2021)
2. Ali, S.I., Mohammed, H.F., Sharaf, H.: A blockchain-based models for student information systems. Egypt. Inf. J. **23**(2), 187–196 (2022)
3. Amelia, S., Ningrum, Q.W.: Application of security system legal documents reviewer letters using blockchain technology. Blockchain Front. Technol. **1**(2), 65–73 (2022)

4. Bessa, F., Martins, L.: A blockchain-based educational record repository. In: Proceedings of the 7th International Workshop on ADVANCEs in ICT Infrastructures and Services (ADVANCE), 1–8 (2019)
5. Bhaskar, V., Tiwari, S.: Blockchain in education management: present and future applications. Int. J. Inf. Learn. Technol. **37**(6), 578–596 (2020)
6. Mohammed, S.I., Ali, H.F., Farouk, H.A., Sharaf, A.A.: A blockchain-based model for student information systems. Egypt. Inf. J. **23**(2), 187–196 (2022)
7. Sathiyaseelan, M., Gobi, V.: Blockchain solution to healthcare record system using hyperledger fabric. Int. J. Comput. Sci. Eng. **11**(3), 1–7 (2023)
8. Shrestha, A.K., Vassileva, E.: User data sharing frameworks: a blockchain-based incentive solution. In 2019 IEEE 16th International Conference on Industrial Engineering and Engineering Management (IEMCON), pp. 1425–1430. IEEE (2019)
9. Shuaib, M., Alam, S., Hasan, S.H.: A blockchain-based framework for secure educational credentials. Turk. J. Comput. Math. Educ. (TURCOMAT) **12**(10), 5157–5167 (2021)
10. Tavares, B.F., Correia, F., Restivo, A.: A survey on blockchain technologies and research. J. Inf. Assur. Secur. **14**, 118–128 (2019)
11. Baig, F., Wang, F.: Blockchain enabled distributed data management-a vision. In: 2019 IEEE 35th International Conference on Data Engineering Workshops (ICDEW), pp. 28–30. IEEE (2019)
12. Ivan, D.: Moving toward a blockchain-based method for the secure storage of patient records. In: ONC/NIST Use of Blockchain for Healthcare and Research Workshop (2016)
13. Arenas, R., Fernandez, P.: CredenceLedger: a permissioned blockchain for verifiable academic credentials. In: IEEE International Conference on Engineering, Technology and Innovation (ICE/ITMC), Stuttgart, Germany, pp. 1–6 (2018)
14. Faber, B., Michelet, G., Weidmann, N., Mukkamala, R.R., Vatrapu, R.: BPDIMS: a blockchain-based personal data and identity management system. In: The 52nd Hawaii International Conference on System Sciences. HISS 2019: HISS 2019, pp. 6855–6864. Hawaii International Conference on System Sciences (HICSS) (2019)
15. Hoang, V.-H., Nguyen, T.-M.-T., Nguyen, V.-C.: Blockchain-based secure data sharing platform for research data rights management over the Ethereum network. In: Proceedings of the 2021 International Conference on Computer Science and Software Engineering (CSSE), pp. 148–153 (2021)
16. Xu, Y., Zhang, F., Lia, X.: Blockchain-based data provenance framework for trustworthy research collaboration. In: The International Conference on Cloud Computing and Big Data (CCBD), pp. 78–89. Springer, Heidelberg (2022)
17. Zhang, X., Li, M., Ma, J.: Decentralized and secure research data sharing platform using interplanetary file system (IPFS) and blockchain. In Security and Communication Networks, vol. 15, no. 1, pp. 1–12. Springer, Heidelberg (2022)
18. Li, Y., Sun, J., Xiang, Y., Zhang, F.: Blockchain-empowered secure and efficient data sharing for scientific research collaboration. Inf. Sci. **593**, 532–548 (2022)
19. Hassan, S., Hamida, K., Ouahman, A.A.: Enhancing research data integrity and traceability using blockchain technology. In the International Symposium on Frontiers of Information Technology (FIT), pp. 123–134. Springer, Heidelberg (2023)
20. Li, J., He, F., Ma, J.: A Privacy-preserving and fine-grained access control scheme for research data sharing on blockchain. In Concurrency and Computation: Practice and Experience. Wiley Online Library (2023)

DeepFake Image, Video and Audio Detection

A. H. Mahima, M. Monica, S. Neha(✉), and Deepti Balaji Raykar

B N M Institute of Technology, Bengaluru 560 070, India
nehasuni107121@gmail.com, deeptibalajiraykar@bnmit.in

Abstract. In the realm of digital media, the proliferation of deepfake technology has brought forth significant concerns regarding its potential for misuse, particularly in areas such as political disinformation and privacy infringement. To counter these threats, we propose a groundbreaking approach to deepfake analysis that encompasses both image and audio detection. This method integrates advanced techniques such as Res-Next CNN for precise frame-level feature extraction from images and MFCC feature extraction for audio, coupled with LSTM-based RNN for comprehensive temporal analysis across both modalities. This model exhibits a remarkable ability to accurately discern between authentic content and deepfakes in both images and audio recordings. Additionally, our methodology emphasizes adaptability and scalability, ensuring its effectiveness across various digital platforms and evolving deepfake techniques in both visual and auditory domains. By continuously refining our model with ongoing advancements in AI and deep learning, we remain steadfast in our commitment to staying ahead of emerging threats posed by malicious actors. In addition to its utility in detecting deepfakes across multiple modalities, our system features a user-friendly interface for swift mitigation of AI-generated manipulations. A key feature of our system is the use of probability scores to indicate the confidence level of the classification. The resultant probability score obtained for various input data ranged between 50% and 90%. This probabilistic approach provides a nuanced measure of the system's confidence in its predictions, offering more detailed insights compared to the binary accuracy metrics commonly used in other studies. By leveraging the power of AI to combat AI, our methodology prioritizes simplicity and reliability, validated through rigorous evaluations on diverse datasets encompassing both image and audio samples. Through these efforts, we safeguard against the harmful impact of deepfakes, thus preserving the integrity of digital media and protecting individuals' privacy and societal well-being.

Keywords: AI-generated manipulations · DeepFake technology · Res-Next CNN · LSTM-based RNN · MFCC feature extraction

1 Introduction

Deepfake videos, epitomizing the pinnacle of digital manipulation, have sparked widespread concern in today's era due to their potential to deceive and manipulate audiences on an unprecedented scale. The proliferation of deepfake technology has engendered a myriad of challenges, notably in exacerbating misinformation and undermining trust in media content. These digitally manipulated videos, often indistinguishable from reality, pose significant risks by fuelling political manipulation, privacy breaches, and the spread of false narratives. However, technology plays a pivotal role in addressing these challenges. Advanced machine learning algorithms, such as deep neural networks, are harnessed to develop robust detection methods capable of distinguishing between genuine and deepfake images audios and videos with increasing accuracy. Additionally, automated content verification tools leverage artificial intelligence to identify inconsistencies and anomalies within media content, aiding in the rapid identification and removal of deepfakes from online platforms.

This research paper presents an innovative deep neural network-based framework and algorithm specifically crafted to discern deepfake videos within the realm of social media. Its primary contribution lies in addressing the escalating difficulty of differentiating between authentic and manipulated videos, a challenge exacerbated by the advancing sophistication of deep learning techniques used in the creation of deepfakes. By introducing a novel algorithm integrating a deep neural model, the paper aims to achieve highly accurate detection of deepfake videos disseminated across social media platforms. The overarching objective is to counteract the rampant spread of fabricated videos, which pose substantial risks to societal integrity, including threats to national security, democratic processes, and individual identities.

This endeavour is motivated by the exponential rise in face forgery within multimedia content over the past two decades, driven by the emergence of transformative technologies like deep learning networks capable of generating visually seamless deepfake videos. The capability to manipulate images and videos using AIpowered tools presents a formidable challenge to the authenticity of media content, potentially fuelling misinformation, undermining trust in journalistic sources, and precipitating broader societal repercussions. Thus, the urgent necessity for effective deepfake detection methodologies has become paramount in safeguarding against the adverse consequences of counterfeit videos across myriad facets of society.

2 Literature Survey

2.1 Introduction

This section delves into prior research concerning deepfake technology, with a focus on the proliferation of deceptive media content, concerns surrounding misinformation, and the potential societal impacts. By scrutinizing existing literature, the objective is to identify gaps, evaluate the efficacy of previous mitigation strategies, and lay the groundwork for our project's distinctive contributions. This review underscores the urgency for innovative approaches to combat the multifaceted challenges posed by deepfake technology in various domains, including politics, social media, and cybersecurity.

2.2 Key Survey Points

- The study [1] focused on creation of deepfakes which relies on the utilization of deep learning encoders and decoders, extensively employed in the realm of machine vision. Encoders function by extracting all features present in an image, which are then used by decoders to generate the fabricated image. Traditionally, generating deepfakes necessitated a substantial number of images and videos for training deep learning models, posing a significant challenge. However, in the current era, acquiring large datasets from social media platforms has become notably easier, leading to the development of more sophisticated deepfake techniques. Many deepfake algorithms leverage TensorFlow, an open-source software library initially developed by Google for internal use in machine learning and deep neural network research. TensorFlow has gained widespread popularity for various machine learning applications since its public release, offering a convenient means to design neural networks with satisfactory performance. Its APIs were compatible with the Python programming language, enabling easy experimentation with different Convolutional Neural Network (CNN) architectures and designs without extensive code modifications.

- A novel algorithm called the Self-Supervised Decoupling Network (SSDN)[2], which integrates similarity decoupling techniques such as Siamese networks or contrastive learning. Furthermore, it utilized Generative Adversarial Networks (GANs) to eliminate compression-related features. Implementation is anticipated to involve Python along with deep learning frameworks like TensorFlow or PyTorch, supplemented by OpenCV for preprocessing and managing image data.
- The investigation in [3] unveiled NA-VGG, a pioneering approach to detecting Deep-Fake face images, leveraging an enhanced VGG Convolutional Neural Network alongside image noise and augmentation methodologies. Initially, the RGB image underwent processing via a SRM filter layer to accentuate image noise characteristics, subsequently serving as input for the network. Further enhancing the dataset, the noise image underwent horizontal/vertical flipping for augmentation purposes. Despite these promising results, challenges loom regarding the model's sensitivity to variations in image noise, limited generalization due to dataset constraints, and reliance on the SRM filter, potentially hindering performance in the face of novel DeepFake techniques.
- In the realm of deepfake video detection, the Certainty-based Attention Network (CAN) emerges as a groundbreaking solution, prioritizing certainty-key frames to make predictions with heightened confidence levels. Comprising two integral components, the CAN incorporated a certainty-based attention map generation segment and a certainty-attentive feature generation module. However, the effectiveness of such detection methods is contingent upon diverse and representative training datasets, with limitations arising from biased or limited data potentially compromising real-world applicability. Moreover, the CAN's adaptability to evolving deepfake techniques may be challenged by rapid advancements in the field, necessitating comprehensive coverage of emerging methods in the training data for robust detection capabilities has discussed in [4].
- The proposed deepfake detection method in [5] leveraged Convolutional Neural Networks (CNNs) to extract frame features and utilizes a custom classifier for identification purposes. Development likely involved Python, TensorFlow, or PyTorch for neural network implementation, with Keras facilitating model construction on these frameworks. Additionally, image and video processing tasks were performed using OpenCV, while GPU acceleration was achieved through the NVIDIA CUDA

Toolkit to expedite computations. For interactive code execution, Jupyter Notebooks or Google Colab provided suitable environments. The effectiveness of CNN-based models is inherently reliant on the diversity and representativeness of the training dataset, with potential compromises in real-world generalization if the dataset is limited or biased.

- Literature [6] presented the Deepfake Detection Model with Mouth Features (DFTMF), employing a deep learning approach to identify Deepfake videos by analyzing lip/mouth movement. Utilizing Convolutional Neural Networks (CNNs), videos were segmented into frames and converted to grayscale images for processing and classification. The DFT-MF model utilized supervised deep learning with CNNs to categorize videos as fake or real based on a threshold number of identified fake frames, determined by variables such as words per sentence, speech rate, and frame rate. The DFT-MF model faced limitations such as it primarily focuses on the mouth region, potentially overlooking manipulations in other facial or contextual cues, and its efficacy heavily relied on accurate lip/mouth movement detection, making it sensitive to variations in facial expressions, lighting conditions, or video quality.
- The literature survey underscores the diverse and innovative methodologies and strategies employed in utilizing deep learning algorithms for the detection of deepfake images, videos, and audio. From advanced neural network architectures to sophisticated ensemble methods, deep learning offers compelling avenues for enhancing detection accuracy and understanding the nuances of manipulated media. Nevertheless, significant challenges persist, particularly in the realms of interpretability and model adaptability, underscoring the imperative for continued research and advancement in this domain.

3 Proposed Model

The proposed solution for deepfake image and video detection involves a comprehensive approach, integrating data analysis, system design, and model development. Thorough analysis highlights the importance of balanced training datasets to mitigate bias and variance in algorithmic predictions [7]. Leveraging deep learning frameworks like PyTorch [8], the system architecture is designed with a focus on feature extraction and classification, utilizing a combination of MTCNN (MultiTask Cascaded Convolutional Networks) for face detection [9], FaceSwap[14] for facial manipulation detection, and ResNeXt for feature extraction from images and videos. Long Short-Term Memory (LSTM) networks are incorporated to capture temporal dependencies in video sequences.

Preprocessing steps involve image segmentation, face detection, and frame extraction, ensuring uniformity and relevance of data inputs. By incorporating ensemble methods and fine-tuning techniques, the model aims to enhance detection accuracy and adaptability to evolving deepfake techniques. The user interface, developed using Django framework [10], offers seamless interaction, allowing users to upload images and videos and receive real-time predictions with confidence scores, thereby facilitating efficient deepfake detection.

The utilization of the librosa library [11] for audio processing within the Python environment. Librosa provides convenient tools for loading audio files and extracting relevant features necessary for our deepfake detection system. In the provided code snippet, librosa is employed to load the audio file and extract MFCC features, which serve as essential inputs for the classification model.

The aim of this proposed solution is to develop a robust deepfake detection system that outperforms existing methods by leveraging advanced techniques and addressing limitations such as biased training data and evolving manipulation techniques.

3.1 System Architecture

The following Fig. 1 portrays the proposed system"s architecture with which the system is built, and hence can be used to detect deepfake images and videos.

In this cloud-based system, users upload both real and fake datasets comprising images and videos to secure cloud storage services such as AWS S3, Google Cloud Storage, or Azure Storage Blobs. Upon upload, the datasets undergo preprocessing steps including noise reduction, image enhancement, and format standardization to ensure consistency and quality. The preprocessed datasets are then split into training, validation, and testing sets, with a data loader component retrieving batches of data during model training and evaluation [12]. For image data, the ResNeXt architecture is employed for feature extraction, capturing discriminative information for classification of real and fake content.

Meanwhile, Long Short-Term Memory (LSTM) networks are utilized for video classification, capturing temporal dependencies in video sequences [13]. Evaluation of both image and video classification models is performed using metrics such as accuracy, precision, recall, and F1-score, with results logged and stored for analysis. The system includes user authentication mechanisms like AWS Cognito or Azure Active Directory, ensuring secure access to data and functionalities. Visualization tools present evaluation results and performance metrics via dashboards, providing users with insights into model effectiveness and performance over time (Fig. 2).

Upon initiating the detection process, the system performs feature extraction and classification using advanced techniques like Res-Next CNN and LSTM-based RNN. Users receive real-time predictions accompanied by confidence scores, providing insights into the likelihood of the content being authentic or deepfake. Based on these results, users can take appropriate actions, such as content removal or further investigation.

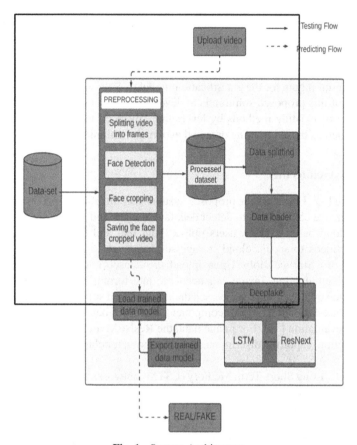

Fig. 1. System Architecture

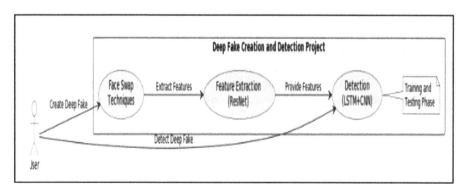

Fig. 2. Usecase diagram

4 Workflow

In our quest to tackle the deepfake menace, we begin by gathering diverse datasets and preprocessing them meticulously. Next, we design a robust model architecture, leveraging advanced techniques like CNNs and LSTMs. With a focus on optimization and validation, we fine-tune our model and evaluate its performance rigorously.

Comparative analysis with existing methods guides our path towards innovation, while seamless deployment and integration ensure real-world impact. This workflow embodies our commitment to combating deepfake proliferation and upholding the integrity of digital content.

4.1 Data Collection and Preprocessing

- Diverse datasets containing real and deepfake videos are collected from sources like Kaggle.
- Preprocessing involves standardizing formats, removing noise, and extracting features like facial landmarks using OpenCV.
- Videos are split into frames, and irrelevant information is discarded, focusing only on facial regions using techniques like face detection.
- Librosa [11] plays a crucial role in our deepfake detection system by providing the necessary tools for audio processing, enabling us to extract informative features essential for accurate classification.

4.2 Model Architecture Design

- The architecture comprises a combination of Convolutional Neural Networks (CNNs) for spatial feature extraction and Recurrent Neural Networks (RNNs) for temporal analysis.
- ResNext CNN models are utilized for feature extraction such as eyes, ears, nose, chin etc. from video frames, leveraging pre-trained models for efficient processing.
- Long Short-Term Memory (LSTM) networks are integrated to capture temporal dependencies between frames, enhancing the model's ability to discern subtle manipulations.

4.3 Model Training and Optimization

- Training the model involves fine-tuning the CNN and LSTM layers using PyTorch framework and TensorFlow, with CUDA support for GPU acceleration.
- Hyperparameters such as learning rate, batch size, and dropout rates are optimized through iterative experimentation.
- Data augmentation techniques like flipping, rotation, and scaling (resizing of the images) are applied to augment the dataset and prevent overfitting.

Table 1. Sample data for image video audio input and probability

Input	Category	Outcome	Probability
Input1	Image	Real	87%
Input2	Image	Fake	61%
Input3	Video	Real	83%
Input4	Audio	Real	–

4.4 Evaluation

The trained model's performance is evaluated using validation datasets, assessing metrics like probability, precision. Statistical tests validate the significance of the results and verify the model's ability to generalize to unseen data using probability (Table 1).

To calculate the probability of a prediction being a deepfake or authentic, our model utilizes the output from the neural network's final layer, which typically employs a softmax or sigmoid function depending on the classification task.

4.5 Comparison with Existing Methods

The proposed model is benchmarked against baseline methods, including traditional machine learning algorithms and existing deepfake detection frameworks.

Comparative analysis considers performance metrics, computational efficiency, and scalability, highlighting the superiority of the proposed approach.

Qualitative assessments examine interpretability, ease of implementation, and potential for real-world deployment.

4.6 Deployment and Integration

The trained model is deployed in real-world scenarios through user-friendly interfaces developed using Django frameworks. Extensive testing ensure compatibility, reliability, and scalability of the deployed system, with provisions for continuous monitoring and updates.

Advantages

Combatting Misinformation
Deepfake detection helps in identifying and mitigating the spread of misinformation and fake news, especially on social media platforms. By identifying manipulated content, users can avoid being misled by false information.

Preserving Authenticity
Deepfake detection helps in preserving the authenticity of digital media by distinguishing between genuine and manipulated content. This is crucial for maintaining trust in online information and media.

Protecting Privacy

Deepfake detection can help protect individuals' privacy by identifying and flagging unauthorized manipulations of their images or videos. This is particularly important in preventing the misuse of personal data for malicious purposes.

Preventing Fraud and Cybercrime

Deepfake detection can assist in preventing various forms of fraud and cybercrime, such as identity theft, phishing attacks, and financial scams. By identifying fake videos or images used for fraudulent purposes, potential victims can be alerted and protected.

Maintaining Trust in Digital Content

Deepfake detection contributes to maintaining trust in digital content, including media, entertainment, and advertising. By ensuring that content is authentic and unaltered, consumers can have confidence in the integrity of the information they consume.

Facilitating Content Moderation

Deepfake detection aids content moderation efforts on online platforms by identifying and removing harmful or inappropriate content. This helps in creating safer and more responsible online environments for users.

Overall, deepfake image and video detection play a crucial role in safeguarding the integrity of digital media, protecting individuals' privacy, combating misinformation, and supporting various aspects of cybersecurity and law enforcement.

5 Results

5.1 Deepfake Image Creation

For the creation of deepfake image a source image and a destination image are used. Some of the features of source images are extracted and swapped with the features of destination image. The destination image will the deepfake image created (Fig. 3).

Fig. 3. Deepfake Image Creation

- Deepfake Image Detection:

Upon processing an image through our deepfake detection system, the output includes a probability score indicating the likelihood of the image being fake or real. This probability score serves as a quantitative measure of confidence in the prediction, with values ranging from 0% to 100%. For instance, if the system assigns a probability of 50% to an image, it signifies a balanced uncertainty regarding its authenticity, suggesting an equal chance of it being either real or fake. A higher probability score, such as 80% or above, indicates stronger confidence in the classification, leaning towards either real or fake, depending on the direction of the probability.

Conversely, a lower score, such as below 50%, suggests a higher degree of uncertainty or ambiguity in the classification. This probabilistic output empowers users to make informed decisions based on the level of confidence in the authenticity of the image, enabling effective identification and mitigation of deepfake content in various contexts, including media forensics, cybersecurity, and content moderation (Figs. 4 and 5).

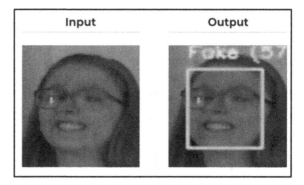

Fig. 4. Fake Image Detection

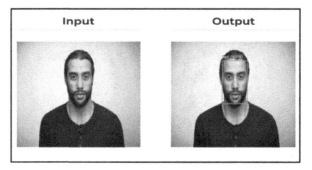

Fig. 5. Real Image Detection

5.2 Deepfake Audio Detection

Upon processing each audio file, the system efficiently extracted relevant features using the librosa library and subsequently employed a Random Forest classifier trained on a diverse dataset encompassing both real and fake audio instances. The classification results demonstrated the system's capability to discern between real and fake audio content with notable accuracy.

Librosa's functionality allows to handle audio files efficiently, ensuring accurate feature extraction while optimizing computational resources. By leveraging librosa, we can streamline the preprocessing pipeline and focus on extracting discriminative features that facilitate the classification of real and fake audio content.

For instance, when testing the system on a sample audio file, it correctly classified the content as 'Real' or 'Fake' based on the extracted features and learned decision boundaries. This successful classification underscores the effectiveness of the employed methodology, leveraging feature extraction techniques and machine learning algorithms to distinguish between authentic and manipulated audio content, thereby contributing to the broader effort of combating the proliferation of deepfake media (Figs. 6 and 7).

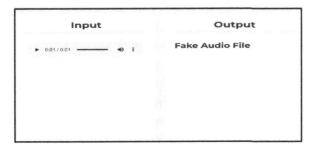

Fig. 6. Fake Audio File Detection

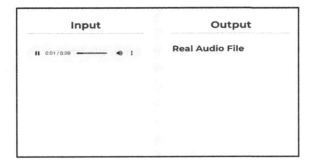

Fig. 7. Real Audio File Detection

5.3 Results and Discussion

The proposed approach to deepfake analysis, integrating advanced techniques such as Res-Next CNN and LSTM-based RNN for both image and audio detection, presents a robust solution to combat the proliferation of deepfake technology. By leveraging diverse datasets and continuously refining our model with advancements in AI and deep learning, we've developed a system capable of accurately discerning between authentic content and deepfakes across multiple modalities.

Unlike traditional studies that primarily focus on accuracy, such as those utilizing deep learning encoders and decoders, SSDN with GANs, NA-VGG with image noise processing, CAN for video detection, and CNN-based frame feature extraction, our approach employs probability as a measure of confidence.

Our system is unique in its comprehensive integration of advanced techniques for deepfake detection across images, videos, and audio. The combination of Res-Next CNN for detailed frame-level feature extraction, LSTM-based RNN for temporal analysis, and Librosa for audio feature extraction ensures robust detection capabilities across all forms of digital media. This multi-modal approach surpasses existing methods by addressing deepfake detection holistically, rather than focusing on a single modality. Compared to existing methods, where they have used accuracy, our approach with probability scores offers superior flexibility and insight. A key feature of our system is the use of probability scores to indicate the confidence level of the Kaggle dataset used. The resultant probability score obtained for various input data ranged between 50% and 90%.

6 Conclusion and Future Enhancement

Several enhancements can elevate deepfake detection system. Integrating multimodal fusion techniques will improve accuracy by combining information from sources like images and audio. Adversarial training methods can fortify the model against sophisticated deepfake generation techniques. Dynamic adaptation to emerging deepfake techniques is vital for sustained effectiveness in real-world scenarios. Human-in-theloop systems can enhance performance by incorporating human feedback into the detection process.

Privacy-preserving techniques should be explored to handle sensitive data during detection. Enhancing the explainability of the model's decisions fosters user trust and understanding. Optimizing the system for real-time detection is crucial for proactive mitigation of harmful content. By exploring these avenues, we can advance our deepfake detection system's capabilities, ensuring its effectiveness against evolving threats.

References

1. Khalil, H.A., Maged, S.A.: Deepfakes creation and detection using deep learning. Department of Mechatronics Engineering, Ain Shams University, Cairo, Egypt (2021)
2. Zhang, J., Ni, J., Xie, H.: Deepfake videos detection using self-supervised decoupling network. School of Computer Science and Engineering, Sun Yat-sen University, China (2021)
3. Chang, X., Wu, J., Yang, T., Feng, G.: DeepFake face image detection based on improved VGG convolutional neural network. School of Cyber Security, Shandong University of Political Science and Law, Jinan (2020)
4. Choi, D.H., Lee, H.J., Lee, S., Kim, J.U., Ro, Y.M.: Fake video detection with certainty-based attention network. Image and Video Systems Lab, School of Electrical Engineering, KAIST, South Korea (2020)

5. Mitra, A., Mohanty, S.P., Kougianos, E.: A Novel Machine Learning based Method for Deepfake Video Detection in Social Media. University of North Texas, USA, Peter Corcoran National University of Ireland, Galway, Ireland (2020)
6. Tayseer, M., Mohammad, J., Ababneh, M., Al-Zoube, A., Elhassan, A.: Digital Forensics and Analysis of Deepfake Videos. Princess Sumaya University for Technology Amman, Jordan (2020)
7. Rössler, A., Cozzolino, D., Verdoliva, L., Riess, C., Thies, J., Nießner, M.: FaceForensics++: learning to detect manipulated facial images. In: Proceedings of the IEEE/CVF International Conference on Computer Vision, pp. 1–11 (2019)
8. Masood, M., Nawaz, M.,Malik, K.M., Javed, A., Irtaza, A.: Deepfakes generation and detection: State-of-the-art, open challenges, countermeasures, and way forward (2021). arXiv: 2103.00484
9. Paszke, A., et al.: PyTorch: an imperative style, high-performance deep learning library. In: Advances in Neural Information Processing Systems, pp. 8024–8035 (2019)
10. Zhang, K., Zhang, Z., Li, Z., Qiao, Y.: Joint face detection and alignment using multitask cascaded convolutional networks. IEEE Signal Process. Lett. **23**(10), 1499–1503 (2016)
11. Holovaty, A., Kaplan-Moss, J.: The definitive guide to Django: Web development done right. Apress (2009). https://www.javatpoint.com/librosa-library-in-python
12. Pedregosa, F., et al.: Scikit-learn: machine learning in python. J. Mach. Learn. Res. **12**, 28252830 (2011)
13. Hochreiter, S., Schmidhuber, J.: Long short-term memory. Neural Comput. **9**(8), 17351780 (1997)
14. Qurat-ul-ain, A.J., Malik, K.M.: Faceswap deepfakes detection using novel multidirectional hexadecimal feature descriptor. In: 2022 19th International Bhurban Conference on Applied Sciences and Technology (IBCAST), Islamabad, Pakistan, pp. 273–278 (2022). https://doi.org/10.1109/IBCAST54850.2022.9990223

Targeted Image Manipulation

S. Sanjay Sen, Sameer Dushyant Patel, A. Bedict, and Golda Dilip(✉)

Department of Computer Science and Engineering, SRM Institute of Science and Technology, Chennai, India
`{ss1531,sp1284,ba6324}@srmist.edu.in, goldadilip@gmail.com`

Abstract. This paper addresses the problem of customization of an image at an object-level, where the aim is to enable users to customize images containing objects unseen during training without explicit supervision. Our approach utilizes a self-supervised method coupled with a semantic embedding space to facilitate object-level manipulation without the need for paired data. Through extensive experiments on various datasets, we demonstrate the effectiveness of our method in accurately customizing images with unseen objects while preserving visual coherence. Our results showcase the potential of targeted image manipulation in diverse applications such as content creation and image editing. In conclusion, this work presents a promising avenue for enabling users to seamlessly customize images containing unseen objects, opening up new possibilities for creative expression and content manipulation.

Keywords: object-level customization · self-supervised method · semantic embedding space · novel image customization · content manipulation

1 Introduction

The rapid advancements in diffusion models have greatly expanded the possibilities for generating and manipulating images using various input methods like text prompts and sketches. This progress has also led to exploration in image manipulation, including altering posture, style, or content based on provided instructions, as well as regenerating specific regions within images guided by textual cues.

This research focuses on the concept of "object teleportation," which involves accurately placing a target object into a specified location within a scene image. Our goal is to regenerate specific regions within a scene image using the target object as a reference. This capability has wide practical applications across fields like image composition, special effects, and virtual try-on. However, existing methods, such as Paint-by-Example and ObjectStitch, often struggle with maintaining consistency, especially for unfamiliar categories. They also require extensive manual intervention or fine-tuning, limiting their real-world applicability.

In the realm of existing work, challenges persist in achieving precision and coherence in object teleportation, particularly for unfamiliar categories or complex scenes. Existing methods often rely on manual adjustments or require significant fine-tuning, making them impractical for real-world use. Our proposed approach aims to address these challenges by utilizing advanced machine learning techniques, including deep neural networks and semantic understanding. By integrating self-supervised learning and semantic embedding spaces, our method simplifies the customization process, allowing users to seamlessly teleport objects into scene images with minimal manual effort. Additionally, our approach empowers users to specify desired locations within a scene, offering unprecedented control in object-level image customization. Overall, our study offers a novel solution to the challenges of object teleportation, facilitating practical and efficient customization of scene images across diverse applications.

2 Problem Statement

In the current landscape of image editing software, a conspicuous gap exists in the realm of object-level image customization, where precise alterations to individual objects within an image are lacking. Conventional editing tools primarily focus on global adjustments, such as colour correction, cropping, and basic retouching, which often fall short in addressing nuanced modifications at the object level. As a consequence, users face limitations when attempting to refine specific elements or aspects within an image, such as enhancing the clarity of a particular object or seamlessly removing unwanted elements. This deficiency underscores a pressing need for innovative solutions harnessing machine learning algorithms to enable granular control over object-level editing, revolutionizing the way users interact with and manipulate images.

3 Related Work

Similar softwares perform local image editing, but with text guidance. InPaintAnything and ObjectStitch are models that perform these functions. Other pieces of similar software make use of Customized Image Generation to images of specified objects with text prompts. Similar models also make use of Image Harmonization to cut the foreground object and paste it on the given background. Zero-shot learning is also made use of in computer vision applications. Similar software applications that delineate objects in within images, make use of CNNs and more traditional methods such as region-based algorithms. Both Paint-by-Example and Graphit operate with a similar input format to ours, utilizing a target image to modify a specific region within a scene image without requiring parameter adjustments. Customized generation has been thoroughly investigated in previous research. Typically, prior studies involve refining a subject-specific text inversion method to depict the target object, enabling generation with diverse text prompts.

Previous studies primarily focus on enhancing local image regions guided by textual input. Blended Diffusion employs multi-step blending within masked regions to achieve more cohesive outputs. Inpaint Anything utilizes SAM and Stable Diffusion to substitute any object in the source image with a text-described target. Paint-by-Example utilizes CLIP image encoding to convert the target image into an embedding for guidance, allowing for the painting of semantically consistent objects onto scene images. Similarly, ObjectStitch proposes a comparable approach, training a content adaptor to align CLIP image encoder outputs with text encoder outputs to guide the diffusion process. However, these methods offer only rudimentary guidance for generations and frequently struggle to generate ID-consistent results for novel, untrained concepts.

Customized or subject-driven generation aims to produce images corresponding to specific objects based on a set of target images and relevant textual prompts. Certain approaches refine a "vocabulary" to describe target concepts, while Cones identifies the relevant neurons for the referenced object. Although these methods can generate high-fidelity images, they lack user control over the scenario and object placement, and their time-intensive fine-tuning process hampers scalability for large-scale applications. More recently, BLIP-Diffusion employs BLIP-2 to align images and text, enabling zero-shot subject-driven generation. Other techniques explore large-scale pretraining to facilitate fine-tune-free subject-driven generation. Fastcomposer associates image representations with specific text embeddings for multi-person generation. Nonetheless, these zero-shot approaches are still in their nascent stages, exhibiting unsatisfactory performance or limited applicability.

A conventional image composition pipeline involves isolating the foreground object and inserting it onto a designated background. Image harmonization techniques are then applied to further refine the pasted region, ensuring more natural lighting and color integration. DCCF introduces pyramid filters to enhance foreground harmonization, while CDTNet utilizes dual transformers for improved processing. HDNet proposes a hierarchical framework that addresses both global and local consistency, achieving state-of-the-art results. However, these methods primarily focus on low-level adjustments, such as modifying the structure, viewpoint, and pose of foreground objects, without considering factors like shadows and reflections (Fig. 1).

Fig. 1. Comparison between existing methods

4 Methodology

This project aims to produce object-scene compositions of high fidelity and diversity based on specified target objects, scenes, and locations. The fundamental concept involves capturing the identity and detailed attributes of the object, then integrating these features into a pre-trained diffusion model to reconstruct the object within the provided scene. To facilitate learning of appearance variations, our approach harnesses extensive datasets comprising both videos and images for comprehensive training. The first part of this process is identity feature extraction - first Background Removal is performed followed by Self-Supervised learning. Following that, Detail Feature Extraction is performed - this includes Collage Representation, High-Frequency Mapping, and Focus Region Visualization. Then, Stable Diffusion is used to project the image into latent space. To train the model, we use Image Pair Collection same object in different scenes and angles. We also make use of Adaptive Timestamp Sampling - we denoise the images at the selected timestamps of a video sample and process them accordingly.

4.1 Removal of Background

We remove the background of the target image with a segmentor and position the object to the center of the image. The image is then fed to the ID Extractor. Both automatic and interactive models of the segmentor is used. This process provides more information on important features.

4.2 Self-supervised Learning

In this study, we observe that self-supervised models demonstrate significant proficiency in retaining highly discernible features. Trained on extensive datasets, these self-supervised models inherently possess the capability for instance retrieval and can effectively map objects into a feature space that remains invariant to augmentations. Specifically, we opt for the most robust self-supervised model available, namely DINO-V2, to serve as the foundational framework for our instance discrimination (ID) extractor.

4.3 Collage Representation

By employing collage as controls, we aim to establish robust priors for the image generation process. Our approach involves stitching the "background removed object" onto predetermined locations within the scene image. While utilizing collage leads to a notable enhancement in generation fidelity, we observe a drawback wherein the generated results exhibit excessive similarity to the provided target, thereby lacking diversity. To address this issue, we investigate the implementation of an information bottleneck mechanism to regulate the influence of the collage in imposing appearance constraints. Specifically, we devise a high-frequency map to represent the object, preserving intricate details while accommodating diverse local variations such as gesture, lighting, orientation, and more.

4.4 High-Frequency Mapping

Starting with an image I, our initial step involves isolating high-frequency areas utilizing high-pass filters, followed by the extraction of RGB colors through the application of the Hadamard product. Additionally, we introduce an eroded mask, Merode, to eliminate information in proximity to the outer contour of the target object. Once the high map is obtained, it is seamlessly integrated into the scene image based on specified positions, subsequently undergoing processing by the detail extractor. This detail extractor, designed akin to a ControlNet-style UNet encoder, generates a range of detail maps characterized by hierarchical resolutions.

4.5 Feature Administration

Once the ID tokens and detail maps are obtained, they are integrated into a pre-trained text-to-image diffusion model to influence the generation process. Specifically, we utilize Stable Diffusion, a model that maps images into latent space and employs probabilistic sampling via a UNet architecture. Regarding the detail maps, we append them to the UNet decoder features at different resolutions. Throughout the training phase, we maintain the pre-trained parameters of the UNet encoder unchanged to retain the underlying priors, while adjusting the UNet decoder parameters to align with the specific requirements of our task.

4.6 Training: Image Pair Compilation

The optimal training data consists of pairs of images depicting "the same object in different scenes," a scenario not readily available in current datasets. Prior studies have addressed this limitation by using individual images and applying augmentations such as rotation, flipping, and elastic transformations. However, these simplistic augmentations fail to adequately capture the diverse poses and viewpoints encountered in real-world scenarios. To address this challenge, our approach in this study involves leveraging video datasets to capture various frames featuring the same object across different instances.

4.7 Time Varying Interval Selection

While video data offers potential for learning appearance variations, its frame qualities often suffer from low resolution or motion blur. Conversely, images provide high-quality details and diverse scenarios but lack dynamic appearance changes. To harness the benefits of both modalities, we introduce adaptive timestep sampling, which tailors the utilization of different data types to distinct stages of denoising training. The conventional diffusion model uniformly samples the timestep (T) across training data. However, our observations indicate that initial denoising steps primarily focus on generating overall structure, pose, and viewpoint, while later steps attend to finer details such as texture and color. Hence, for video data, we prioritize sampling early denoising steps (large T) during training to better capture appearance changes. Conversely, for images, we elevate the likelihood of late steps (small T) to enhance learning of fine details (Fig. 2).

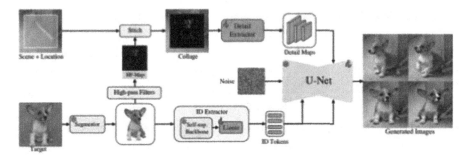

Fig. 2. Architectural Diagram

5 Results and Discussion

In this section, we present the results and subsequent discussion of our study on Targeted Image Manipulation. The process of obtaining results involved several key steps, each crucial in assessing the effectiveness and performance of the proposed framework.

Firstly, we conducted experiments using diverse datasets, encompassing a wide range of object categories and scene types. These datasets were meticulously curated to provide a comprehensive evaluation platform for the project. During training, the model was exposed to a combination of video and image data, enabling it to learn rich representations of objects and scenes. Qualitative evaluation involved visual inspection and subjective assessment of generated images. We assessed the quality, fidelity, and diversity of the generated images (Table 1).

Table 1. User study on the comparison between our project and existing reference-based alternatives.

Fidelity (↑)		Diversity (↑)		
Paint-by-Example	2.71	2.10		**3.04**
Graphit	2.65	2.11		2.84
Ours	**3.04**	**3.06**		2.88

The results of our experiments indicated significant improvements in performance metrics compared to baseline methods. The project demonstrated superior accuracy, robustness, and generalization capabilities, showcasing its effectiveness (Table 2).

Additionally, we evaluated the ID extractor using different backbones. In this table "G" refers to global token, "P" refers to patch tokens and "Seg" refers to the removal of the target object from the background image (Table 3).

Overall, the results and discussion underscored the efficacy and potential of the project in contributing to the advancement of computer vision and visual content manipulation.

Figure 3 shows the basic UI of the project. The user is given multiple options and gets to select the preferred background and the preferred reference.

Table 2. Quantitative study on the core components of our project

	Our Score (↑)	DINO Score (↑)
Baseline	73.8	31.5
DINO−V2 (with Seg)	80.4	63.2
High-frequency Map	81.5	64.8
Adaptive Timestep Sampling	82.1	67.8

Table 3. Evaluation of multiple backbones of ID extractor

	Our Score (↑)	DINO Score (↑)
VGG	71.6	27.7
Ours	73.7	31.5
Dino-V2-(G)	73.0	35.4
Dino-V2-(G + P)	79.9	64.1

Fig. 3. UI

Figure 4 depicts a particular case where the user has slelected a desired background and a desired reference picture. The selected images have been loaded onto the respective tabs.

Figure 5 is the output received after the project has been run. It showcases the capabilities and boundaries of targeted image manipulation.

Figure 6 gives a better look at the controls available to the user after the initial processing has been complete.

Targeted Image Manipulation 49

Fig. 4. Background and reference selection

Fig. 5. Output

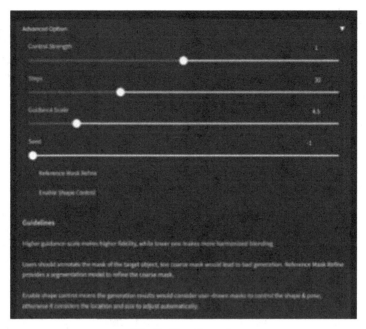

Fig. 6. User Controls

6 Conclusion

In summary, the study presents a groundbreaking approach to targeted image manipulation, offering a versatile solution applicable across various domains. Leveraging a diffusion-based generator, the model demonstrates the capability to perform object teleportation, seamlessly integrating target objects into scene images. This innovation is underscored by the utilization of discriminative instance discrimination (ID) extractors and frequency-aware detail extractors, which define the characteristics of the target object with precision. Through comprehensive training on diverse datasets encompassing both image and video data, the model achieves remarkable performance in customizing images, presenting a viable solution to region-to-region mapping tasks.

Moreover, the proposed approach holds immense practical utility in applications ranging from image composition to special effects rendering and virtual try-on. By enabling precise and seamless object placement within scene images, the model offers users unprecedented control and flexibility in image customization. Furthermore, the model's adaptability to diverse customization tasks underscores its potential to address a wide range of user needs and preferences.

While the study focuses on object-level customization, the implications extend beyond the realm of image editing. The integration of such capabilities into various domains, including fashion design, interior decoration, and digital art, opens up new avenues for creative expression and innovation. By tailoring the model's capabilities to specific application domains, users can harness its potential to unlock new opportunities and push the boundaries of creative exploration.

In conclusion, targeted image manipulation represents a significant advancement in the field of image manipulation, offering a powerful tool for users to realize their creative visions. As technology continues to evolve, further refinements and advancements in this area promise to shape the future landscape of image customization, offering endless possibilities for creative expression and innovation.

7 Future Scope

In considering the future trajectory of targeted image manipulation, several avenues emerge for further exploration and advancement. One significant avenue involves delving into more intricate model architectures finely tuned to the unique demands of object-level customization. This entails investigating novel network structures, attention mechanisms, and memory-augmented models aimed at enhancing both the fidelity and adaptability of this project to diverse customization tasks.

Furthermore, the integration of multimodal inputs stands out as a promising direction for future development. By incorporating additional modalities such as textual descriptions, audio cues, or contextual information, this project could offer users a more comprehensive and intuitive means of specifying their customization preferences. Research efforts in this area could focus on devising effective fusion techniques to seamlessly integrate multimodal data, thereby enriching the customization process.

Interactive user interfaces represent another fertile ground for exploration. Future endeavors may center on the development of interactive interfaces that facilitate real-time feedback and collaboration between users and the customization system. Such interfaces could incorporate interactive visualization tools, user-guided editing features, and intuitive controls to empower users with greater control and understanding of the customization process.

Moreover, the adaptation of this project to specific application domains holds significant promise for future innovation. Tailoring the capabilities of this project to domains such as fashion design, interior decoration, or digital art could unlock new opportunities for creative expression and innovation. By understanding the unique requirements and challenges of different domains, this project could be optimized to deliver tailored solutions to meet the particular requirements of users in these domains.

Lastly, ethical and societal implications deserve careful consideration in the future development of this project. As the technology advances, it is crucial to address issues related to privacy, fairness, and cultural sensitivity. Research efforts in this area can help ensure that this project is developed and deployed responsibly, taking into account the broader societal impact of its use. By navigating these challenges and opportunities, this project has the potential to shape the future landscape of object-level image customization in profound and meaningful ways.

References

1. Jampani, V., Aberman, K., Pritch, K.: Dreambooth: optimizing text-to-image diffusion models for generation driven by subjects. In: IEEE Conference on Computer Vision and Pattern Recognition (2023)

2. Agrawala, L.: Integrating dependent control into diffusion theories of text to image. arXiv.org
3. Yipeng, P., Abdal.: ImagetoStyleGAN++: ways to modify integrated pictures. In: IEEE Conference on Computer Vision and Pattern Recognition
4. Shechtman, J.Y.Z., Zhang, R., Nupur.: Customization of text-to-image dispersion over many concepts. In: Institute of Electrical and Electronics Engineers Conference on CVPR (2023)
5. Sun, J., Ren, S., Zhang, X., He, K.: Deep residual learning for the identification of images. In: Institute of Electrical and Electronics Engineering Conference on Pattern Analysis and Machine Vision (2016)
6. He, K., Zhang, X., Ren, S., Sun, J.: Deep residual learning for image recognition. In: Institute of Electrical and Electronics Engineering Conference on CVPR (2016)
7. Lowe, D.G., et al.: Brown: identifying panoramas. In: Presented at the 2003 in IEEE Conference on Computer Vision (2003)
8. Wu, Z., Lin, S., Lau, R.W., Zhao.: Simplifying positioning for self-supervised representational training. In: the Association for the Advancement of Artificial Intelligence Conference on AI Proceedings
9. Girdhar, D.R., Lee.: Meta-Pix: Adapting Few-Shot Videos for Different Targets
10. Zeynep, Y., Naeem, M.F., Massimiliano.: ZeroShot compositional learning in an open world. In IEEE Conference and Computer Vision Foundation Conference on Proceedings on Pattern Recognition & Computer Vision (2021)
11. Kumari, N., Zhang, B., Zhang, R., Shechtman, E., Zhu, J.Y.: Customizing text-to-image diffusion with multiple concepts. In: IEEE Conference on Computer Vision and Pattern Recognition (2023)
12. Cong, W., et al.: DoveNet: enhancing deep image harmonization through domain verification. In: IEEE Conference on Computer Vision and Pattern Recognition (2020)
13. Sunkavalli, K., Johnson, M.K., Matusik, W., Pfister, H.: Multi-scale image harmonization. ACM Trans. Graph. (2010)
14. Wang, L., et al.: Learning to detect salient objects with image-level supervision. In: IEEE Conference on Computer Vision Pattern Recognition (2017)
15. Wang, W., Feiszli, M., Wang, H., Tran, D.: Unidentified video objects: a benchmark for dense, openworld segmentation. In: International Conference on Computer Vision (2021)
16. Xiao, G., Yin, T., Freeman, W.T., Durand, F., Han, S.: Fastcomposer: tuning-free multi-subject image generation with localized attention. arXiv:2305.10431 (2023)
17. Xie, S., Zhang, Z., Lin, Z., Hinz, T., Zhang, K.: Smartbrush: text and shape guided object inpainting with diffusion model. In: IEEE Conference on Computer Vision Pattern Recognition (2023)
18. Yu, X., et al.: Mvimgnet: a large-scale dataset of multi-view images. In: IEEE Conference on Computer Vision Pattern Recognition (2023)
19. Yu, T., et al.: Inpaint anything: segment anything meets image inpainting. arXiv:2304.06790 (2023)
20. Yang, L., Fan, Y., Xu, N.: Video instance segmentation. In: International Conference on Computer Vision (2019)

Efficient Deep Fake Image Detection Using Dense CNN Architecture

P. Chitra[✉], N. Venkateswarlu, V. Prem Kumar, and O. Abhishek Rithik

Department of Computer Science and Engineering (Emerging Technologies),
SRM Institute of Science and Technology, Vadapalani, Chennai, Tamil Nadu, India
`{chitra1p,vn3652,pv5830,or1839}@srmist.edu.in`

Abstract. Deepfakes innovative manipulations of aesthetic web content utilizing deep understanding methods have actually arised as a substantial risk increasing problems regarding false information as well as personal privacy violations. Their influence covers different domain names from social networks to political unsupported claims, highlighting the immediate demands for durable discovery devices. This research study deals with the danger of deepfake with a thorough examination right into binary category techniques. Concentrated on determining genuine from adjusted pictures, this research utilizes varied datasets to educate and also review methods. Taking advantage of typical artificial intelligence formulas as well as Convolutional Neural Networks (CNNs) the here and now method highlights function removal for spatial reliances essential in picture evaluation. The ready methods will certainly be examined based upon accuracy in sight of their efficiency in setting apart in between real as well as artificial visuals. Speculative outcomes show the version's efficiency in finding adjusted pictures throughout numerous situations showcasing its capacity in dealing with the deepfake obstacle as well as carrying out 97% accuracy in identifying deep- fakes on undetected information.

Keywords: Deep learning · CNN · Efficient Net · Depthwise Seperable convolutions (DSCB) · GAN

1 Introduction

Expert system (AI) has actually come to be a transformative pressure, affecting different elements of our lives, from computer system vision as well as speech handling to the layout of intricate multi-agent systems. Nevertheless within this world of progression exists an expanding worry: the development of deepfakes.

Deepfakes are meticulously developed making use of innovative deep understanding methods that adjust aesthetic products, primarily pictures as well as motion pictures. The difference in between construction plus fact is coming to be a growing number of over cast because of these progressively natural changes. This capability to perfectly mix genuine and also artificial components increases substantial moral as well as social worries postulating a straight danger to info protection, private personal privacy as well as the actual structure of count on the electronic globe. The possible damage caused by deepfakes is diverse together with extensive. Destructive stars can manipulate this innovation to set up sophisticated plans of false information as well as false information controling popular opinion and also weakening rely on genuine resources of details. Deepfakes can be militarized to harm track records, grow discord within neighborhoods as well as also affect the result of political elections.

This research study explores this vital obstacle, recommending an unique technique labelled Efficient Deep Fake Image Detection utilizing Dense CNN Architecture. This approach intends to effectively compare genuine coupled with controlled photos splitting pictures right into "Fake" and also "Real" groups. 2 datasets are suggested under "Real" CelebA as well as FFHQ, each with 5000 pictures; these datasets most likely include high-resolution face photos and also star pictures specifically. On the other hand expert system versions like StarGAN, StyleGAN, StyleGAN2, SFHQ, Stable Diffusion and also Face Synthetics generate artificial pictures that drop under the "Fake"classification. These versions supply in between 1000 along with 5000 artificial pictures [21] in an initiative to duplicate actual human faces. This differed dataset building and construction is necessary for artificial intelligence research study particularly in the locations of picture synthesis, control, coupled with discovery, considering that it offers a wide variety of synthetic photos for contrast plus evaluation making use of numerous AI techniques.

2 Literature Review

M. M El-Gayar et al. [1] presented a novel hierarchical structure integrating multi-scale image properties and graph neural networks to discern real from deepfake videos. They fused convolutional neural networks and graph convolutional networks using three strategies and noted challenges such as high computational requirements, limited generalizability, and a narrow focus on mouth features. Despite achieving high accuracy rates on datasets like FaceForensics++ and Celeb-DF, further exploration is needed, particularly for tasks requiring relational reasoning and in diverse domains.

Chuangchuang Tan et al. [2] created Neighboring Pixel Relationships (NPR) to record structural irregularities in generative networks based on CNN for deepfake identification. They emphasized difficulties such as adversarial vulnerability, interpretability problems, and pre-training dependency. Even while models like EfficientNetV2M and InceptionResNetV2 have achieved great accuracy and precision, these issues still need to be resolved for more broadly applicable detection.

Chunlei Peng, et al. [3] utilised facial image symmetry to describe regional relationships in the Symmetric Spatial Attention Augmentation based vision Transformer (SSAAFormer) for deepfake detection. They highlighted difficulties such as limited explainability, resource-intensive inference, and vocal track separation issues. Even with average precision rates and moderate accuracy, cross-domain detection problems still need to be addressed in order to increase efficiency and dependability.

Dragos Tantaru et al. [4] introduced a novel approach to deepfake detection, focusing on identifying manipulated regions instead of solely classifying images. By treating it as a weakly-supervised localization task, they compared various methods using a shared architecture and analyzed factors influencing method performance. Despite facing challenges in incorporating generation methods adequately and noting variations in baseline model performance, further efforts were deemed necessary to enhance model robustness, particularly in speech-trained scenarios.

Tal Reiss et al. [5] introduced FACTOR, a practical approach for deepfake factchecking without relying on temporal data, showcasing its efficacy in critical attack scenarios like face swapping and audio-visual synthesis. The research highlighted FACTOR's factual basis and its ability to generalize to unseen attacks, despite limitations in unconditional deepfakes. Specialized encoders for non-standard facts were employed, although effectiveness was tied to generative model limitations.

Srijan Das et al. [6] presented a novel approach, Self-Supervised Auxiliary Task (SSAT), to enhance the training of Vision Transformers (ViTs) with limited data. By integrating a secondary self-supervised task alongside the main task, SSAT improved performance compared to conventional SSL pre-training and fine-tuning methods. The study demonstrated SSAT's effectiveness across 10 datasets and emphasized its potential in reducing the carbon footprint of training processes. Challenges such as limited masking ratio exploration and methodology reliance on specific manipulation techniques were noted, along with the dependency on the DFDC dataset for pretraining.

Rui Shao et al. [7] introduced the research problem of Detecting Sequential DeepFake Manipulation (Seq-DeepFake), addressing an important challenge. However, the paper lacked quantitative impact comparison and comprehensive discussion on generalization capabilities. There was an oversight on environmental impact and efficiency, with limited exploration of real-world applicability. The absence of future directions in the conclusion was noted.

Zhimin Sun et al. [8] addressed the challenge of identifying the source of manipulated faces (deepfakes) by proposing a new benchmark (OW-DFA) and introducing a new framework (CPL) to improve attribution accuracy, especially for unknown forgery techniques. However, the research lacks comprehensive analysis of CPL failure cases and discussion on its generalization. There's a brief mention of interpretability and security implications, with missing direct comparison with existing methods. Ethical considerations are also incompletely addressed. Testing on ImageNet, FaceForensics++, and Celeb-DF datasets, with CPL achieving an accuracy of 71.89%.

Chao Shuai et al. [9] proposed a two-stream network with three functional modules for improved deepfake detection, incorporating collaborative learning and multistream, multi-scale feature handling, along with a Semi-supervised Patch Similarity Learning strategy. However, the paper lacks detailed analysis of failure cases in the conclusion and makes ambiguous claims regarding generalization to unseen forgeries. Computational cost and scalability discussions are omitted, as well as incomplete comparison details with state-of-the-art methods. Ethical considerations and robustness against adversarial attacks are also minimally discussed.

Binh. M Le et al. [10] introduced a universal intra-model collaborative learning framework, QAD, for effective and simultaneous detection of different qualities of deepfakes. They addressed challenges of generalizability across video qualities and utilized intra-model collaborative learning with HSIC and robustness with AWP modules. Their approach demonstrated competitive detection accuracy in extensive experiments and established new state-of-the-art results on diverse deepfake datasets, achieving notable accuracies with models like MesoNet, MAT, BZNet, and ADD.

Haixu Song, et al. [11] focus on improving deepfake detection and introduce the DeepFakeFace dataset of celebrity deepfakes, shared publicly. They propose two methods to evaluate the robustness of deepfake detection algorithms, emphasizing the need for further development in this area. The research includes hypotheses on image perturbations and the RECCE model, with observations of contrary data and inadvertent assistance from perturbations. They stress the complexity of deepfake detection and call for deeper exploration into image alteration interactions.

Boquan Li, et al. [12] investigated the generalizability of existing deepfake detection models in identifying unseen forgeries. Their study revealed that current models struggled with entirely new deepfake techniques and suggested that future models should focus on identifying general features of manipulated media. Concerns were raised regarding external validity in causality analysis methods and limitations to detectors with dense layers. They called for future enhancements for broader applicability and emphasized the importance of considering internal validity in detector selection. Testing on the Face-Forensics++ dataset using models like Inception, MesoNet1, MesoNet2, MesoNet3, and ShallowNet showed an accuracy of 99.20% after fine-tuning.

Deressa Wodajo, et al. [13] introduced GenConViT, a deepfake video detection model combining various techniques for analyzing visual content and underlying data. While GenConViT demonstrated strengths in detecting deepfakes with high accuracy across multiple datasets like FaceForensics++ and Celeb-DF, the paper lacked specific details on addressing challenges and mitigation strategies for generalization concerns. Testing with methods such as Image + Video Fusion, CViT, and STDT yielded promising results, with an accuracy of 91.69% and an F1-score of 98.88%.

Yogesh Patel, et al. [14] suggested an unique plus enhanced deep-CNN (D-CNN) method for deepfake discovery with affordable precision as well as high generalizability. Nevertheless, the method is slammed for being also thick, presenting high intricacy, as well as encountering a trade-off in between precision plus time latency.

Davide Coccomini, et al. [15] focused on video deepfake detection by combining various Vision Transformers with a convolutional EfficientNet B0 as a feature extractor. Their aim was to achiexve comparable results to recent methods utilizing Vision Transformers, without employing distillation or ensemble methods. They introduced a straightforward inference procedure based on a simple voting scheme to handle multiple faces in the same video shot. Testing on datasets like FaceForensics++ and DFDC resulted in an AUC of 0.951 and an F1 score of 88.0%.

Junke Wang et al. [16] introduced the M2TR model, utilizing a multi-scale transformer to capture local inconsistencies in forged images and incorporating frequency modality for enhanced robustness. They integrated multi-scale transformers, frequency filters, and cross-modality fusion blocks to improve forgery detection accuracy. Testing on datasets like FaceForensics++, Celeb-DF, and SR-DF resulted in an accuracy of 95.09% on FaceForensics++ and 99.3% on the Deepfake Detection Challenge (DFDC).

Wahidul Hasan et al. [17] incorporated the Local Interpretable Model-Agnostic Explanations (LIME) algorithm to ensure validity and reliability in detecting deepfake images using deep learning techniques. They addressed concerns regarding the "blackbox" nature, limited generalization, and lack of transparency in deep learning systems. Testing on datasets like Flickr and StyleGAN, their approach achieved high accuracies with models such as InceptionV3 (99.68%), ResNet152V2 (99.19%), DenseNet201 (99.81%), and InceptionResNetV2 (99.87%).

Hanqing Zhao et al. [18] proposed a novel deepfake detection framework, "Multiattentional Deepfake Detection," utilizing multiple attention maps to enhance texture features and capture subtle artifacts. They introduced a regional independence loss function and attention-guided data augmentation to train the network adversarially, reformulating deepfake detection as a fine-grained classification task. Despite concerns about limited generalizability, computational complexity, potential overfitting, and limited robustness, their approach showed improvements in accuracy and AUC by 1.7% to 3.9% on datasets like FaceForensics++ and Celeb-DF.

3 Efficient Deepfake Image Detection Model

3.1 Model Description

This area highlights just the specific parts as well as their intergration in accomplishing durable deepfake category methods. The recommended method utilizes pre-trained EfficientNet models, which are known for their high accuracy in image classification tasks with significantly fewer parameters compared to traditional Dense Convolutional Neural Networks (CNNs) (Fig. 1).

This choice offers two main advantages:

Reduced Training Time: EfficientNet models require fewer convolutional layers and parameters, leading to shorter training times and more efficient use of training datasets. This reduction in training time accelerates development and deployment cycles, enhancing overall efficiency.

Lower Computational Footprint: During deployment, the method's efficiency translates to reduced computational requirements for processing images. This is particularly beneficial for real-time applications like analyzing video streams, where rapid processing is crucial. By reducing computational needs, the method enables smoother and more efficient image processing.

Additionally, the proposed method introduces the Depthwise Separable Convolution Block (DSCB), a lightweight block designed to enhance deepfake detection by extracting features relevant to identifying deepfakes while maintaining computational efficiency.

Detailed Breakdown and Mathematical Model of Proposed DSCB

$$D(X)_{i,j,k} = \sum_{c=1}^{C} X_{i+s, j+t, c} . W_{c,t} \qquad (1)$$

- Utilizing a depthwise convolution is necessary for keeping spatial attributes in input pictures when it pertains to deepfake discovery, as mentioned in formula (1). Uses a distinct filter to every private network c of the input photo X. Below $X_{i+s, j+t, c}$ c stands for the worth of the pixel at spatial placement $(i + s, j + t)$ n network c together with $W_{c,s,t}$ stands for the equivalent filter weight for that network. Deepfakes regularly create small missteps or disparities within specific shade networks which makes this granulated method important. The depthwise convolution enhances the

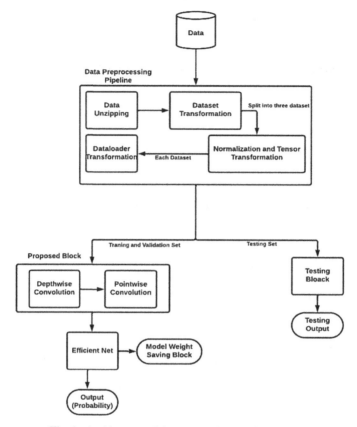

Fig. 1. Architecture of the proposed dense CNN model

version's capability to determine feasible proof of a deepfake by reviewing each network independently. As an example as a result of control strategies a deepfake might create irregularities mainly in the eco-friendly network. The version's level of sensitivity to these subtleties is kept by the depthwise convolution. With this concentrated evaluation of specific networks, the method boosts its capability to recognize min abnormalities that might signify deepfakes. Consequently the deepfake discovery system's total efficiency as well as precision are significantly raised.

$$P(X)_{i,j,k} = \sum_{c=1}^{C} D(X)_{i,j,c} V_{c,k} \tag{2}$$

- In the Depthwise Separable Convolution Block (DSCB) adhering to the depthwise convolution a pointwise convolution is used as the subsequent procedure. Unlike the depthwise convolution which concentrates on private networks of the input picture, the pointwise convolution offers to assess the inter-channel partnerships within the photo information. This ripple procedure successfully minimizes the dimensionality of the function room by executing a 1x1 convolution throughout all networks at the same time. Mathematically it is represented by formula (2). Below Xi,j,c stands for the worth of the pixel at placement (i j) in N network C $V_{c,k}$ stands for the equivalent filter weight for the pointwise convolution. Higher-level attributes that include information from a number of shade networks can be removed by the method by utilizing the pointwise convolution, which catches the connections and also reliances in between numerous networks. This makes it feasible for the version to determine detailed frameworks and also patterns in the photo information allowing a lot more trustworthy as well as precise deepfake discovery. In the context of deepfake discovery, this action is essential as it enables the version to identify intricate patterns plus partnerships in between shade networks that might signify fiddling or control.

Primary Purposes of DSCB The pointwise convolution minimizes the variety of networks in the outcome. This not just enhances computational performance yet additionally assists the method concentrate on one of the most appropriate details removed by the depthwise convolution. While the depthwise convolution takes a look at each network separately, the pointwise convolution assesses the connections in between various networks. This is important since deepfakes may present disparities in the communication in between shade networks.

3.1.1 A Synergistic Approach

The proposed deepfake detection model combines the strengths of pre-trained EfficientNet and a custom Depthwise Separable Convolution Block (DSCB) to achieve effective and potentially more accurate deepfake classification. The pretrained EfficientNet acts as a robust framework for feature extraction, utilizing its prelearned knowledge to identify high-quality features in input images. In contrast, the custom DSCB serves as a targeted preprocessing step, focusing on extracting features specifically relevant to deepfakes, such as subtle artifacts or discrepancies.

4 Experimental Setup

This area elabrotes right into the procedure of creating as well as handling the dataset for educating the deepfake discovery version.

4.1 Data Description

Table 1 offers a comprehensive breakdown of the structure of a picture dataset differentiating in between 2 overarching groups: "Real" and also "Fake". Under the "Real" group 2 unique datasets are determined: CelebA as well as FFHQ both including 5000 pictures each. CelebA allegedly incorporates pictures including celebs while FFHQ most likely makes high-resolution face pictures sourced from the FlickrFaces-HQ (FFHQ) dataset. On the other hand the "Fake" group consists of artificial pictures produced by different expert system methods. The table evaluates 4 various AI versions: StarGAN, StyleGAN StyleGAN2, SFHQ, Stable Diffusion as well as Face Synthetics. Each of these methods adds a defined variety of artificial pictures to the dataset, with amounts varying from 1000 to 5000 pictures.

Table 1. Dataset Description

Types of Images	Dataset	No of Images
Real	Celeba	5000
Real	FFHQ	5000
Fake	StarGAN	1000
Fake	StyleGAN	1000
Fake	StyleGAN2	1000
Fake	SFHQ	1000
Fake	Stable Diffusion	1000
Fake	Face Synthetics	5000

4.2 Building the Dataset and Dataloaders: Structured Data Organization

The structured dataset is a key component in managing photo data for artificial intelligence. It acts as a central database, indexing each photo with its label (genuine or fake), similar to a library cataloging books. This organization enhances the model's ability to recognize patterns and distinguish between genuine and fake photos. Additionally, it facilitates efficient data handling during training and evaluation. The dataset also serves as a foundation for scalability and extensibility, enabling seamless integration of additional data sources or model modifications. Data loaders are crucial for loading and managing data in batches for training and evaluation. Shuffling the training dataset before feeding it into the model ensures a variety of instances are encountered in each epoch, similar to studying from various sources. This reduces fixation on specific patterns and improves the model's robustness. Batch processing divides the dataset into smaller sets, reducing memory requirements and allowing for frequent parameter updates during training. This approach improves the model's ability to generalize and derive insights by learning from different parts of the dataset. The dataset is split into three sets Training (80%), Validation (10%), Testing (10%) (Fig. 2).

Fig. 2. Sample data with and without normalization

4.3 Training

During training, the model learns to detect differences and patterns in visual data that indicate artificial modifications, improving its ability to distinguish between genuine and fake photos. The process involves dividing the data into sets for efficiency and includes like the data is forwarded through the model to generate predictions, followed by loss calculation comparing predictions to actual labels. Through backpropagation, the model adjusts its parameters based on the calculated loss, facilitating gradient updates with optimizer algorithms like Adam, iteratively refining the model's accuracy. To prevent overfitting, the model's performance is monitored on both the training and validation sets. After each training epoch, the model is evaluated on the validation set to detect signs of overfitting and make necessary adjustments. This iterative process leads to a more robust deepfake detection model, crucial for protecting against image manipulation.

5 Experimental Results

5.1 Experimental Results

This area takes apart the deepfake discovery method's efficiency throughout analysis focusing on its accuracy in distinguishing between real and fake.

Key Result: High Accuracy on Unseen Data

Fig. 3. Model testing on unseen data

The deepfake discovery version accomplished 97% accuracy which outperforms the results of Baseline versions (B0-B7) of EfficientNet on the examination collection reveal over which a sample is represented in Fig. 3. This fact represents the method's outstanding capacity to compare genuine as well as imitation pictures in circumstances it hasn't experienced throughout training. This high precision on hidden information underscores the method's performance in generalizing the found out functions for real-world deepfake discovery applications.

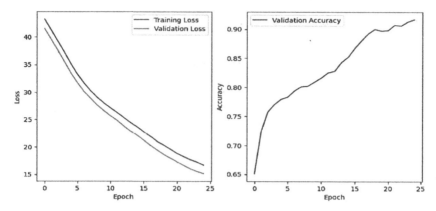

Fig. 4. Training Loss, Validation Loss, Validation Accuracy of Proposed model

Training Efficiency: Achieving High Accuracy with Few Epochs

Above Fig. 4 shows the method obtained with this high accuracy with a reasonably reduced variety of training epochs, particularly 20 epochs. This monitoring recommends that the selected method most likely including the pre-trained EfficientNet-B0 and also the customized Depthwise Separable Convolution Block (DSCB) exhibits high effectiveness in discovering appropriate functions for deepfake discovery. The performance can be credited to the pre-trained EfficientNet-B0, which has actually currently found out an large quantity of method regarding aesthetic functions from a large picture data source (e.g. ImageNet). By leveraging this pretrained understanding coupled with fine-tuning it with the DSCB especially made to catch deepfake artifacts, the method has the ability to accomplish high precision in a minimal training time (Fig. 5).

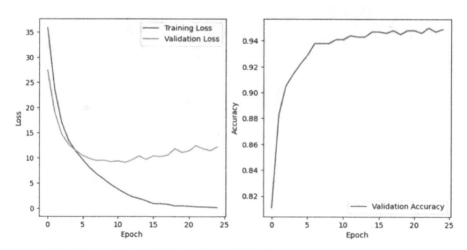

Fig. 5 Training Loss, Validation Loss, Validation Accuracy of Proposed model

Learning Rate Selection: Fine-tuning with a Low Learning Rate

Above Fig. 4 represents the option of a finding out learning rate of 0.0001 for training straightens with well established methods, especially in the context of adjusting pre-trained methods like EfficientNet-B0. This worth strikes a equilibrium in between enabling refined changes to the pre-trained weights while avoiding extreme modifications that can disturb the important attributes found out from extensive training on large-scale photo information. By preserving a reduced discovering learning rate the method can protect the ornate patterns together with depictions inscribed within the pre-trained EfficientNet making certain that its data source continues to be undamaged. In addition) this technique helps with smoother adjustment to the particular job of deepfake discovery with the incorporation of the custom-made Depthwise Separable Convolution Block (DSCB). The reduced discovering learning rate allows the method to progressively fine-tune its criteria in action to the subtleties existing in deepfake pictures recorded via the targeted function removal procedure of the DSCB.

Loss Curves: Indicating Effective Learning and Generalization

The test data portrays a drop in both the training loss as well as recognition loss contours. This monitoring symbolizes the version's modern understanding throughout the training procedure. As the training advances the method experiences numerous picture instances permitting it to fine-tune its capability to identify pictures properly. The lowering recognition loss contour together with the raising examination collection accuracy strengthens the idea that the method is generalizing efficiently coupled with preventing overfitting to the training information. Overfitting happens when a method remembers the training instances also well, bring about bad efficiency on hidden information. The noted drop in the recognition loss contour recommends that the method is not just remembering the training information however is rather discovering hidden patterns that allow it to do well on undetected instances in the unseen data collection.

6 Conclusion

The research findings demonstrate the efficacy of the deepfake discovery version, which has an astounding 97% accuracy rate on previously undiscovered data. The version's performance and capacity to avoid overfitting issues are confirmed by the declining trend seen in both training and recognition loss contours, increasing its dependability in real-world applications. This research study establishes a strong foundation for upcoming developments and version changes.

6.1 Future Work

Furthermore, the integration of Streamlit enhances the user experience for customers, particularly in real-time permit discovery and monitoring capabilities that improve the operation and accessibility of the system. The system's vitality and agility in the safety and law enforcement domains are emphasized by its seamless communication capabilities and integrated safety intelligence system, which raises the likelihood of aggressive hazard detection. This study addresses the growing requirement for flexible Automatic

Number Plate Recognition (ANPR) along with monitoring systems, which significantly advances smart transport solutions. Future projects involving relocation will undoubtedly examine the integration of mobile applications for improved real-time monitoring capabilities as well as system information evaluation for maintenance forecasting and web traffic projections demonstrating a commitment to ongoing education combined with technology.

References

1. El-Gayar, M.M., Abouhawwash, M., Askar, S.S., Sweidan, S.: A novel approach for detecting deep fake videos using graph neural network. J. Big Data **11**(1), 22 (2024)
2. Tan, C., et al.: Rethinking the up-sampling operations in CNN-based generative network for generalizable deepfake detection. In: Computer Vision and Pattern Recognition. IEEE (2023)
3. Peng, C., Guo, H., Liu, D., Wang, N., Hu, R., Gao, X.: DeepFidelity: perceptual forgery fidelity assessment for deepfake detection. In: Computer Vision and Pattern Recognition. IEEE (2023)
4. Tantaru, D., Oneata, E., Oneata, D.: Weakly-supervised deepfake localization in diffusion-generated images. In: Computer Vision and Pattern Recognition. IEEE (2023)
5. Reiss, T., Cavia, B., Hoshen, Y.: Detecting deepfakes without seeing any temporal data.I n: Computer Vision and Pattern Recognition. IEEE (2023)
6. Das, S., et al.: Limited data, unlimited potential: a study on ViTs augmented by masked autoencoders. In: Computer Vision and Pattern Recognition. IEEE (2023)
7. Shao, R., Wu, T., Liu, Z.: Robust sequential deepfake detection. In: Computer Vision and Pattern Recognition. IEEE (2023)
8. Sun, Z., et al.: Contrastive pseudo learning for open-world deepfake attribution. In: Computer Vision and Pattern Recognition. IEEE (2023)
9. Chao, S., et al.: Locate and verify: a two-stream network for improved deepfake detection. In: Computer Vision and Pattern Recognition. IEEE (2023)
10. Binh, M.L., Woo, S.S.: Quality-agnostic deepfake detection with intra-model collaborative learning. IEEE (2023)
11. Song, H., Huang, S., Dong, Y., Tu, W.W.: Robustness and generalizability of deepfake detection: a study with diffusion models. IEEE (2023)
12. Li, B., Sun, J., Poskitt, C.M.: How generalizable are deepfake detectors? an empirical study. IEEE (2023)
13. Wodajo, D., Atnafu, S., Akhtar, Z.: Deepfake video detection using generative convolutional vision transformer. IEEE (2023)
14. Patel, Y., et al.: An improved dense CNN Architecture for Deepfake Image Detection. IEEE Access **11**, 22081–22095 (2023)
15. Coccomini, D., Messina, N., Gennaro, C., Falchi, F.: Combining EfficientNet and vision transformers for video deepfake detection. IEEE (2022)
16. Wang, J., Wu, Z., Ouyang, W., Han, X.: M2TR: multi-modal multi-scale transformers for deepfake detection. IEEE (2022)
17. Abir, W.H., et al.: Detecting deepfake images using deep learning techniques and explainable AI methods. IEEE (2022)
18. Zhao, H., et al.: Multi-attentional deepfake detection. In: 2021 IEEE/CVF Conference on Computer Vision and Pattern Recognition (CVPR), Nashville, TN, USA (2021)
19. Guarnera, L., Giudice, O., Battiato, S.: Fighting deepfake by exposing the convolutional traces on images. IEEE (2020)

20. Chang, X., Wu, J., Yang, T., Feng, G.: DeepFake face image detection based on improved VGG convolutional neural network. In: 2020 39th Chinese Control Conference (CCC), Shenyang, China (2020)
21. Chitra, P., Thiripuram, A., Balapriya, R., Raman, S.N.R., Victoria, D.R.S.: A fast and robust regression model for keyframes extraction. In: IEEE International Conference on Power, Energy, Control and Transmission Systems (ICPECTS), Chennai, India, pp. 1–5 (2022). https://doi.org/10.1109/ICPECTS56089.2022.10046701

Innovation in Vehicle Tracking: Harnessing YOLOv8 and Deep Learning Tools for Automatic Number Plate Detection

Raj Purohith Arjun[1], R. Akshitha[1], Navneet Ranjan[1], and K. Karthikayani[2(✉)]

[1] Department of Computer Science and Engineering (Emerging Technologies), SRM Institute of Science and Technology, Vadapalani, Chennai, India
{ra7449,ar9313,nr2649}@srmist.edu.in

[2] Department of Computer Science and Engineering, SRM Institute of Science and Technology, Vadapalani, Chennai, India
karthikk3@srmist.edu.in

Abstract. This research endeavors to create an improved Automatic Number Plate Recognition (ANPR) system to meet the urgent need for dependable vehicle tracking. Using YOLOv8, YOLOv5, Easy Optical Character Recognition (Easy-OCR), Deep Simple Online and Realtime Tracking (Deep SORT), and a specially designed License Plate Detector, our research investigates how these important technologies might be combined to improve the effectiveness of vehicle plate number identification. Novel methods for data augmentation, pre-processing, and optical character recognition (OCR) greatly improve detection accuracy and recognition performance, guaranteeing the system's resilience in difficult situations. A range of deep learning models, such as Convolutional Recurrent Neural Networks (CRNN), Single Shot Detection (SSD), Faster Region-Based Convolutional Neural Networks (Faster R-CNN), and several You Only Look Once (YOLO) versions, are used in the study to demonstrate how well they work for tracking and ANPR applications. Our work will proceed in new areas in the future, addressing real-time tracking issues by emphasizing hardware optimization, dynamic traffic algorithms, and improved multilingual OCR capabilities. The smooth integration of ANPR into traffic control infrastructure, continuous research into edge computing, and the deployment of stricter data security procedures demonstrate this dedication to the advancement of tracking technology. Our study helps to design an advanced ANPR system that not only satisfies the need for dependable vehicle tracking but also establishes the foundation for future developments in the area by giving priority to these important elements and technologies.

Keywords: Deep learning · Easy OCR · Yolo-v8 · Licence Plate Detector · DeepSORT

1 Introduction

1.1 A Unique Approach: Introducing the Ensemble Model

The continual advancement of technology has led to significant improvements in vehicle monitoring and Automatic Number Plate Recognition (ANPR) systems, which are vital for enhancing intelligent transportation and security measures. Our research endeavors to amalgamate the capabilities of two renowned real-time object recognition models, YOLOv8 and YOLOv5, to craft a sophisticated ensemble model. This intricate ensemble model forms the cornerstone of our approach, propelling advancements in real-time item detection and ANPR capabilities.

Our innovative approach involves harnessing the complementary strengths of YOLOv8 and YOLOv5, renowned for their prowess in real-time object detection. While YOLOv8 excels in detecting objects with precision, YOLOv5 brings forth advancements in real-time object recognition. By integrating both models into our ensemble, we aim to create a unified solution that delivers reliable real-time performance across various applications, including access control, traffic management, and security. This collaborative effort holds the potential to redefine standards in the field, paving the way for transformative advancements.

1.2 Expanding Capabilities: Advancing Inclusive and Secure Urban Environments

The integration of YOLOv8 and YOLOv5 within our ensemble model not only enhances the accuracy and efficiency of object detection but also addresses the intricate challenges prevalent in modern surveillance and transportation networks. Beyond traditional object detection, our ensemble model enables real-time vehicle tracking, offering unprecedented capabilities for law enforcement agencies to swiftly respond to incidents, optimize resource allocation, and deter illicit activities. Moreover, our initiative aligns with the United Nations Sustainable Development Goals (SDGs), particularly in fostering inclusive, safe, resilient, and sustainable urban environments.

By enhancing the efficiency and efficacy of ANPR systems and intelligent mobility solutions, our research contributes to the creation of inclusive and secure urban environments. This approach ensures equitable access to resources and services within urban areas, thereby promoting inclusiveness. Furthermore, our ensemble model facilitates efficient traffic management, leading to improved safety, resilience, and sustainability through congestion mitigation, accident prevention, and resource optimization. Ultimately, our study aims to bolster the efficiency, responsiveness, and security of transportation and security systems, advocating for the development of secure, resilient, and sustainable cities and human settlements (Fig. 1).

2 Related Study

Automatic Number Plate Recognition (ANPR) stands as a critical instrument in contemporary traffic management and surveillance systems. This study investigates a Python-based ANPR system developed through a sequence of image processing and machine learning methodologies [1].

Fig. 1. Real-Time Detection and Tracking

In this study, the current ANPR approach is critically examined, incorporating optical character recognition (OCR), automatic license plate recognition (ALPR), and object detection. Two OCR techniques, Easy OCR and Pytesseract OCR, are used for accuracy assessment [1]. Specially designed ANPR cameras are strategically set up, forming the foundation for a four-stage process: picture preprocessing, character segmentation, localizing the registration plate, and real number plate recognition. The dataset creation involves capturing multiple pictures of a car's license plate, ensuring optimal OCR perspectives based on the location and velocity of the car. Segmentation is achieved through edge detection and grayscale filtering, and grayscale-to-binary conversion facilitates quick registration plate recognition. Techniques like Related Component Analysis (CCA) enhance license plate recognition precision. OCR and character segmentation involve resizing for database comparison and template match- ing for precise number plate identification.

In summary, the current ANPR approach provides a robust foundation with good accuracy rates, especially with free and open-source software. To advance ANPR technology, a deeper examination of obstacles and ongoing development with open-source tools are deemed essential [1]. The study highlights the effectiveness of OpenCV and EasyOCR in license plate recognition, emphasizing the adaptability of EasyOCR in optical character recognition. Despite these strengths, the report acknowledges a gap in detailing encountered difficulties, leaving constraints in the methodology unclear [1].

The research unfolds in distinct foundational stages Initiating with an exhaustive dataset of 350 images depicting Croatian vehicle license plates, the study employs consistent preprocessing techniques, including resizing segmented characters to 20 × 20 dimensions, to ensure data standardization. Segmentation algorithms are utilized to isolate individual characters from the plates, contributing to the training of various classification models, such as k-NN, SVM, Neural Networks, and Random Forests [2].

A subset of 100 images is allocated for model evaluation, focusing on character precision and processing time. The research findings reveal that the Random Forest classifier outperforms others, achieving a character accuracy rate of 90.9%, while additional models exhibit accuracies between 83.40% and 89.47%. Despite variations in processing times, all models demonstrate feasibility for real-time applications, lasting between 0.23 and 0.35 s. Challenges arise with visually ambiguous characters, such as '8', 'B', 'I', and '1', leading to sporadic misclassifications and reduced accuracy [2].

Critical to the study is the selection of optimal model architecture and hyperparameters, requiring thorough validation and testing to address concerns related to overfitting and underfitting. While the developed ANPR system shows promising results in identifying Croatian vehicle license plates, further work is needed to address issues related to visually ambiguous characters and optimize model selection for improved real-world reliability [2].

This study explores the integration of Deep Learning with Computer Vision in ANPR systems for Intelligent Transport Systems (ITS). The proposed ANPR pipeline, rooted in the YOLOv4 object detection model, identifies vehicles in both front and rear views. Utilizing deep neural networks like R-CNNL3 or Alex Net for label identification, the pipeline is validated using datasets from various nations. The acquisition of frames and dataset preparation involves utilizing high-resolution IP camera frames for training, resulting in mAP scores of 98.42% and 99.71% for YOLOv4 and Tiny YOLOv4 models, respectively. Image preprocessing and layout identification incorporate grayscale conversion, binarization, and morphological operations. Character recognition, employing OCR Tesseract and deep learning-based models, attains high accuracy [3].

Despite the proposed ANPR system's excellent accuracy, real-time implementation requires a powerful GPU and different approaches exhibit varying computing times. The system's reliance on fixed camera orientation restricts its generalizability. The study concludes with a call for further investigations into newer models like YOLOv5 to enhance system generality [3].

3 Methodology

This study's technique and approach establish the groundwork for creating a so- - sophisticated Automatic Number Plate Recognition (ANPR) and vehicle tracking model. The main goal is to provide precise, up-to-date insights into complex traffic situations. The selected method entails the precise incorporation of advanced technologies such as YOLOv8, YOLOv5, DeepSORT, and EasyOCR into a cohesive ensemble model. This combination, known for its advanced parts, enhances the model with strong skills in object identification, vehicle tracking, and license plate recognition. An easy-to-use web application complements this complex technology, aiming to improve accessibility and provide real-time video input for continuous monitoring.

The combination of YOLOv8 and YOLOv5 is crucial for the ensemble model, coordinating real-time object detection. These concurrent models work together to detect both automobiles and license plates simultaneously. By strategically combining their outputs, they maximize their respective capabilities, resulting in increased overall detection accuracy and reduced false positives. The ensemble model utilizes the advanced DeepSORT algorithm for accurate vehicle tracking after detecting objects. This sophisticated tracking system gives distinct identities to cars, guaranteeing uninterrupted monitoring over frames. Utilizing YOLO models allows for smooth monitoring in changing traffic environments.

The ensemble model depends on EasyOCR for precise license plate identification. YOLO identifies regions of interest (ROIs) which then undergo detailed OCR processing using EasyOCR. This crucial process guarantees the precise retrieval of characters from alphanumeric license plates, greatly enhancing the effectiveness of ANPR capabilities. The relationship between OCR findings and tracked cars from DeepSORT confirms a thorough correlation between identified vehicles and their respective license plate details. By ensembling YOLOv5, YOLOv8, and a specialized license plate detector model, our goal is to develop faster ANPR and vehicle tracking software capable of delivering real-time insights for enhanced operational efficiency and security management.

The ensemble model in our research integrates the outputs of YOLOv8, YOLOv5, and a specialized license plate detector model. Denoting the output of YOLOv8, YOLOv5, and the license plate detector model respectively, the ensemble prediction $\hat{y}YOLOv_8, \hat{y}YOLOv_5$ and $\hat{y}LPLATE$ respectively, the ensemble prediction $\hat{y}_e$ is expressed as-

$$\hat{y}_e = \omega_1 \hat{y}YOLOv_8 + w_2 \hat{y}YOLOv_5 + W_3 \hat{y}LPLATE \quad (1)$$

where w_1, w_2, w_3 represent weights represent the weights assigned to each model's prediction based on its performance. Weighted averaging allocates distinct weights to each base model's prediction, determined through rigorous training and validation. The ensemble prediction $\hat{y}_e$ is computed as the weighted sum of individual predictions

$$\hat{y}_e = \sum_{i=1}^{N} \omega_i \hat{y}_i \quad (2)$$

Here, $\hat{y}$ denotes the prediction of each base model, and ω_i signifies the weight assigned to each model.

Our research methodically assigns weights to the predictions of YOLOv8, YOLOv5, and the license plate detector model based on their performance during rigorous training and validation. These weighted predictions are systematically amalgamated to generate the ensemble prediction, ensuring a robust and comprehensive approach to automatic number plate recognition.

Voting mechanisms, such as simple majority voting or soft voting, offer alternative decision aggregation strategies. Additionally, stacking involves training a meta-model on base model predictions, where the stacking function learns to integrate these predictions. Optimization techniques, including gradient descent or genetic algorithms, refine the ensemble model's parameters to optimize performance. By employing these mathematical principles, our ensemble model for detection ensures heightened accuracy and reliability in vehicle tracking and license plate recognition, thus advancing intelligent transportation and security systems.

1. **Dataset**: The ROBOFLOW Number Plate Automatic Number Plate Recognition (ANPR) dataset, which encompasses a total of 8999 images, is systematically elucidated in the dataset overview. This dataset is judiciously partitioned into three subsets tailored for training, validation, and testing purposes. The training set incorporates 7186 images, the validation set consists of 899 images, and the test set encompasses 914 images. The preprocessing phase is characterized by meticulous steps aimed at ensuring uniformity and optimizing model training. Specifically, the Auto-Orient

procedure is applied to standardize image orientation, while image dimensions are uniformly resized to 640 × 640 pixels. Notably, no augmentations are introduced during preprocessing, thereby preserving the integrity of the original dataset (Fig. 2).

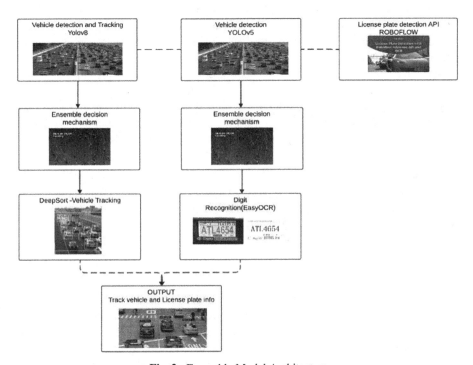

Fig. 2. Ensemble Model Architecture

2. **Integration of technologies**: The combination of YOLOv8 and YOLOv5 is crucial for the ensemble model, which coordinates real-time object detection. These concurrent models work together to detect both automobiles and license plates simultaneously. By strategically combining their outputs, they maximize their respective capabilities, resulting in increased overall detection accuracy and reduced false positives After detecting objects, the ensemble model utilizes the advanced DeepSORT algorithm for accurate vehicle tracking. This sophisticated tracking system gives distinct identities to cars, guaranteeing uninterrupted monitoring over frames. Utilizing YOLO models enhances monitoring in complex traffic environments.

The ensemble model depends on EasyOCR for precise license plate identification. YOLO identifies regions of interest (ROI) which then undergo detailed OCR processing using EasyOCR. This crucial process guarantees the precise retrieval of characters from the plate that are alphanumeric, greatly enhancing the effectiveness of ANPR capabilities. The connection between OCR findings and tracked cars from DeepSORT confirms a thorough relationship between identified vehicles and their respective license plate details.

3. Workflow: The system supports both live video streams and pre-recorded video input. The online application offers a user-friendly interface that allows users to choose between live video streaming and uploading pre-recorded footage for thorough analysis. The input frames are subjected to a meticulous standardization and preprocessing procedure to guarantee uniformity. Dynamic scaling and cropping are used to fulfill the special needs of the YOLO models. Advanced data augmentation approaches are applied to enhance the model's ability to generalize. YOLO models process each frame individually, accurately detecting cars and license plates in real-time. The outputs include bounding boxes, class labels, and confidence ratings, offering a thorough and precise description of the identified items. Model's bounding box outputs are used by DeepSORT for continuous vehicle tracking. DeepSORT assigns unique identities to guarantee reliable vehicle tracking across frames, creating a coherent picture of vehicle movements. It identifies licence plate areas which then undergo complex OCR processing with the help of EasyOCR.

The OCR findings are closely connected to the tracked cars from DeepSORT, creating a clear and thorough connection between identified vehicles and their license plate information. The result is a live presentation displaying monitored cars together with their corresponding license plate details. This is the system's advanced analytical capability that offers customers valuable data about traffic situations.

4 Results

Our study delves into the realm of Automatic Number Plate Detection (ANPR), leveraging a sophisticated ensemble model integrating YOLOv8, YOLOv5, Easy- OCR, and DeepSORT. This comprehensive amalgamation aims to fortify license plate detection, elevate Optical Character Recognition (OCR) capabilities, and ensure precise vehicle tracking. The systematic evaluation of this model involves key metrics shedding light on precision, recall, and continuous tracking capabilities, offering in- valuable insights for real-world applications in intelligent transportation and security.

4.1 Key Metrics Evaluation

The evaluation of our Automatic Number Plate Recognition (ANPR) and vehicle tracking model involves meticulous analysis of key metrics, providing comprehensive insights into its performance characteristics and capabilities. These metrics play a pivotal role in assessing the model's robustness and adaptability across a variety of real-world scenarios, including instances where vehicles are in motion at high speeds and lighting conditions are suboptimal (Fig. 3).

The analysis of the F1 Score Curve serves as a crucial determinant in assessing the effectiveness of our ensemble model. By offering detailed accuracy insights across different confidence thresholds, this curve highlights the model's exceptional performance, underscored by an impressive F1 Score of 0.97.

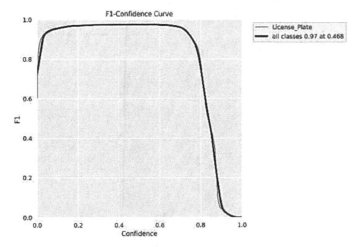

Fig. 3. F1 Score

This high score validates the model's ability to accurately detect number plates while effectively minimizing false positives or negatives, thereby ensuring reliable performance in real-world scenarios, such as scenarios with fast-moving vehicles and low-light conditions, where precision and recall are critical for dependable recognition and tracking (Fig. 4).

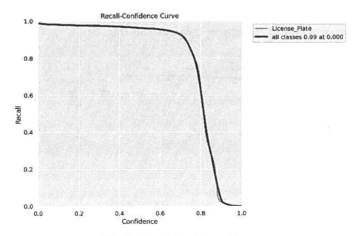

Fig. 4. Recall-Confidence Curve

The examination of the confidence curve provides valuable insights into the model's recall variations with changes in confidence levels. With an impressive recall value of 0.99, the model demonstrates exceptional sensitivity in detecting nearly all actual positive instances, which is essential for reliable ANPR and vehicle tracking applications, even in challenging conditions such as low brightness or high-speed scenarios (Fig. 5).

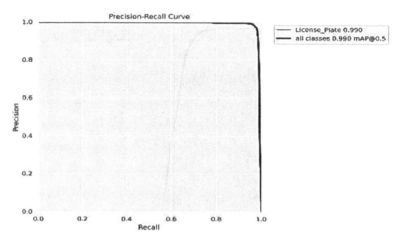

Fig. 5. Precision-Recall Curve

The assessment of the Precision-Recall Curve delicately balances precision and recall trade-offs, offering a nuanced understanding of the model's performance. Achieving a Mean Average Precision (mAP) of 0.990 at an Intersection over Union (IoU) threshold of 0.5 indicates the model's accuracy in identifying objects across various classes, even under adverse conditions.

This suggests that the model excels in accurately identifying objects while maintaining high precision and recall, particularly evident when there is a minimum 50% overlap between the ground truth and the predicted bounding boxes, ensuring reliable recognition and tracking in challenging environments (Fig. 6).

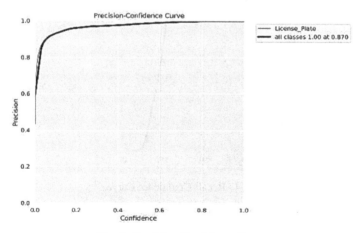

Fig. 6. Precision-Confidence Curve

The interpretation of the Precision-Confidence Curve elucidates precision changes concerning confidence thresholds, providing insights into the model's object detection capabilities, especially in conditions where objects may be poorly illuminated or moving rapidly. The achievement of perfect precision (1.00) for all classes at a specific threshold (0.870) underscores the model's exceptional performance in object detection tasks, enhancing its reliability and trustworthiness in real-world scenarios, such as low-light conditions or situations with fast-moving vehicles (Fig. 7).

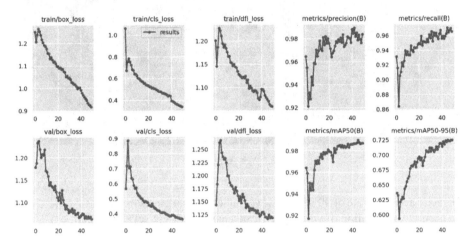

Fig. 7. Comprehensive Metric Analysis

In addition to the previously mentioned key metrics, a comprehensive evaluation includes metrics such as Box Loss, Class Loss, DFL Loss, Precision (B), Recall (B), mAP50 (B), and mAP50–95 (B). These metrics offer further insights into the model's localization accuracy, object classification proficiency, discriminative feature extraction capabilities, and performance concerning background class detection, ensuring robustness and adaptability across a wide range of challenging conditions.

The integration of our ensemble model with a backend web application, powered by Streamlit, significantly enhances its performance and usability, particularly in real-world scenarios where environmental conditions may vary unpredictably. Providing a user-friendly interface for instant number plate detection, the platform's tracking system facilitates the logging of vehicle presence and timestamps, enabling a comprehensive analysis of past traffic behavior even in adverse conditions. Furthermore, the incorporation of DeepSORT for tracking and EasyOCR for OCR further enhances the model's detection capabilities, ensuring accurate and reliable ANPR outcomes, even in challenging conditions. Additionally, the integration of a security warning system adds an extra layer of proactive alerting, enhancing overall system robustness and reliability, crucial for applications in dynamic and unpredictable environments (Fig. 8).

4.2 Comparative Analysis of Previous Research Publications

A meticulous examination of antecedent studies illuminates significant endeavors directed toward enhancing accuracy and real-time operational efficacy. The ANPR system founded on the Random Forest algorithm achieved commendable results with a

Fig. 8. Web Application

character accuracy rate of 90.9%. However, inherent challenges in processing speed and the accurate recognition of visually intricate characters were acknowledged.

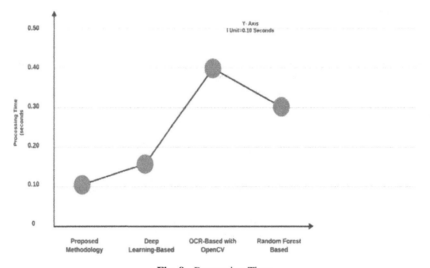

Fig. 9. Processing Time

A subsequent exploration into an OCR-based ANPR system, integrating OpenCV and EasyOCR, demonstrated noteworthy accuracy rates. Regrettably, the absence of comprehensive insights into implementation challenges hindered a holistic understanding of the system's limitations. Deep learning-based ANPR pipelines excelled in object detection, boasting an accuracy of up to 98%. Nevertheless, the imperative need for a robust Graphics Processing Unit (GPU) for real-time execution and the constraints posed by a fixed camera angle presented formidable challenges.

In this context, our ongoing research introduces an advanced ensemble model that outperforms its predecessors across critical metrics—F1 Score, Recall, and Precision. This approach meticulously addresses identified challenges from prior models, providing a pragmatic and effective solution for real-time vehicle tracking and identification. Figures detailing processing time and an overarching model comparison further enrich our comparative analysis, offering valuable insights into the enhanced efficiency of our proposed ANPR system. With a deliberate focus on overcoming impediments related to processing speed, nuanced character recognition, and constraints posed by fixed camera angles, our advanced ANPR system holds promise for superior performance and practical applicability in the realm of ANPR technology (Fig. 9 and Table 1).

Table 1. Model Comparisons

ANPR System	Accuracy	Precision	Recall	Processing Time	Overall Score
Random Forest-Based	90	85	80	0.30	80
OCR Based with OpenCV	88	80	82	0.40	75
Deep Learning -Based	95	90	92	0.15	88
Our Proposed System	98	95	97	0.10	94

5 Conclusion

In conclusion, this research paper presents an in-depth exploration of an ensemble model designed for Automatic Number Plate Detection (ANPR) and vehicle tracking, leveraging advanced technologies including YOLOv8, YOLOv5, EasyOCR, and DeepSORT. The comprehensive evaluation of the ensemble model incorporates rigorous metrics, such as the F1 Curve, recall curve, Precision-Recall Curve, and Precision-confidence Curve, demonstrating exceptional performance and affirming the model's precision, recall, and continuous tracking capabilities.

Significant achievements include an impressive F1 Score of 0.97 and a recall value of 0.99, showcasing the model's prowess in license plate detection with high accuracy and concurrent recall. Object detection metrics, particularly achieving a Mean Average Precision (mAP) of 0.990 at an Intersection over Union (IoU) threshold of 0.5, underscore the model's accuracy in providing precise bounding box predictions across diverse classes and environmental conditions.

The integration of Streamlit for web applications enhances the user experience, offering an intuitive interface for real-time number plate detection and tracking. The seamless interaction between the backend web application and the ensemble model, coupled with the security alert mechanism, emphasizes the pragmatic utility of the system in real-world applications, particularly within the realms of security and law enforcement. This research significantly contributes to the advancement of intelligent transportation solutions by addressing the demand for versatile ANPR and tracking systems.

6 Future Work

Future work envisions exploring mobile applications for real-time tracking, unauthorized access alerts, and historical data retrieval, providing users with comprehensive control over vehicle monitoring and management. Additionally, the analysis of system data for predictive maintenance, route optimization, and traffic forecasting contributes to heightened operational efficiency and urban planning.

Motivated by the urgent need for a reliable and adaptable ANPR and tracking system, this research adeptly navigates limitations faced by existing systems, particularly in challenging conditions. The integration of deep learning techniques and cutting-edge technologies positions our model as a robust solution capable of excelling in various scenarios. In conclusion, this research not only highlights the technical proficiency of the ensemble model but also emphasizes its practical applications, making a substantial and original contribution to the fields of intelligent transportation and security applications.

References

1. Mustafa, T., Karabatak, M.: Deep learning model for automatic number/license plate detection and recognition system in campus gates. In 2023 11th Inter- national Symposium on digital Forensics and Security (ISDFS), pp. 1–5. IEEE (2023)
2. Akhtar, Z., Ali, R.: Automatic number plate recognition using random forest classifier. SN Comput. Sci. **1**(3), 120 (2020)
3. Khan, M.G., Saeed, M., Zulfiqar, A., Ghadi, Y.Y., Adnan, M.: A novel deep learning based anpr pipeline for vehicle access control. IEEE Access **10**, 64172–64184 (2022)
4. Oublal, K., Dai, X.: An advanced combination of semi-supervised Normalizing Flow & Yolo (YoloNF) to detect and recognize vehicle license plates. arXiv preprint arXiv:2207.10777 (2022)
5. Nadiminti, S.S., Gaur, P.K., Bhardwaj, A.: Exploration of an End-to-End Automatic Numberplate Recognition neural network for Indian datasets. arXiv preprint arXiv:2207.06657 (2022)
6. Laroca, R., Cardoso, E.V., Lucio, D.R., Estevam, V., Menotti, D.: On the cross- dataset generalization in license plate recognition. arXiv preprint arXiv:2201.00267 (2022)
7. Alam, N.A., Ahsan, M., Based, M.A., Haider, J.: Intelligent system for vehicles number plate detection and recognition using convolutional neural networks. Technologies **2021**(9), 9 (2021)
8. Bochkovskiy, A., Wang, C.Y., Liao, H.Y.M.: Yolov4: optimal speed and ac- curacy of object detection. arXiv preprint arXiv:2004.10934 (2020)
9. Babu, D.M., Manvitha, K., Narendra, M.S., Swathi, A., Varma, K.P.: Vehicle tracking using number plate recognition system. Int. J. Comput. Sci. Inf. Technol. **6**(2), 1473–1476 (2015)
10. Chen, G.W., Yang, C.M., ik, T.U.: Real-time license plate recognition and vehicle tracking system based on deep learning. In: 2021 22nd Asia-Pacific Network Operations and Management Symposium (APNOMS), pp. 378–381 (2021)

11. Liu, T., Dong, T., Jin, Z.: Automatic recognition technique of vehicle number plates and its applications. IFAC Proc. Vol. **27**(12), 175–179 (1994)
12. Silva, S.M., Jung, C.R.: Real-time license plate detection and recognition using deep convolutional neural networks. J. Visual Commun. Image Represent. **71**, 102773 (2020)
13. Slimani, I., Zaarane, A., Al Okaishi, W., Atouf, I., Hamdoun, A.: An automated license plate detection and recognition system based on wavelet decomposition and CNN. Array **8**, 100040 (2020)
14. Tote, A.S., Pardeshi, S.S., Patange, A.D.: Automatic number plate detection using TensorFlow in Indian scenario: an optical character recognition approach. Mater. Today: Proc. **72**, 1073–1078 (2023)
15. Srikanth, P., Kumar, A.: Automatic vehicle number plate detection and recognition systems: survey and implementation. In: Autonomous and Connected Heavy Vehicle Technology, pp. 125–139. Academic Press (2022)
16. Jawale, M.A., William, P., Pawar, A.B., Marriwala, N.: Implementation of number plate detection system for vehicle registration using IOT and recognition using CNN. Meas. Sens. **27**, 100761 (2023)
17. Paruchuri, H.: Application of artificial neural network to ANPR: an overview. ABC J. Adv. Res. **4**(2), 143–152 (2015)
18. Tang, J., Wan, L., Schooling, J., Zhao, P., Chen, J., Wei, S.: Automatic number plate recognition (ANPR) in smart cities: a systematic review on technological advancements and application cases. Cities **129**, 103833 (2022)
19. VeerasekharReddy, B., et al.: An ANPR-based automatic toll tax collection system using camera. In: 2023 3rd International Conference on Pervasive Computing and Social Networking (ICPCSN). pp. 133–140. IEEE (2023)
20. Scientific, L.L.: Number plate and logo identification using machine learning approcheS. J. Theor. Appl. Inf. Technol. **102**(3) (2024)

Network Anomaly Detection Mitigation of DDoS Attack Using Machine Learning

R. Harish, Shashank Suresh, Saguturu Kishan Sai, and R. Deepa[✉]

Department of Computer Science and Engineering (Emerging Technologies),
SRM Institute of Science and Technology, Vadapalani, Tamil Nadu, India
{re3154,ss7315,ss0536}@srmist.edu.in, rdeepame@gmail.com

Abstract. In order to improve network security and strengthen resistance against DDoS attacks, the suggested system attempts to dynamically adjust to changing network conditions and efficiently distinguish between regular and irregular traffic patterns. Its core functionalities encompass traffic flow initialization, feature extraction, dataset creation, intrusion detection, and mitigation strategies. During the initialization phase, the system collects traffic flow data from SDN switches, configures monitoring parameters, and establishes baseline profiles. Feature extraction entails the selection of pertinent characteristics from the data, followed by transformation and encoding for subsequent analysis. The dataset is then formed, with each flow entry categorized as normal or abnormal based on observed behavior. The detection module utilizes classification methodologies such as KNN and Random Forest to pinpoint intrusions, leveraging flow-based detection for its scalability and operational efficiency. In response to identified threats, the mitigation module implements proactive security measures and dynamic response strategies. Non-functional requirements stipulate detection and analysis timeframes, ensuring prompt responses to security incidents. In a comparative analysis, Random Forest exhibits superior accuracy compared to KNN, showcasing its efficacy in handling intricate datasets and delivering precise predictions. Overall, the proposed system integrates diverse components and methodologies to enhance network security and effectively mitigate DDoS attacks. The technological and operational issues got resolved to achieve high performance, including improved DDoS attack detection and mitigation using network anomaly detection. The suggested work makes it clear that the systems' efficacy and dependability have improved.

Keywords: Traffic flow initialization · Feature extraction · Dataset creation · Intrusion detection · Mitigation strategies · SDN switches · Monitoring parameters · Baseline profiles

1 Introduction

Network-based anomaly detection is pivotal in cybersecurity, aiming to identify irregular patterns in network traffic indicative of potential security threats. As organizations increasingly rely on digital networks, robust anomaly detection systems become imperative for safeguarding sensitive assets. This process entails continuous real-time monitoring and analysis of network traffic to establish a baseline of normal behavior. Any deviations from this baseline, such as sudden spikes in traffic or unauthorized access attempts, may signal security breaches. Techniques like statistical analysis, machine learning, and behavioral modeling are employed to automatically detect and respond to these anomalies, thereby enhancing overall cybersecurity measures. The pervasive integration of network-based services and the escalating dependence on digital infrastructure have ushered in a new era of information accessibility, dissemination, and processing. However, this transformation has also introduced a host of cyber threats, with Distributed Denial of Service (DDoS) attacks being a primary concern. These attacks, aimed at disrupting network availability by flooding it with malicious traffic, pose significant risks, including financial losses, reputational damage, and threats to public safety. Traditional methods of DDoS detection and mitigation often fall short in addressing the dynamic and expansive nature of modern attacks. To counter these challenges, this project proposes a novel approach to network-based anomaly detection and mitigation, specifically targeting DDoS threats. By leveraging machine learning techniques, the proposed system aims to swiftly and accurately identify anomalous network behavior indicative of DDoS attacks in real-time. Proactive measures are then applied to mitigate their impact, safeguarding network resources and ensuring uninterrupted service availability.

2 Literature Review

A Novel Approach for Detection of DoS/DDoS Attack [1] in Network Environment using Ensemble Machine Learning Model, introduce an innovative method for identifying DoS/DDoS attacks in network environments. Utilizing an ensemble machine learning model, they analyze the "mynetwork.csv" custom dataset. However, a potential drawback noted is the risk of overfitting due to exceptionally high accuracy rates. Additionally, the study does not incorporate a deep learning model, suggesting avenues for future exploration in enhancing detection methodologies.

Machine Learning Based Classification Model [2] for Network Traffic Anomaly Detection focusing on network traffic anomaly detection through a machine learning-based classification model. Presented at CICDdos 2019, their study utilized Naive Bayes, KNN, and SVM algorithms. However, the model's scope was restricted to identifying low-rate DDoS attacks. Despite this limitation, their research represents a significant step forward in the development of techniques for detecting and mitigating network anomalies, laying the groundwork for future advancements in the field.

Anomaly based real time prevention of Distributed Denial of Service attacks [3] on the web using machine learning approach devised a real-time prevention system aimed at countering Distributed Denial of Service (DDoS) attacks on web platforms, leveraging an anomaly-based approach powered by machine learning. Their methodology, which harnessed the K-Means Algorithm alongside the Microsoft Kaggle's Malware Prediction dataset, demonstrated notable efficacy, particularly in addressing Application DDoS attacks. While excelling in preemptive measures, the system fell short in its ability to actively mitigate attacks once they occurred. Nevertheless, their research constitutes a significant contribution to the ongoing endeavors in crafting proactive solutions to combat DDoS threats. It offers valuable insights into potential enhancements for real-time attack response strategies, underscoring the importance of continuous refinement and innovation in the realm of cybersecurity to safeguard digital infrastructures against evolving threats.

Improving Transferability of Network Intrusion Detection [4] in a Federated Learning Setup. The research titled "Improving Transferability of Network Intrusion Detection in a Federated Learning Setup", geared towards augmenting the transferability of deep learning models for network intrusion detection within a federated learning framework. By leveraging the comprehensive CIC-IDS dataset, the study meticulously explored various federated learning techniques to enhance model performance. However, while the research showcased promising advancements in transferability, it lacked the capability to actively counteract attacks post-occurrence. Furthermore, a note of caution was sounded regarding the potential risks associated with very high accuracy rates, which could potentially signal overfitting issues within the model. Notably, the study abstained from utilizing a deep learning model in its methodology, suggesting avenues for future research to explore the integration of deep learning techniques to further enhance the efficacy and robustness of intrusion detection systems within federated learning environments.

Effective and Efficient DDoS Attack Detection Using Deep Learning Algorithm [5], Multi-Layer Perceptron explored innovative approaches to detecting DDoS attacks, employing the MLP deep learning algorithm, K-Means Algorithm, and Extended CIC dataset. Their study highlighted deep learning's efficacy in identifying Application DDoS attacks, enhancing network security. However, while proficient in preemptive detection, the method lacked post-attack mitigation capabilities, underscoring the challenge of real-time response to cyber threats. Nonetheless, their research offers valuable insights, driving proactive cybersecurity advancements.

Anomaly Detection in IoT Networks [6]: A Review and Comparative Analysis delved into the realm of "Advances in Deep Learning for Anomaly Detection in Networks," harnessing the power of deep learning algorithms. Their investigation centered on employing the NSL-KDD dataset for network anomaly detection. However, despite its efficacy, the approach encountered challenges related to scalability, particularly in the context of large-scale networks. This limitation underscores the importance of further research and development efforts aimed at enhancing the scalability of deep learning-based anomaly detection systems to effectively address the evolving landscape of network security threats across diverse and expansive network infrastructures.

Advances in Deep Learning for Anomaly Detection in Networks [7], conducted a comprehensive methodology review and comparative study focusing on IoT network traffic traces. While their research provides valuable insights into network traffic analysis, a notable drawback is the limited attention given to emerging IoT threats. By expanding the project paper to address this limitation, researchers can enhance the relevance and applicability of their findings in the rapidly evolving landscape of IoT security. Incorporating a deeper analysis of emerging threats and vulnerabilities specific to IoT devices and networks would strengthen the research's contribution to mitigating cybersecurity risks in IoT environments.

A Comparative Study of Machine Learning Techniques [8] for Network Anomaly Detection In their study titled "A Comparative Study of Machine Learning Techniques for Network Anomaly Detection," undertook a comprehensive comparative analysis of various machine learning techniques. Employing the UNSW-NB15 dataset for network anomaly detection, they meticulously evaluated the performance of different models. However, their findings highlighted a notable deficiency in the diversity of the assessed models. This observation underscores the importance of incorporating a wider range of machine learning algorithms and methodologies in future research endeavors to ensure a more comprehensive understanding of their effectiveness and applicability in addressing network security challenges.

Detecting DDoS Attacks Using Machine Learning Techniques [9] in Cloud Computing Environments. Their methodology employed semi-supervised learning and cloud traffic analysis techniques. They utilized the CloudSim simulation platform along with a custom cloud dataset for their research. Evaluation metrics included the detection rate and false positive rate. The performance of their approach was noted to adapt well to dynamic cloud environments. However, a drawback mentioned was its limitation to specific cloud configurations.

DDoS attack detection in cloud computing [10] environments by employing feature engineering and flow-based analysis. Utilizing datasets from AWS CloudTrail and NetFlow, their method exhibited improved detection accuracy compared to traditional methods. Yet, its drawback lay in its dependency on accurate feature engineering, which could pose challenges in scenarios with evolving attack patterns or diverse network configurations (Fig. 1).

3 Methodology

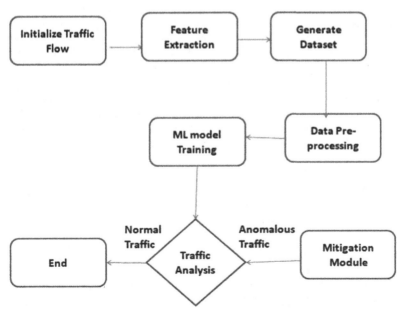

Fig. 1. Architecture Diagram

3.1 Initialize Traffic Flow

3.1.1 Data Collection

This module is responsible for initiating the process of collecting traffic flow data from the flow tables of Software-Defined Networking (SDN) switches within the network infrastructure. It establishes communication channels with the SDN controller to retrieve flow statistics using the OpenFlow protocol.

3.1.2 Initialization of Monitoring Parameters

The module configures filters to exclude harmless traffic, allowing focus on potential threats. Setting thresholds and defining metrics ensures accurate anomaly detection. Proactive monitoring aids in timely security incident response, bolstering network security. Proper parameter initialization establishes a robust traffic monitoring system, enhancing cyber threat defense.

3.1.3 Establishment of Baselines

In part of the initialization process, this module may establish baseline profiles of normal network behavior based on historical traffic data or predefined models. These baselines serve as reference points for identifying deviations that may signify anomalous activity, such as DDoS attacks.

3.2 Feature Extraction

3.2.1 Input Data

One of the most important parts of the network-based anomaly detection system is the feature extraction module, which processes the raw traffic flow data that is gathered from SDN switches. The flow collector module collected this data, which includes a wealth of details about network traffic flows, such as protocol, ports, source and destination IP addresses MAC addresses, packet sizes, timestamps, and other relevant information. The feature extraction module creates a thorough snapshot of network activity by consuming this raw data, setting the stage for further analysis and anomaly identification.

3.2.2 Feature Generation

The feature extraction module carefully chooses the most relevant characteristics from the given dataset in order to achieve effective anomaly detection. This entails a careful feature selection procedure in which information that is redundant or useless is eliminated in order to concentrate only on elements that are important for detecting DDoS attacks. A vast range of factors, including as packet rates, payload sizes, communication patterns, temporal information, and different statistical measurements, can be included in the features selected for extraction. The module's ability to prioritize the selection of pertinent features speeds up the analysis process that follows, improving the system's capacity to identify and counteract DDoS attacks effectively.

3.2.3 Selection of Relevant Features

The ESP32 WROOM microcontroller and environmental sensors, such as the MQ-137 for ammonia and MQ-7 for carbon monoxide, are used in real-time data processing by the alert system to improve efficiency. With the use of an LED, it monitors pollution levels in relation to health standards and issues color-coded air quality warnings (from moderate to dangerous), urging users to take immediate precautionary measures like turning on air purifiers or packing to evacuate. Furthermore, by integrating the system with mobile apps and home automation systems, users will be able to better regulate indoor air quality as notifications and updates on air quality are sent to them directly.

3.3 Generate Dataset

3.3.1 Data Collection

In The first step in generating the dataset is to collect traffic flow data from the flow tables of SDN switches. This data is collected by the flow collector module and typically includes information such as source and destination IP addresses, MAC addresses, ports, protocols, packet sizes, timestamps, and other relevant attributes of network traffic flows.

3.3.2 Feature Extraction

Once the raw traffic flow data is collected, the feature extraction module processes the data to extract relevant features that capture essential characteristics of network traffic. These features may include packet rates, payload sizes, communication patterns, temporal information, and other statistical measures.

3.3.3 Labeling

Each flow entry in the dataset is labeled as either "normal" or "abnormal" based on its behavior. Anomalous behavior indicative of DDoS attacks is labeled as "abnormal," while normal traffic behavior is labeled as "normal." Labeling is essential for supervised machine learning algorithms, as it provides the ground truth for training the model to distinguish between normal and abnormal network traffic.

3.4 Data Preprocessing

3.4.1 Data Transformation

Data transformation involves converting the raw data into a more suitable format for analysis or modeling. This may include scaling numerical features to a common range, transforming variables to meet the assumptions of statistical tests, or encoding categorical variables into numerical representations.

3.4.2 Normalization

Data Normalization and standardization are techniques used to rescale numerical features to a common scale. Normalization scales the features to a range between 0 and 1, while standardization scales the features to have a mean of 0 and a standard deviation of 1. These techniques ensure that all features.

3.4.3 Encoding Categorical Variables

Categorical variables represent qualitative data and need to be encoded into numerical representations for analysis or modeling. This can be done using techniques such as one-hot encoding, label encoding, or binary encoding, depending on the nature of the categorical variables.

3.4.4 Data Splitting

Finally, the preprocessed data is typically divided into training, validation, and testing sets. The training set is used to train the model, the validation set is used to tune hyper parameters and evaluate model performance during training, and the testing set is used to assess the final performance of the trained model on unseen data.

3.5 Detection Module

3.5.1 Classification of Intrusion Detection

To identify malicious activity or policy violations, intrusion detection entails monitoring network traffic or system operations. The source of the data being examined (such as network flows or individual packets) and the selected detection strategy are two important criteria that determine how intrusion detection technologies are classified.

- **Information Source**: One method of detecting intrusions is by examining individual packets or network flows. Packet-based detection looks closely at each individual packet, whereas flow-based detection analyzes aggregated traffic flows between network endpoints.
- **Detection Method**: Signature-based and anomaly-based are the two main types of detection techniques.
- **Signature-based Detection**: In this technique, patterns in system activity or network traffic are detected and compared to a database of known threat signatures. It may have trouble spotting new or invisible attacks, but it is good at spotting known threats (Fig. 2).

2. **Anomaly-based Detection:** This method searches for departures from typical behavioral patterns. It creates a reference point for typical behavior and marks any departures from it as possible intrusions. Anomaly-based detection frequently makes use of machine learning techniques to automatically detect these Discrepancies (Fig. 3).

3.6 Mitigation Module

An essential part of the architecture for network security, the mitigation module is made to react quickly and efficiently to malicious behaviour that is identified. Its main goal is to lessen the effects of cyberattacks and stop additional damage from occurring to the network, its components, and the data it contains. Now let's explore the Mitigation module in more detail:

Network Anomaly Detection Mitigation of DDoS Attack 87

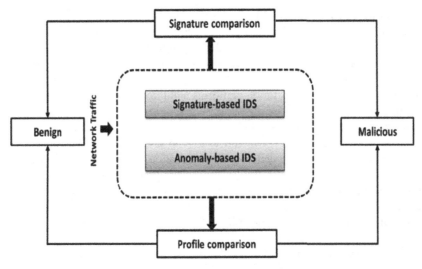

Fig. 2. Signature Based Detection

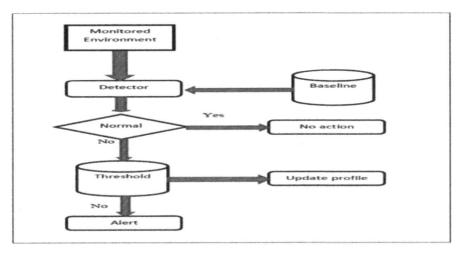

Fig. 3. Anomaly Detection

- **Goal and Range:** The mitigation module collaborates with other modules, including intrusion detection, prevention, and response systems, to function as a part of an all-encompassing.cybersecurity.architecture. The aim of this system is to promptly detect and eliminate any potential risks, hence reducing the likelihood of network outages, data.breaches,or.illegal.access. The scope of the module includes a variety of attack types, such as malware infections, insider threats, DDoS (Distributed Denial of Service) attacks, and vulnerability exploitation.

- **Workflow and Integration:** When suspect or malicious activity is discovered, the mitigation module easily interfaces with detection systems to receive alerts and notifications. The module responds to an alert by initiating predefined mitigation procedures that are customized to the particular threat and the security policies of the network. By providing real-time threat data and automating response activities, integration with security orchestration platforms and threat intelligence feeds improves the module's capabilities.
- **Proactive Security Measures:** One of the main features of the mitigation module is its proactive security strategy, which tries to stop assaults before they have a major negative impact. Implementing network segmentation, vulnerability patching, access controls, and security best practices like the least privilege and least astonishment principles are examples of proactive methods. The module lowers the attack surface and increases the network's resistance to cyberattacks by proactively correcting potential security flaws and vulnerabilities.

3.7 Comparative Analysis: K-Nearest Neighbours (KNN) vs. Random Forest

3.7.1 KNN (K-Nearest Neighbours)

$$d(x, y) = \sqrt{\sum_{i=1}^{n}(xi - yi)^2} \qquad (1)$$

Relevance of Output: KNN's accuracy of 98% suggests that it works well for basic categorization jobs. Due to its dependency on distance computations for predictions, it may perform worse with huge feature spaces and complex datasets.

3.7.2 Random Forest

$$\text{MSE} = 1/N \sum_{i=1}^{N}(yi - \hat{y}i)^2 \qquad (2)$$

N-Sample size, Yi -Actual value, ŶI -Predicted value.

Output Relevance: With a better accuracy of 99% Random Forest demonstrated its resilience in managing intricate datasets with high-dimensional feature spaces and non-linear correlations. It works very well at recognizing complex patterns and producing precise forecasts categorization jobs. Due to its dependency on distance computations for predictions (Table 1).

Table 1. Acuracy of two models

Classifier	Accuracy(%)
KNN	98
Random Forest	99

Network Anomaly Detection Mitigation of DDoS Attack 89

```
*** Creating network
*** Adding controller
*** Adding hosts:
h1 h2 h3 h4 h5 h6 h7 h8 h9 h10 h11 h12 h13 h14 h15 h16 h17 h18
*** Adding switches:
s1 s2 s3 s4 s5 s6
*** Adding links:
(h1, s1) (h2, s1) (h3, s1) (h4, s2) (h5, s2) (h6, s2) (h7, s3) (h8, s3) (h9, s3) (h10, s4) (h11, s4)
 (h12, s4) (h13, s5) (h14, s5) (h15, s5) (h16, s6) (h17, s6) (h18, s6) (s1, s2) (s2, s3) (s3, s4) (s
4, s5) (s5, s6)
*** Configuring hosts
h1 h2 h3 h4 h5 h6 h7 h8 h9 h10 h11 h12 h13 h14 h15 h16 h17 h18
*** Starting controller
c0
*** Starting 6 switches
s1 s2 s3 s4 s5 s6
```

Fig. 4. Network Topology

4 Result

Network Structure: Outline devices and connections. Device Types: Identify switches, routers, hosts, etc. Connections: Describe interconnections briefly. Layout: Specify physical/logical arrangement. Segments: Highlight subnetworks. Traffic Flow: Mention traffic direction. Bandwidth: Note bandwidth allocations. Redundancy: Include redundant paths/devices. Scale: (Fig. 4).

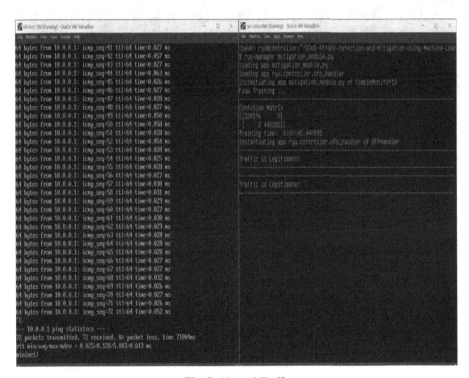

Fig. 5. Normal Traffic

Regular Patterns: Consistent data flow within expected parameters Authorized Users: Traffic generated by authorized users or devices. Typical Protocols: Usage of standard protocols like TCP/IP, HTTP, etc. Expected Volume: Traffic volume within normal operational levels. No Anomalies: Absence of unusual or suspicious activity. Smooth (Fig. 5).

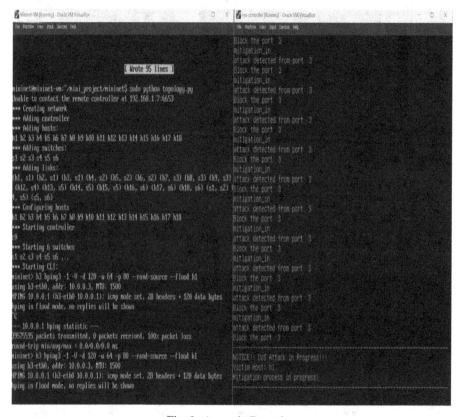

Fig. 6. Anomaly Detection

Abnormal Traffic Patterns: Anomaly detection identifies deviations from normal network behaviour. Unusual Traffic Volume: Detects sudden spikes or unusually high traffic levels. Atypical Protocol Usage: Identifies abnormal usage of network protocols (e.g., SYN flood) (Fig. 6).

System Comparison with Existing Model

After designing the attack detection and mitigation method, rigorous testing and evaluation were conducted to assess its effectiveness. The evaluation process involved analyzing the performance of various machine learning (ML) algorithms and comparing different models used within the framework. The table below presents the accuracies of the different models (Table 2):

Table 2. Comparison of Models

Model	Accuracy (%)
Random Forest	98
SVM	88
KNN	82

From the table, it's evident that the random forest model achieves the highest accuracy in categorizing normal and malicious traffic. To evaluate the performance of the model, a test was conducted involving online traffic classification and attack mitigation. For mitigating attacks.

5 Conclusion

In this research endeavor, a sophisticated approach utilizing a random forest machine learning algorithm is devised to construct a model capable of autonomously identifying and mitigating Distributed Denial of Service (DDoS) attacks within Software-Defined Networking (SDN) environments. The model operates by routinely gathering all traffic flow entries, meticulously extracting intrinsic flow features, and augmenting them with additional relevant features. Employing a detection module endowed with five distinct criteria, each flow is meticulously categorized as either normal or anomalous. Upon detecting an attack, immediate action is taken to prevent its source from inflicting further harm.

6 Future Work

Enhanced Feature Extraction: Investigate advanced techniques for feature extraction from network traffic data. This could involve the exploration of deep learning architectures, such as convolutional neural networks (CNNs) or recurrent neural networks (RNNs), to automatically learn hierarchical representations of network traffic patterns. Dynamic Model Adaptation: Develop adaptive machine learning models that can dynamically adjust their parameters and decision boundaries in response to changing network conditions and evolving attack strategies. This could involve the use of reinforcement learning techniques to enable models to continuously learn and improve over time.

References

1. Rao, G.S., Subbarao, P.K.: A novel approach for detection of DoS/DDoS attack in network environment using ensemble machine learning model. Int. J. Recent Innov. Trends Comput. Commun. **11**(9), 244–253 (2023)
2. Kumar, S., Viinikainen, A., Hamalainen, T.: Machine learning classification model for network based intrusion detection system. In: 2016 11th International Conference for Internet Technology and Secured Transactions (ICITST), pp. 242–249. IEEE, December 2016

3. Kishore, P.K., Ramamoorthy, S., Rajavarman, V.N.: ARTP: anomaly based real time prevention of Distributed Denial of Service attacks on the web using machine learning approach. Int. J. Intell. Netw. **4**, 38–45 (2023)
4. Ghosh, S., Jameel, A.S.M.M., Gamal, A.E.: A study on transferability of deep learning models for network intrusion detection. arXiv preprint arXiv:2312.11550 (2023)
5. Ahmed, S., et al.: Effective and efficient DDoS attack detection using deep learning algorithm, multi-layer perceptron. Future Internet **15**(2), 76 (2023)
6. Osho, O., Hong, S.: A survey paper on machine learning approaches to intrusion detection. Int. J. Eng. Res. Technol. (IJERT) **10**, 94–102 (2021)
7. Ade, J.V.: Ensemble learning methods for DDoS attack detection in cloud environments: a comprehensive review (2024)
8. Mihoub, A., Fredj, O.B., Cheikhrouhou, O., Derhab, A., Krichen, M.: Denial of service attack detection and mitigation for internet of things using looking-back-enabled machine learning techniques. Comput. Electr. Eng. **98**, 107716 (2022)
9. Musa, N.S., Mirza, N.M., Rafique, S.H., Abdallah, A., Murugan, T.: Machine learning and deep learning techniques for distributed denial of service anomaly detection in software defined networks-current research solutions. IEEE Access (2024)
10. El Sayed, M.S., Le-Khac, N.A., Azer, M.A., Jurcut, A.D.: A flow-based anomaly detection approach with feature selection method against DDoS attacks in SDNs. IEEE Trans. Cogn. Commun. Netw. **8**(4), 1862–1880 (2022)

Leveraging the Trio of 5G, MIMO, and IoT for Proactive Vehicular Crash Prevention in VANET

S. Bharathi[1(✉)], S. Sathiyapriya[2], K. Nivethika[2], S. Sivachitralakshmi[1], and P. Durgadevi[1]

[1] Department of Computer Science and Engineering, SRM Institute of Science and Technology, Chennai, India
`barla.indian@gmail.com`
[2] Department of Computer Science and Engineering, VelTech Rangarajan Dr. Sagunthala R&D Institute of Science and Technology, Chennai, India

Abstract. Road transport is extensively used worldwide for the transportation of people and many kinds of goods. However, this mode of transportation is prone to accidents and faces several continuous problems, such as the frequent loss of life and property in the case of an accident. To successfully tackle these issues, it is advisable to implement a self-regulating event detection system that makes use of wireless communication, 5G technologies, and the Internet of Things. The Internet of Things (IoT) is a technological advancement that raises the level of connection between machines and people. The online system improves communication between cars and infrastructures, transmits data, and forecasts event happenings using networks such as eCall, OneM2M, and mobile internet access. The advancement of 5G technology is occurring fast in conjunction with the development of Internet of Things (IoT), Artificial Intelligence (AI), and intelligent platforms for essential communications. 5G technology is the most recent iteration of technology that operates on the Ultra High Spectrum Band (UHSB). This design employs a communication system that connects pedestrians, vehicles, roads, and cloud infrastructure to provide reliable and efficient transmission of vehicle locations and temperatures. Intelligent transportation systems depend on the sharing of information and the ability to forecast incidents in order to prioritize safety. This study focuses on the detection of accidents utilizing 5G networks, integrated mobile broadband, and MIMO wireless systems. Lastly, we will address modern technology, challenges, and ongoing and potential research opportunities.

Keywords: 5G · IOT · MIMO · Intelligent transportation systems (ITS) · VANET

1 Introduction

Road transportation systems are used worldwide. The benefit of this in the 21st century stems from the culmination of technological breakthroughs and inventions. If successfully used by professionals, this would considerably enhance the process of identifying accidents.

Drivers and pedestrians who use the route. The research described in [1] utilizes an ultrasonic detector for the purpose of detecting traffic accidents, whereas other investigations rely on the infrared sensor of a smartphone. Various modern techniques exist for the automatic identification of traffic incidents and are now being used. Scientific study has several ideas for different automated accident alert systems. These include various machine learning techniques, mobile applications, the Global System for Mobile Communications (GSM), the Global Positioning System (GPS), vehicle ad hoc networks, and other technologies. Optical wireless communication (OWC), which operates independently of radio frequency (RF)-based wireless technology, may be a feasible choice for upcoming communication networks like 5G and 6G systems. A variety of wireless technologies are accessible for communication between vehicles and vehicles (V2V and V2l). Among these, dedicated short range communication (DSRC) is the most widely used technology. DSRC is utilized for medium to short range (up to 1000 m) communication, as its name implies. For DSRC, the Frequency Communications Commission (FCC) has set aside 75 MHz of bandwidth at a frequency of 5.9 GHz. Additionally, the 75 MHz bandwidth is split up into seven 10 MHz channels. One of these channels is a control channel that is only used for safety, while the other six are operational frequencies.

Vehicular adhoc networks (VANET) and multiple-input multiple-output (MIMO) are two distinct technologies that have been presented and investigated by separate, mostly distinct research communities. On the one hand, the fields of information theory and wireless communications have led the way in MIMO research, with the majority of their work focusing on point-to-point links. Recently, the issue of multi-user and mobile MIMO has come to light [1–3], although VANETs and their distinct problems and use cases have not received any special attention. However, a combined effort from several groups, including the wireless communications and networking, mobile computing, and automotive research communities, has contributed to the advancement of VANET research. This is explained by the fact that it is inherently multidisciplinary, bringing mobile computing and upcoming wireless network technologies directly to the needs of developing automotive applications.

The solutions offered by this technology are unmatched in the 5G and other related industries because to its extensive coverage, high-speed data transmission, and minimal latency. The Internet of Things (IoT) is a cutting-edge network technology that allows for the connecting of all items, as explained in reference [3].

This study investigates the detection of road accidents using wireless technologies, namely 5G and the Internet of Things. The study proposes innovative techniques for the swift identification of eCalls, OneM2M, and Unmanned Aerial Vehicle using the 5G ultra high spectrum band (UHSB) and Multiple Input and Multiple Output (MIMO) technology. The primary contribution of this work is to present a new approach for utilizing technology to identify and promptly inform the relevant authorities about accidents. The subsequent section of this article is organized in the following manner, Sect. 2 provides an overview of the research that are relevant to the topic. Section 3 analyzes the incidence of incidents across different IoT networks. Section 4 demonstrates the identification of accident incidents using 5G technology. The Conclusion is presented in Sect. 5.

2 Literature Review

The transport business places the highest value on safety and recognizes the significant risk of road accidents. Therefore, it is crucial to utilize technology to effectively mitigate these issues. Developing automated methods for detecting traffic accidents and reducing the response time from an accident occurrence and the arrival of first responders are two efficient strategies for decreasing the fatality rate in traffic incidents. Current approaches utilize the inherent automated accident detection and reporting mechanism included inside cars [4]. The primary significance of built-in automated accident detectors in automobiles is their independence from human intervention in triggering alerts to the appropriate authorities. The system's functionality must be of high quality, built and installed by default in the vehicle, and undergo regular servicing and maintenance. This ensures that the system operates effectively, promptly, and accurately in terms of timing and location, as depicted in Fig. 1. In the event of an accident, the system will detect it and transmit a notification.

Another research in [5] indicated that speed is the primary factor contributing to accidents. Speed is well recognized as a primary factor contributing to automobile accidents. The Internet of Things (IoT) design serves as a system capable of promptly detecting accidents and transmitting the relevant information to both the first aid center and safety professionals. Timely access to accident information might have potentially resulted in the preservation of several lives by enabling emergency workers to respond more effectively.

An application was launched in [6] that used an accelerometer to identify risky driving and trigger vehicle alarms. Following an accident or immediately thereafter, it is utilized as a vehicle crash or rollover detection system. The accelerometer receives the signal that is utilized to determine the severe accident. The vibration sensor in this study will identify the signal and transmit it to the ATMEGA 8A controller whenever a vehicle experiences an accident or flips over. Given the continuous advancements in technology, it is imperative to revamp emergency services by incorporating state-of-the-art technologies [7].

The drawback of this method is its high maintenance cost, which is necessary for optimal efficiency. Wireless technology will enhance the detection and emergency services.

The collision avoidance system is a wireless communication system that can be used to detect road accidents and aid in rescue operations. In a study referenced as [8], various literature on collision avoidance techniques was examined, with a specific focus on communication challenges within the realm of automotive cyber-physical systems. The study proposes the use of affective computing and affective emotions, along with a novel taxonomy, to better understand the principles of Vehicle Collision Prevention Systems (VCPS).

The importance of accident identification cannot be overstated, since safety is the paramount concern of the industry. The utilization of wireless communications will enhance the promptness and precision in detecting accidents. Table 1 provides an overview of various studies that have been conducted in the field of literature.

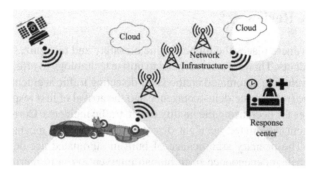

Fig. 1. Accident detection and Reporting

Table 1. Analysis of Accident detection in various Literature

Ref	Accident Exposure	Restrictions	Evaluation metric	Simulators
[1]	Ultrasonic detector	Constraints imposed by CPUs on different devices	Demonstrates optimal performance in diverse urban environments and various nature settings	Infrared sensors (IR sensors), smartphones, airbags, and mobile apps
[6]	Utilizing a motion sensor for an automobile alarm system	Experienced internet connectivity issues and loss of GSM signal due to an accident	Enhanced acoustic sensor with improved sensitivity and precision	GSM texting. GPS and an MCU ATmega 8A
[9]	Video recordings captured by surveillance cameras installed on roadways and highways Indicate the potential occurrence of a vehicular collision	The approach is unable to identify automobiles that are extensively damaged to the point where none of the relevant vehicle parts are available	Demonstrates promising outcomes when assessed on two distinct datasets of impaired vehicles, which differ according to of the picture's resolution, its distance from the camera, and the number of objects present in the image	Damaged Cars Dataset-1 (DCD-1) and Damaged Cars Dataset-2 (DCD-2)

(*continued*)

Table 1. (*continued*)

Ref	Accident Exposure	Restrictions	Evaluation metric	Simulators
[10]	IoT system model-GMM and CART	Dependent on continuous Internet connectivity	Enhanced in terms of both recall and accuracy	Acceleration, overturn, detachment, overturn, combustion, crash site, and velocity
[11]	Intelligent traffic light system	The delay in the queue-based method of GSM messaging	Completely motorized	GSM messaging
[12]	Accelerometer designed for tiny electromechanical devices using an ultrasonic sensor	Not universally applicable	Take prompt and expeditious measures to preserve the life of the sufferer	Several communication protocols, such as cellular, LoRaWAN, and NB-IoT, are available

3 Detecting Incidents Across Various IoT Networks

The Internet of Things (IoT) has seen significant advancements over the past century in order to efficiently identify road accidents. The primary objective of utilizing IoT is to establish a precise connection with first responders in the event of an accident. Upon detecting an accident, the Internet of Things (IoT) will promptly activate a network of interconnected devices, whether it be between humans and machines or between machines themselves. The authors assert that the Internet of Things (IoT) is a network of interconnected computing equipment, objects, animals, or individuals that are assigned unique IDs and has the ability to transmit data across a network without the need for human-to-human or human-to-computer contact. A network is comprised of tangible components that have the ability to gather and exchange digital data.

The IoT architecture facilitates the rapid identification and alerting of automobile collisions. This can be achieved by integrating sophisticated sensors with an embedded microcontroller within the vehicle. The microcontroller can be programmed to detect accidents promptly and trigger an immediate response by alerting emergency personnel. By incorporating these functionalities, the system can effectively address a significant number of accident scenarios. The text is referenced by the number 13.

In their paper, the author discussed the existence of several systems and networks that may be employed for early detection and prompt reactions in the case of an accident. These systems utilize wireless technology, Internet of Things (IoT), and 5G to detect accidents and subsequently minimize their occurrence. Hence, the process by which IoT devices share data entails pre-processing, which includes modifying the structure of records and categorizing them. This information is then communicated to other stations

3.1 Electronic Call (E-call)

The main purpose of an eCall system is to automatically identify automotive accidents and promptly tell emergency services. This technology plays a crucial role in saving lives and reducing the severity of casualties by providing essential information to responders and significantly reducing their response time [15].

Figure 2 demonstrates that in the event of a severe traffic incident, eCall is an emergency call that may be initiated either manually by car passengers or automatically through the activation of sensors within the vehicle. When activated, the in-vehicle eCall system establishes a direct voice connection with the relevant Public Safety Answering Point (PSAP) via the 112 emergency number [16].

During a crash, the eCall system, which is a call system installed in vehicles, establishes a communication channel using GSM/CDMA technology. The Vehicular Network then activates the eCall System. An accident can be determined by analyzing the activation of the airbag or the initiation of the fuel pump shutdown [17].

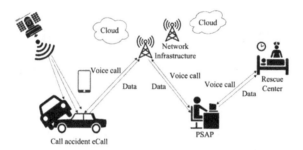

Fig. 2. Electronic call for Accident detection

Separate research elucidated the development of an advanced algorithm designed to accurately identify the kind and location of accidents. A distress call is made to the operator from the vehicle after a collision. If the occupants of the car fail to answer, the operator has the ability to retrieve an SMS including the information described in the previously described sections. The operator can promptly dispatch the necessary rescue troops based on the gathered information. Short Message Service (SMS) has the capability to incorporate other information beyond the data collected by the sensor, such as the proprietor's name, the vehicle's make and model, and the number of the license plate.

3.2 One M2M

Machine-to-machine (M2M) networking and the usage of massive data generated by M2M communication can significantly improve road safety [18]. It is important to highlight that IoT may be included to monitor alarms for automatic accident detection. As per the planned system, it is mandatory for every vehicle to have an M2M device installed in its on-board unit. This unit is responsible for the identification and assessment of accidents, as well as the utilization of wireless technologies to transmit accident data to an M2M web application. In addition, the utilization of lightweight M2M standards protocols enables the M2M server to allocate the necessary resources for the rescue effort and distribute the information to other M2M devices, therefore increasing the chances of survival and providing additional resources to the victims [19]. Nevertheless, human factors significantly influence the causation of road accidents.

The primary factors contributing to this issue encompass driving ailments such as weariness, drowsiness, and other persistent problems. The development of IoT technology aimed to tackle these issues.

This framework model pertains to the Internet of Things (IoT) standard known as oneM2M. Hence, it has the potential to establish communication with additional applications or systems in order to identify accidents, and suitable measures will be taken to notify the relevant authorities, including rescue agencies and medical institutions [20].

3.3 Unmanned Aerial Vehicle

Real-time surveillance of crashes is a crucial factor in road transportation. The utilization of unmanned aerial vehicles (UAVs), such as drones, will be highly advantageous in emergency services, including firefighting and other urgent situations. The integration of artificial intelligence (AI) with 5G and 6G wireless networks will enable the immediate identification of accidents.

Accident detection equipment in [7] promptly provide an emergency alarm upon sensing an accident. Finally, the drone detects the signal and accurately determines the exact location of its target. The mistake may range from little to significant. Nevertheless, prompt emergency rescue aid is necessary. Drones will consequently reach the designated destination promptly and distribute medical aid. Additionally, it will broadcast video to control centers and medical institutions, as seen in Fig. 3.

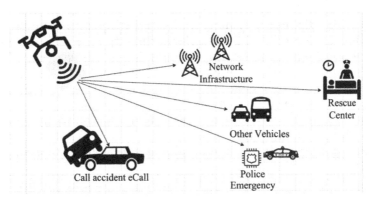

Fig. 3. UAV to transmit to the control centre and facilities.

4 5G-Enabled Accident Detection

The 5th generation technology is capable of efficiently detecting accidents and facilitating effective communication. The authors in [21] suggested that 5G connectivity is revolutionizing road transportation by providing exceptional communication speed, hence altering the idea of networking. Utilizing 5G technology is crucial in reducing the significant number of fatalities and catastrophic injuries resulting from accidents.

4.1 Integration of Mobile Broadband

In their study, the authors focused on forecasting the development of 5G technology and the major obstacles that lie ahead for mobile networks. They also examined the challenges faced by mobile network providers in managing the increasing traffic demand and detecting accidents in mobile broadband services in the future. Additionally, it showcases that 5G would provide a latency-free access data rate comparable to that of fiber optic connections.

Despite the existence of several accident detection technology, the mortality rate continues to increase. The delayed response to catastrophic accidents can be attributed to imprecise accident detection methods and inadequate communication procedures. Traffic management systems commonly utilize either automated or non-automatic approaches to detect issues.

Automated incident detection (AID) processes, including a range of sensors and techniques such as image processing, are commonly employed. In addition, research is conducted on vehicle probes that analyze traffic data to quickly identify events and utilize algorithms based on detectors and sensors for event detection [23].

The advent of 5G network technology will introduce a new era in mobile wireless communication. The terminal must possess the capability to combine various streams from diverse wireless technologies, as 5G devices will have the ability to concurrently connect to many wireless technologies [24].

IMT-Advanced (IMT-A) systems are already being utilized, and research on 5th-Generation (5G) mobile communication technology is presently underway. Additionally, 5G will have the capability to optimize services and enhance user awareness. It will significantly enhance energy and cost efficiency, surpassing current levels by more than a hundred times. This advancement will enable us to fully achieve the goal of 5G, which is to have information readily accessible and everything connected at our fingertips.

Smart transportation is a transport system that utilizes a network infrastructure to enhance productivity, safety, and efficiency. 5G, in conjunction with AI algorithms, is a key technology that enables smart transportation systems. Within a specialized transportation network that prioritizes safety and efficiency, linked automobiles and Roadside Units (RSUs) will offer accident avoidance and route optimization services [25].

4.2 Accident Detection Through MIMO (Multiple Input and Multiple Output)

MIMO, or Multiple-Input Multiple-Output, is a crucial technology used for accident detection and facilitating rapid reaction in transmission methods. Multiple antennae are utilized for both transmitting and receiving signals to detect traffic accidents. MIMO technology enables concurrent transmission of data, resulting in a higher data transfer rate. Increasing the number of antennas can enhance both transmission and reception capabilities [24]. Additionally, array antenna technology holds great potential for utilization in 5G systems.

MIMO systems employ multiple antennas to transmit and receive data. Massive MIMO networks are designed specifically to mitigate interference among users and cells, resulting in enhanced throughput and greater reliability [26]. The MIMO arrangement, with an array of antennas, may also be utilized for accident detection.

However, the performance of MIMO arrangement is mostly reliant on the utilization of broad antennas to identify accidents and connect with rescue centers.

The author in [27] suggests that the usage of millimeter-wave (mmWave) massive multiple-input multiple-output (MIMO) technology would have a significant impact on cellular vehicle-to-infrastructure situations, enabling gigabits-per-second (Gbps) communication. The authors analyze the mmWave massive MIMO vehicular channel by characterizing it with metrics such as route loss and root-mean-square delay spread. The concept of a worldwide network of interconnected physical objects, enabling universal connection for all things, regardless of location or time, has generated enthusiasm for the Internet of Things (IoT). In order to manage the substantial surge in traffic associated with 5G, it is necessary to utilize mm-Wave frequencies, deploy dense base stations (small cells in close proximity), and employ advanced antennas. The integration of massive MIMO, millimeter-wave, and dense small networks proved successful in facilitating the advancement of IoT applications on 5G networks. Massive MIMO utilizes an extensive number of antennas to achieve high array gain and reduce total transmit power. This is achieved through disruption reduction and spatial multiplexing, where the antennas use very low transmit power. This strategy effectively lowers the total power used to transmit while increasing the array gain. The integration of IoT applications into 5G networks is facilitated by the presence of dense tiny cells and mm-Wave technology [3].

Additionally, several additional articles have analyzed the complexities around 5G and 4G technologies.

The introduction of machine learning prediction models for route loss modeling of 4G networks was presented in [28–33]. An exact characterization of route loss is crucial for ensuring stability in signal propagation inside 5G networks. To obtain high-quality and dependable signals, it is suitable to utilize 5G technology for identifying accidents in road transport systems.

5 Conclusion

This study has investigated the role of technology in identifying accidents and the effective methods of communicating with the appropriate rescue personnel. This study just examines the Internet of Things (IoT) and 5G, disregarding other connected frameworks put out by several academics. The integration of eCalls, UAV, OneM2M, MIMO, and mobile broadband into the framework will improve the accuracy and efficiency of accident detection and communication, ultimately leading to a decrease in the number of deaths and injuries. While future developments in wireless technology may improve road safety, they are unlikely to entirely reduce the occurrence of accidents and deaths.

Conflicts of Interest. The authors declare no conflicts of interest regarding the publication of this paper.

References

1. Teyeb, I., et al.: Towards a smart car seat design for drowsiness detection based on pressure distribution of the driver's body. In: ICSEA 2016: The Eleventh International Conference on Software Engineering Advances (2016)
2. World Health Organization, Global Status Report on Road Safety 2018: Summary (No. WHO/NMH/NVI/18.20), World Health Organization (2018). https://www.who.int/publications-detail-redirect/9789241565684
3. Zhang, X., et al.: A survey on deep learning based brain computer interface: Recent advances and new frontiers. arXiv preprint arXiv 1905, 04149 (2019)
4. Blankertz, B., et al.: Optimizing spatial filters for robust EEG single-trial analysis. IEEE Signal Process. Mag. **25**(1), 41–56 (2007)
5. Gao, Z., et al.: EEG-based spatio–temporal convolutional neural network for driver fatigue evaluation. IEEE Trans. Neural Networks Learn. Syst. **30**(9), 2755–2763 (2019)
6. Qafzezi, E., et al.: Coordination and management of cloud, fog and edge resources in SDN-VANETs using fuzzy logic: a comparison study for two fuzzy-based systems. Internet Things **100169**, 1–12 (2020). https://doi.org/10.1016/j.iot.2020.100169
7. Qu, X., et al.: Complex network analysis of VANET topology with realistic vehicular traces. IEEE Trans. Veh. Technol. **69**(4), 4426–4438 (2020)
8. SAE International, Taxonomy and definitions for terms related to driving automation systems for on-road motor vehicles. SAE Int. (J3016) **3**(21), 10–12 (2016)
9. Dong, Y., et al.: Driver inattention monitoring system for intelligent vehicles: a review. IEEE Trans. Intell. Transp. Syst. **12**(2), 596–614 (2010)
10. Yao, K.P., et al.: Real-time vision-based driver drowsiness/fatigue detection system. In: 2010 IEEE 71st Vehicular Technology Conference. IEEE (2010)
11. James, B.: Emergency vehicle proximity Alert system, U.S. Patent No. 16/656,622, April 2020. https://patents.google.com/patent/US20060227008/en
12. Gromera, M., et al.: ECG sensor for detection of driver's drowsiness. In: 23rd International Conference on Knowledge-Based and Intelligent Information & Engineering Systems. Elsevier B.V. (2019). https://doi.org/10.1016/j.procs.2019.09.366
13. Bharathi, S., Durgadevi, P.: A comprehensive investigation on role of machine learning in 6G technology. In: Goyal, V., Gupta, M., Mirjalili, S., Trivedi, A. (eds.) Proceedings of International Conference on Communication and Artificial Intelligence. LNNS, vol. 435. Springer, Singapore (2022). https://doi.org/10.1007/978-981-19-0976-4_4
14. Camilleri, J., et al.: Driver Assistance System for Vehicle, U.S. Patent No. 10,623,704 B2, April 2020. https://patents.google.com/patent/US8636393B2/en
15. Bharathi, S., Durgadevi, P.: An improved machine learning algorithm for crash severity and fatality insight in VANET network. In: Hemanth, J., Pelusi, D., Chen, J.I.Z. (eds.) Intelligent Cyber Physical Systems and Internet of Things, ICoICI 2022. Engineering Cyber-Physical Systems and Critical Infrastructures, vol. 3. Springer, Cham (2023). https://doi.org/10.1007/978-3-031-18497-0_50

16. Bharathi, S., Durgadevi, P.: An Intensive Investigation of Vehicular Adhoc Network Simulators. IoT, Cloud and Data Science (2023). https://doi.org/10.4028/p-715gbh
17. Paul, J., Jahan, Z., Lateef, K.F., Islam, M.R., Bakchy, S.C.: Prediction of road accident and severity of Bangladesh applying machine learning techniques. In: IEEE 8th R10 Humanitarian Technology Conference (R10-HTC), pp. 1–6 (2020). https://doi.org/10.1109/R10-HTC 49770.2020.9356987
18. Lee, J., Yang, J.H.: Analysis of driver's EEG given take-over alarm in SAE level 3 automated driving in a simulated environment. Int. J. Automot. Technol. **21**(3), 719–728 (2020)

Envision, Enhanced, Envisage (EEE) IoT Device Prediction Using Neural Network

S. Menaka[✉][iD], Saswat Biswal, Pulibandla Sri Surya Teja, and Mothukuri Koushik

SRM Institute of Science and Technology, Ramapuram, Chennai 600089, India
{menakas1,sr1225}@srmist.edu.in

Abstract. Frequently the cause of security issues, Internet-of-things (IoT) devices stand to gain a great deal from automated management. Strong device identification is necessary for this in order to implement the proper network security measures. To tackle this challenge, we investigate methods for precisely identifying IoT devices by observing their network behavior and utilizing strategies that have been suggested by other researchers. A unified architecture for identifying and managing Internet of Things (IoT) devices. Leveraging deep metric representation learning, our approach analyzes network communication data to automatically identify both known and unauthorized IoT devices. We extract relevant features from network traffic data, capturing patterns specific to different device types. These features are used as input for the deep metric learning model. Our model learns a compact and discriminative representation for each device based on the extracted features. By minimizing intra-class variations and maximizing inter-class separations, it enhances device identification accuracy. The architecture defines a unified decision boundary that accommodates diverse IoT devices. This boundary ensures robustness against variations in device behavior and network conditions. In summary, our unified architecture combines domain knowledge with deep learning techniques, providing an effective solution for IoT device identification. Its scalability and adaptability make it suitable for large-scale IoT deployments. The architecture achieves high accuracy, exceeding 99%, in identifying various types of devices, including traffic from smartphones and computers. By automating the device identification process, our solution contributes to better management and security of IoT networks.

Keywords: IoT Device · Deep Metric Representation · Domain Knowledge

1 Introduction

The Internet of Things (IoT) is characterized by the connection and interaction of 'smart objects' devices or items equipped with embedded sensors, data processing capabilities, and communication systems allowing for the delivery of various applications and services that would not be feasible without it. When sensors, actuators, information, and communication technologies come together to form the Internet of Things, huge volumes of data are produced [1]. To enable somewhat precise control and decision-making, this data must be sorted through. The integration of learning-enabled components is achieved through the use of Deep Learning (DL) and Deep Neural Networks (DNNs) is a common method of implementing smart decision capability in IoT. Using wireless signals to passively identify IoT devices for Physical Layer authentication and Non-cryptographic Device Identification (NDI) is one common use of DNNs in the Internet of Things. Both DL and DNNs are useful for wireless device identification in a variety of contexts. However, when new devices (as new classes) emerge, DNN models in these applications must constantly innovate to accommodate operational changes. The term "lifelong learning" (IL) refers to this type of constantly changing program. Traditional methods need to retrain DNNs on a regular basis. Retraining DNNs on a regular basis is necessary for [2] conventional techniques. DNNs are trained using a combination of historical and current device inputs in this paradigm, after being initiated from the beginning. Non-Incremental Learning (Non-IL) schemes ensure optimal accuracy, but when more devices are added, they can become significantly more memory-intensive and require longer training periods. Consequently, IL that strikes a fair balance between accuracy, memory usage, and training effectiveness is required. IoT devices pose various security challenges, particularly in domestic installations. To address this, it is recommended to allocate minimal or no memory for storing previous data in the ever-growing IoT ecosystem [3]. Active and especially automated administration of these devices might be beneficial in fending off such dangers. But to ensure the correct policies, actions, and updates are applied, automating these operations is essential need effective device identification. Because IoT devices must communicate across a network to function, it is only reasonable in this setting to identify them by analyzing their analyzing network activity at the home router is effective because devices cannot conceal their behavior. From a privacy, scalability, and independence standpoint, conducting network activity analysis locally eliminates reliance on manufacturer-provided cloud services at the home router is reliable. Moreover, the MUD standard is beginning to receive some supporting the reporting of findings from such investigations is crucial. In the rapidly evolving IoT landscape, smart objects IoT devices equipped with sensors, data processing, and communication capabilities interconnect and interact to provide automated services that would not be feasible otherwise. Around 2020, trillions of IOT devices are predicted to join the worldwide network. An extensive variety of innovative applications and services, including those in the fields of health, transportation, energy and utilities, and other fields, are made possible by the Internet of Things, which is becoming a ubiquitous part of daily life [4]. Big data analytics also makes it possible to go from IoT to real-time control. However, as connection grows, there are perils connected with the IoT. For instance, strategies that claim to be rogue IoT devices defined as those that have corrupted genuine devices or claimed a false identity have exposed the IoT to several hazards that might have serious repercussions. Rogue IoT devices can carry out a range of attacks, such as impersonating trusted entities to gain access to sensitive resources, hijacking legitimate devices to participate in distributed denial of service (DDoS) attacks, and more. In wirelessly linked IoT, the issue of rogue devices becomes much more dangerous Since it is easier to intercept, falsify, and spread network communications, network operators view identifying known or unknown devices and detecting compromised ones as the initial step in safeguarding the Internet of Things (IoT) from threats posed by rogue devices [5]. The system's wider implementation is hampered by privacy and security issues, despite its many benefits.

The main contributions of the research are as follows:

Deep Metric Representation Learning: The study introduces a novel approach for Internet of Things device identification, deep metric representation learning can be applied to effectively differentiate and classify devices based on their unique behavioral patterns.

Feature Extraction and Selection: Relevant features are extracted from network communication data, enhancing device discrimination.

Unified Decision Boundary: The architecture accommodates diverse IoT devices, ensuring robustness and scalability.

2 Related Work

In this section, we'll examine DFL and IoT technology as they now stand. The aspirations for their merger are also examined.

2.1 Deep Federated Learning

The decentralized, Privacy-preserving training of Deep Neural Networks (DNNs) is a core focus of the Federated Learning (FL) subfield of Distributed Federated Learning (DFL). In this framework, participating devices such as smartphones and Internet of Things (IoT) devices retain local models and collaborate to develop a shared global model without exposing personal data. A central server orchestrates communication, gathers model updates from devices, and facilitates the training of complex models on large, distributed datasets. This approach is particularly valuable for applications where data sensitivity or decentralization is a concern, such as healthcare, finance, and autonomous systems, as it allows for model training without compromising privacy or security. Figure 2 illustrates the DFL system architecture. To minimize data transmission, each IoT device performs local data processing before sending updates to the central server over a secure and reliable network. The central server then aggregates these updates, using them to refine the global model via federated learning techniques. This method ensures that local model updates from distributed devices are merged to improve the overall model, reducing data transfer and preserving privacy [9]. Without transferring its local data to other clients or the server, each client trains a model. The model parameters are then shared by the clients and the server. All shared parameters must be processed by the server, which then creates a model using those values. Following that, the clients are shown the finished model [11]. When the model is trained, it can be sent back to the IoT devices for local interpretation.

In light of this, IoT devices may produce forecasts in local proximity to their current location without transmitting information to a regional server. The essential server updates the model on a regular basis using the latest information from the IoT devices. Continuous learning is one way to achieve this; here new data are incorporated into the model to make modifications possible.

2.2 Internet of Things

Many different kinds of items and things are expected to be connected to the Internet of Things by taking use of its omnipresent sensing and processing capabilities. This facilitates enterprises' ability to provide their clients services and applications. An important part of DFL is played by IoT devices.

DFL offers a way to train machine learning (ML) models using data produced by Internet of Things (IoT) devices, which are often dispersed and have constrained processing capabilities. These devices can collaborate and share their local models, allowing them to collectively contribute to the training of a global model without revealing any private information. In a healthcare setting, for instance, Internet of Things devices like wearables and home monitoring systems can gather and analyze vast amounts of health data. By employing Distributed Federated Learning (DFL), healthcare organizations can derive valuable insights into patient health while maintaining patient privacy, as models are trained on this data in a manner that preserves confidentiality.

DFL is also a good fit for large-scale Internet of Things systems because of its decentralized structure, which it enables the generation, processing, and updating of data at the control, hence updating the regional model. By doing this, businesses can use the combined intellect of all Internet of Things devices to train models on this data without compromising privacy that work well in real-world scenarios and are robust and scalable.

2.3 DFL's Vision for IoT

The following are a few of the difficulties that prompted the research of DFL:

These are significant and auspicious. The architecture is able to extract complex features from IoT device data by incorporating DFL approaches. This allows the architecture to recognize minute patterns and traits that are essential for precise identification. One ambition is to use DFL to improve the architecture's anomaly detection skills so that it can recognize abnormal behavior suggestive of security breaches or issues with IoT devices. DFL also creates pathways for ongoing learning and modification, enabling the architecture to change in tandem with new threats and IoT technologies. Moreover, DFL enables cross-domain knowledge transfer, meaning that insights from one domain might improve the identification process in another. In the end, this architecture's synergy between DFL and domain knowledge anticipates a time when IoT device identification is not only precise but also resilient, adaptive, and able to handle changing IoT landscape difficulties [12]. Costs associated with maintenance: Centralized machine learning systems need a lot of care and attention, which can be expensive and time-consuming. In order to overcome this issue, DFL offers collaborative learning across a variety of devices.

Problems with scalability: Large numbers of IoT devices can be too much for centralized machine learning systems. DFL addresses this issue by enabling collective learning across multiple devices, which helps to distribute the computational workload. Through the ability to collaborate on learning across numerous devices, a greater variety of data may be accessed.

Power Consumption: Making use of machine learning might be difficult since IOT devices sometimes have restricted power resources [13]. By enabling local machine learning on IOT devices, DFL resolves the difficulty.

Absence of real-time processing: Many Internets of Things applications may demand real-time data processing, which centralized machine learning systems may not be able to provide. DFL manages this difficulty, without requiring regular contact with a centralized server, by allowing IOT devices.

3 Proposed Model

Our proposed model simplifies the classification of IoT devices by employing consisting of only an input layer and an output layer. This streamlined method contrasts with prevailing, more complex methods. Here are the key details:

1. **Input Layer:**

    For both experiments, the input to our model is a 28×28-pixel grayscale image (equivalent to 784-pixel values). Alternatively, if byte values are provided directly, the input layer contains 784 neurons.

2. **Output Layer:**

    - In the first experiment, which involves classifying nine IoT device classes along with one non-IoT class, the output layer consists of ten neurons (as illustrated in Fig. 3).
    - The second experiment focuses on detecting unauthorized IoT devices that are not included in the whitelist, concentrating on the traffic of nine IoT device classes.
    - For this experiment, we train nine models, each time excluding one IoT class from the training set. The traffic from the excluded IoT class is treated as that of an unknown IoT device to demonstrate feasibility.
    - As a result, the output layer in the second testing contains 8 neurons.

3. **Weight Initialization and Activation Functions:**

    - We initialize weights using a normal distribution for both input and output layers.
    - The input layer employs the ReLU (Rectified Linear Unit) activation function.
    - Output layer uses softmax activation.

4. **Optimization and Evaluation:**

    - We employ the Adam optimizer.
    - The loss function is categorical cross-entropy.
    - Accuracy serves as the estimation metric for the justification set.

5. **Hidden Layers:**

    - We experimented with middle unseen layers, varying their constraints.
    - However, the outcomes remained consistent.

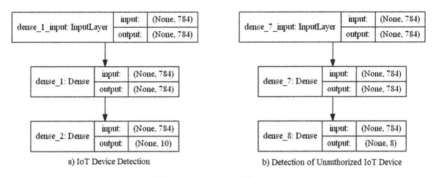

Fig. 1. Model Architecture

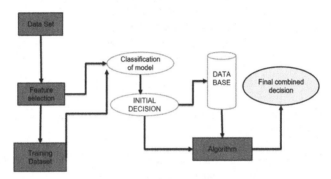

Fig. 2. An Architecture diagram of deep learning model

A. Feature Map Visualization

- The usage of centralized machine learning systems might be complicated by unreliable connection, which can affect IoT devices. By allowing local machine learning on IoT devices, DFL solves this problem by eliminating the requirement for regular contact with a centralized server.
- Data integrity is a significant problem in many Internets of Things applications, and centralized machine learning algorithms may compromise it. In order to address this issue, [14] DFL enables collaboration in learning across devices without requiring data to be sent to a central location.
- Ineffective or sluggish models may arise from centralized machine learning systems' potential inefficiency in utilizing resources. DFL facilitates more effective resource use by distributing the computational burden.
- One potential limitation of centralized machine learning systems in dynamic Internet of Things contexts is their inability to quickly adjust to changing needs. Local ML processing on IoT devices is made possible by DFL, making it easier to adjust to evolving needs.
- Vulnerabilities related to security: The efficacy of the models and the integrity of the data may be jeopardized in centralized machine learning systems by security

threats. One way DFL addresses this issue is by facilitating cross-device collaborative learning, which can lower the possibility of security lapses.

Fig. 3. Challenges of DFL

4 Based on Learning Objectives

In every ML process, including DFL, learning objectives are essential. The process of matching the right DFL technique to an IoT application's specific purpose is facilitated by classifying DFL according to learning objectives. Learning objectives may be categorized into three basic categories: semi-supervised, unsupervised, and supervised. In general, the selection among these various learning objectives is influenced by the specifications and features of the specific IoT software, as well as the quality and visibility of the data produced by the distributed devices.

4.1 Supervised DFL

A distributed ML method called federated learning permits some people to work together and create a model without exchanging any data, which falls under the category of Internet of Things (IoT) applications where the main goal is to classify or predict based on available labeled data. To decrease the communication burden in FL, J. Konecný et al. present a number of techniques, including compaction, induced conformational learning, the suggested method, and important recommendations. Along with providing a thorough summary of the most recent FL research, they also investigated the effectiveness of various communication-efficient strategies across several benchmarks. The authors proposed a Federated Learning (FL) architecture that integrates local model updates and global model aggregation to effectively train a Deep Neural Network (DNN) on data distributed across multiple devices. The recommended strategy reduces the calculation cost and network bandwidth of traditional FL approaches by using quantization and scarification.

4.2 Unsupervised DFL

Unsupervised DFL becomes crucial when labeled data is insufficient or unavailable. This is because a variety of applications depend on the discovery of hidden patterns or abnormalities, yet Internet of Things devices frequently produce enormous volumes of unlabeled data.

Suman et al. presented a novel unsupervised FL method for training ML models with unbalanced datasets dispersed over several devices [4]. The authors presented a clustering-based method that leverages device-specific data commonality to train the local models. By adding up all of the local models' results, the global model is created. Since the suggested method does not need the sharing of labeled data, it is a great choice when managing sensitive data or data distributed across multiple devices. This approach utilizes significantly less data transfer between devices, [17] the authors' method proved to be competitive in performance when compared to existing supervised FL algorithms.

4.3 Semi-supervised DFL

In Internet of Things applications where labeled data is scarce but unlabeled data is abundant, semi-supervised DFL offers a balance between supervised and unsupervised learning. It extracts valuable insights while optimizing resource utilization at hand. A two-step process that utilizes both labeled and unlabeled data was presented by Ywang et al. [5] to train machine learning algorithms in a parallel setting across several Internet of Things sensors. Initially, the raw data is labeled using a self-training technique to enhance the predictive capability of the local models. In the subsequent step, the labeled data are used to refine the global model. Through a reduction in the requirement for labeled data, the suggested technique lessened the dangers related to sharing private info. The authors showed, with the use of a benchmark dataset, that their method preserves user anonymity and improves upon both supervised and unsupervised methods [13]. The authors also discuss the efficiency of regularization techniques in improving semi-supervised learning, including homogeneous generalized linear models, immersive adversarial training and entropy reduction.

5 Result and Discussion

The approach for a Deep Metric Representation Learning-Based Domain Knowledge Driven Unified Architecture for IoT Device Recognition begins with a comprehensive understanding of IoT device characteristics and communication protocols [7]. A thorough literature review identifies existing approaches and gaps in the field. Data collection entails gathering diverse datasets of IoT device data, which are then preprocessed and feature-engineered to capture relevant characteristics [6]. Deep metric representation learning architecture, infused with domain knowledge, is designed and trained on the prepared data. Model performance is evaluated using appropriate metrics and validated on unseen datasets. Integration with domain knowledge enhances model interpretability and trustworthiness. Optimization and fine-tuning ensure the model's efficiency and scalability for deployment in IoT environments. Finally, documentation and reporting communicate the methodology's findings and implications, facilitating its implementation and further research in the domain.

5.1 Approach

In the area of network traffic classification, two primary methods are commonly utilized: the rule-based approach and the statistical and behavioral feature approach. Both methods, however, have certain limitations. The rule-based approach relies on port numbers, which have become unreliable due to the use of dynamic port assignments. Meanwhile, the statistical and behavioral feature approach requires domain knowledge for effective feature identification and preprocessing.

In this research, we propose a novel method based on representation learning, with a specific focus on identifying IoT devices within network traffic. Additionally, we aim to detect connected IoT devices that are not part of the organization's whitelist (i.e., unknown devices). Our approach is motivated by the unique characteristics of the data transmitted by IoT devices, which differ from that of computers and smartphones, including shorter data lengths and specific patterns. While our scope is limited to IoT devices operating over the TCP protocol, we believe our method can be extended to other communication protocols such as Bluetooth, ZigBee, and CoAP. This work addresses the limitations identified in earlier studies conducted by other researchers, as discussed in the previous section.

5.2 Data Preprocessing

During the preprocessing phase, we convert network traffic data (captured in pcap format) into gray-scale images. Our main focus is on the payloads exchanged during TCP sessions between IoT devices (as illustrated in Fig. 1). Importantly, this data processing step is consistent across both of our experiments. The conversion process consists of the following steps:

1. Extract Payloads: We isolate the payloads from the TCP sessions.
2. Image Conversion: These payloads are then converted into grayscale images.

5.2.1 Dataset Description

To assess the effectiveness of our proposed model, we trained a single-layer fully connected neural network. Our dataset was obtained from the IoT Trace Dataset. Below are the key details:

Total TCP sessions: 218,657

- IoT device sessions: 110,849
- Non-IoT device sessions: 107,808.

5.2.2 Device Selection

- Given our deep learning approach, we focused on devices with completed 1,000 TCP sessions.
- This selection ensures robust training and reliable results.

5.2.3 Data Split

The reduced dataset was split into three subsets:
Training, Validation, Test.

A random selection of 10% of the data was made for both the validation and test sets.

For additional information, please see Table 1, which details the subset of the IoT Trace Dataset used in our experiments after the preprocessing phase (Fig. 4).

Table 1. Dataset in Experimental Analysis

S.NO	Device Details	MAC Value	Device	TrainingSet	ValidationSet	TestSet	Total
1	IoTdevices	N/A	IoT	20055	2486	2231	24772
2	Belkin-Wemo	ec:1a:59:83:28:11	IoT	7333	915	818	9066
3	Belkin	ec:1a:59:79:f4:89	IoT	2779	353	312	3444
4	Insteon	00:62:6e:51:27:2e	IoT	2197	281	247	2725
5	Samsung	00:16:6c:ab:6b:88	IoT	3305	417	370	4092
6	Amazon-Echo	44:65:0d:56:cc:d3	IoT	2923	370	328	3621
7	PIX-STAR	e0:76:d0:33:bb:85	IoT	31219	3864	3472	38555
8	Netatmo	70:ee:50:03:b8:ac	IoT	5715	715	638	7068
9	WithingsAura	00:24:e4:20:28:c6	IoT	926	124	105	1155
10	Netatmo	70:ee:50:18:34:43	IoT	1914	246	215	2375

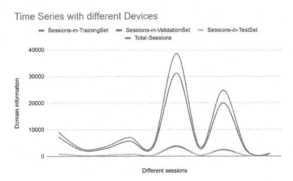

Fig. 4. Session with Different Devices

To estimate our model's performance on the Test, we employed the following evaluation metrics:

Accuracy assesses the overall correctness of the predictions made by the model.

Precision measures the proportion of true positive predictions relative to all positive predictions.

Recall indicates the proportion of true positive predictions among all actual positive instances.

The F1 score provides a balance between precision and recall, offering a single metric that accounts for both false positives and false negatives.

$$A = \frac{TP + TN}{TP + FP + FN + TN}, \quad P = \frac{TP}{TP + FP}$$

$$R = \frac{TP}{TP + FN}, \quad F1 = \frac{2 \times P \times R}{P + R}$$

where:
TP-True Positives
FP-False Positives
TN-True Negatives
FN-False Negatives

Unknown Devices	Minimum Number of Epochs	Thresholds	Test Set Accuracy (%)
Amazon Echo	9	0.97	98.9
Samsung SmartCam	27	0.99	97.9
Belkin Wemo switch	5	0.77	99.3
Netatmo Welcome	18	0.99	98.3
Insteon camera	8	0.92	98.8
Withings Aura smart sleep sensor	6	0.80	99.8
Netatmo weather station	3	0.76	99.8
PIX-STAR photoframe	3	0.87	99.8
Belkin Wemo motion sensor	3	0.90	99.0

The accuracy and loss values recorded on the validation set during the training of our multiclass classifier are significant. We trained the model over 25 epochs with a batch size of 100, achieving the highest accuracy of 99.85% and corresponding loss after just seven epochs. To address the issue of overfitting, we opted to retrain the model for exactly seven epochs. The test set results were impressive, with an accuracy of 99.84%, which either exceeded or equaled that of existing approaches. We also present the confusion matrix for each IoT device as well as the non-IoT device. In this experiment, our dataset included traffic data from nine IoT devices. We trained nine individual multiclass classifiers, each excluding one class (designated as an unknown IoT device) during the training phase. By determining the minimum number of epochs needed to achieve maximum accuracy on the validation set, we ensured the reliability of our findings.

6 Conclusion

The proposed architecture offers a practical solution for precisely identifying IoT devices across diverse settings. By incorporating domain-specific information and leveraging deep metric representation learning techniques, it achieves impressive accuracy even in the presence of noise and heterogeneity. Real-world deployment feasibility, flexibility, and scalability further enhance its value. Additionally, the work emphasizes the importance of domain knowledge in addressing IoT-specific challenges [16]. Leveraging deep learning techniques, the model learns compact and discriminative representations for each IoT device. This work advances the field of IoT device identification. It underscores the need to combine technical expertise with domain-specific insights for successful problem-solving. This approach reduces intra-class variations while maximizing inter-class separations, thereby improving identification accuracy. To address privacy concerns, GDPR-Compliant DFL architecture is introduced. Overall, this research advances IoT device identification and underscores the need for informed architectural choices.

References

1. Lin, J., Yu, W., Zhang, N., Yang, X., Zhang, H., Zhao, W.: A survey on Internet of Things: architecture enabling technologies security and privacy and applications. IEEE Internet Things J. **4**(5), 1125–1142 (2017)
2. Lin, H., Kaur, K., Wang, X., Kaddoum, G., Hu, J., Hassan, M.M.: Privacy-aware access control in IoT-enabled healthcare: a federated deep learning approach. IEEE Internet Things J. **10**(4), 2893–2902 (2023)
3. Ravichandran, P., Saravanakumar, C., Rose, J.D., Vijayakumar, M., Lakshmi, V.M.: Efficient multilevel federated compressed reinforcement learning of smart homes using deep learning methods. In: Proceedings of International Conference on Innovative Computing, Intelligent Communication and Smart Electrical Systems (ICSES), pp. 1–11, September 2021
4. Suman, M., Arulanantham, G.: Efficient differentiation of biodegradable and non-biodegradable municipal waste using a novel MobileYOLO algorithm. Traitement du Signal **40**(5), 1833–1842 (2023). https://doi.org/10.18280/ts.400505
5. Yin, B., Yin, H., Wu, Y., Jiang, Z.: FDC: a secure federated deep learning mechanism for data co laborations in the Internet of Things. IEEE Internet Things J. **7**(7), 6348–6359 (2020)
6. Joshi, R.S., Varun, M., et al.: Multi-task improve domain generalization in EEG-based emotion classification using feature fusion learning model. In: 2023 14th International Conference on Computing Communication and Networking Technologies (ICCCNT), Delhi, India, pp. 1–9 (2023). https://doi.org/10.1109/ICCCNT56998.2023.10307526
7. Gayathri, A., et al.: To improving the performance of identification and segregation of liquid and solid from municipal waste using Adam Optimization Algorithm. In: 2023 International Conference on Self Sustainable Artificial Intelligence Systems (ICSSAS), Erode, India, pp. 1515–1520 (2023). https://doi.org/10.1109/ICSSAS57918.2023.10331841
8. Kaspour, S., Yassine, A.: A federated learning model with short sequence to point mechanism for smart home energy disaggregation. In: Proceedings of IEEE Symposium on Computers and Communications (ISCC), pp. 1–6, June 2022
9. . Zhang, T., Zhang, X., Ke, X.: Quad-FPN: a novel quad feature pyramid network for SAR ship detection. Remote Sens. **13**(14) (2021)
10. Arik, S.O., Pfister, T.: Proto Attend: attention-based prototypical learning. J. Mach. Learn. Res. **21**(1) (2022)
11. Ram, B.R., Gowtham, N.V.S., Reddy, G.V.M., et al.: Data transformation, modelling and prediction of customer churn using deep learning. In: 2023 14th International Conference on Computing Communication and Networking Technologies (ICCCNT), Delhi, India, pp. 1–6 (2023). https://doi.org/10.1109/ICCCNT56998.2023.10306384
12. Bach, S., Binder, A., Montavon, G., Klauschen, F., Müller, K.-R., Samek, W.: On pixel-wise explanations for non-linear classifier decisions by layer-wise relevance propagation. PLos One **10**(7) (2015)

13. Gayathri, A., et al.: An accuracy of identifying recyclable objects and the number of objects identified from municipal waste without occlusion using computer vision techniques. In: Nayak, R., Mittal, N., Kumar, M., Polkowski, Z., Khunteta, A. (eds.) Recent Advancements in Artificial Intelligence. ICRAAI 2023. Innovations in Sustainable Technologies and Computing. Springer, Singapore (2024). https://doi.org/10.1007/978-981-97-1111-6_5
14. Abbas, Z., Ahmad, S.F., Syed, M.H., Anjum, A., Rehman, S.: Exploring deep federated learning for the Internet of Things: a GDPR-compliant architecture. IEEE Access (2023)
15. Rahman, S.A., Tout, H., Talhi, C., Mourad, A.: Internet of Things intrusion detection: centralized, on-device, or federated learning? IEEE Network **34**(6), 310–317 (2020). https://doi.org/10.1109/MNET.011.2000286
16. Khan, L.U., Saad, W., Han, Z., Hossain, E., Hong, C.S.: Federated learning for Internet of Things: recent advances, taxonomy, and open challenges. IEEE Commun. Surv. Tutorials **23**(3), 1759–1799, thirdquarter (2021). https://doi.org/10.1109/COMST.2021.3090430
17. Sun, W., Lei, S., Wang, L., Liu, Z., Zhang, Y.: Adaptive federated learning and digital twin for industrial Internet of Things. IEEE Trans. Industr. Inf. **17**(8), 5605–5614 (2021). https://doi.org/10.1109/TII.2020.3034674
18. Khan, L.U., Alsenwi, M., Yaqoob, I., Imran, M., Han, Z., Hong, C.S.: Resource optimized federated learning-enabled cognitive Internet of Things for smart industries. IEEE Access **8**, 168854–168864 (2020). https://doi.org/10.1109/ACCESS.2020.3023940

Image Tampering Detection Using Deep Learning

S. S. Nagamuthu Krishnan[✉], Saran Chowdam, Sandeep Badarla, and C. S. Nithin Tejesh

Department of CSE, Srinivasa Ramanujan Centre, SASTRA Deemed to be University, Kumbakonam, India
`nagamuthukrishnan@src.sastra.edu`

Abstract. The field of image forensics employs a range of techniques to confirm the integrity of images. Specific approaches are used to address particular problems. In this paper, our focus is on detecting copy-move forgery by using deep learning. To determine whether an image has been tampered with, we utilize a customized CNN model architecture and train it with images after performing error-level analysis on them within the images. Additionally, we use traditional method like feature extraction like the SIFT detector which extracts key-points invariant to any changes and by using DBSCAN algorithm we pinpoint areas of copy-move forgery.

Keywords: CNN · DBSCAN · SIFT Detector · Copy-move · Error Level Analysis · Image Forensics · Deep learning

1 Introduction

In this rapidly developing era of AI, image forgery has become effortless and straightforward. Nowadays, we have plenty of photo editors that can easily manipulate the images, making it almost impossible for a normal human eye to detect the forgery. One of the most commonly used methods for image manipulation is the "copy-move" technique, where a part of an area in an image is copied and pasted onto the same image, as shown in Fig. 1.

In the past, individuals relied on traditional methods such as block-based matching and feature extraction to detect copy-move forgery. However, today's society has transitioned towards more advanced techniques, such as deep learning. This approach trains a model with an extensive amount of datasets, allowing the model to predict the outcome. The accuracy of this model is dependent on various factors, including the type of dataset and algorithm utilized. This paper will explore the implementation of these techniques.

In this paper a CNN model that has been customized to determine whether an image has been tampered with or not has been used which predicts whether an image is tampered or not by analyzing the error level it produced during the error level analysis. In addition to that traditional method like feature extraction is used to pinpoint the areas of the copy-move forgery to extract those features SIFT detector is used. These extracted key-points are invariant to changes like rotation and brightness. After extraction, by using DBSCAN algorithm clustering is formed which results in forging areas falling into same cluster. The primary objective is to ensure that the model can be trained with less computation and should provide accurate results. This model has been named as Copy-move Forgery Detection (CMFD) model, and its architecture is displayed in Fig. 2.

Fig. 1. White car copied to the opposite lane [18]

It can be noticed that remarkably different and improved approaches are used in the ELA method as opposed to the given existing techniques in the above papers. While Abdullah M. Mousa used a block-based matching approach to focus on the contrast of pixel values of divided image blocks, the ELA method combines deep learning with traditional techniques. Such a combination allows for a systematic approach to detecting possible tampering before model training, promoting the model's capacity to detect important image characteristics. The combination of ELA preprocessing with deep learning and traditional techniques avails a more robust and comprehensive solution to detect tampered images, potentially achieving better accuracy and reliability.

2 Existing Methods to Detect the Copy-Move Forgery Detection

Various methods have been proposed to detect copy-move forgery in images. Let's discuss a few of them. Abdullah M. Mousa proposed a block-based matching approach in his paper [1]. He divided the image into blocks and used a kd-tree to match similar blocks with similar pixel values. Yohanna Rodriguez-Ortega, Diego Renza, and M. Ballesteros proposed a customized architecture and transfer learning model using CNN in their paper [2]. They aimed to prove that pre-trained architecture is better than customized architecture models, but the transfer learning model which uses VGGNet-16 took more time for training than the customized one. In our project, we also used a customized architecture model to decrease the training time while pre- serving its accuracy. We also combined various datasets into a single dataset and trained the model with it to

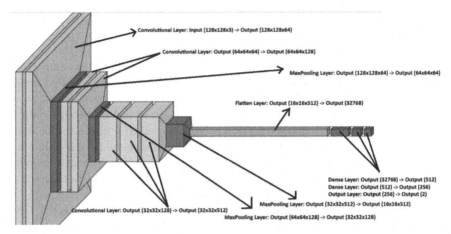

Fig. 2. The Architecture of the CMFD model [4]

improve the generalization of the model. Finally, Yaqi Liu, ChaoXia, XiaobinZhu, and Shengwei Xu proposed the SuperGlue method that uses SIFT detection in their paper [3]. This method extracts distinct key points and descriptors and matches them with another image that contains the same objects. It is a powerful and robust method.

In a research paper [4] proposed by Mamta, Anuradha Pillai, and Deepika Punj, they used the DBSCAN clustering algorithm and VGGNet-16 model to detect copy-move forgery in images. The algorithm clusters similar pixels that contain similar values to detect the copy-move forgery. In another research paper [5], SVM was used for the same purpose but required more training data to produce accurate results.

In yet another research paper [6], Marcello Zanardelli, Fabrizio Guerrini, Riccardo Leonardi, and Nicola Adami studied various research papers and listed every image forging method technique. As our research is focused on copy-move forgery detection, we focused on that particular section where they used different pre-trained architecture models like VGGNet, ResNet, AlexNet, etc. They also used benchmark datasets to train the model and achieved different accuracies. However, using pre-trained models increased the training time and complexity of the process to detect copy-move forgery in an image.

In another research paper [7], traditional and modern methods were used to detect tampered images. The authors used block-based forgery detection along with VGG-Net-16. They extracted features from images as blocks and trained the model with them. This method was followed to detect copy-move forged areas in an image. These are the existing methods that are mostly used to detect copy-move forged areas in an image.

3 Comparing Proposed Method to Detect the Copy-Move Forgery Detection with Existing Method

A method to identify tampered images has been developed, which is simple and reliable. To do this, Error Level Analysis (ELA) is used on each image. The process starts by compressing the image into JPEG format with a quality of 85. This type of image format can cause significant changes to be observed in tampered images. Next, the pixel values of the original image are subtracted from the JPEG format, and the maximum pixel value of the difference is determined. This value is then used to calculate the scaling factor for the brightness of the image. The calculation is done by dividing the value 255 by the maximum pixel value, allowing for easy detection of any differences in the images (as shown in Fig. 3). Like JPEG format uses lossy compression means it discards some of the image data while compression, compression is performed on each 8 × 8 pixel block by using direct cosine transform (DCT). The low level error places will be shown in colors like blue and green where as high error level places will be shown in red or yellow This process is repeated for all images in the dataset, and they are converted into a 128 × 128 × 3 array containing their pixel values. These arrays will be used to train the model. The prediction depends upon the error level of the image.

The model consists of several hidden layers and takes an input size of a 128 × 128 × 3 image. It generates an output of two values that represent the probabilities of classes 0 and 1, where 0 denotes a tampered image and 1 represents an untampered image. Prior to training the model, labels are assigned to each image, such as 0 for tampered and 1 for untampered images. By comparing the output class probabilities, the model can identify whether or not the image has been tampered with, thereby predicting whether an image is tampered with or not.

Fig. 3. The original image and the image we got after performing the error-level analysis [18]

To detect tampered regions in an image, traditional methods such as SIFT detector and DBSCAN are utilized. Initially, key points and their respective descriptors are detected in the image. This process involves performing Scale-space peak selection, which is a complex operation. Gaussian blur is applied on every octave, as shown in Fig. 4. For every subsequent octave, half the size of the first octave is taken. Usually, around 8 octaves are taken to carry out this operation.

$$G(x, y, \sigma) = \frac{1}{2\pi\sigma^2} e^{-\frac{x^2+y^2}{2\sigma^2}}$$

Fig. 4. Gaussian blur is applied to each pixel [19]

The scale operator is denoted by σ while x and y represent the positions of the pixel. After this step, the DOG (Difference of Gaussian kernel) is applied, where the first step is performed with a different sigma value and then subtracted from the first sigma value results (shown in Fig 5). Key points are identified by comparing them with the neighboring pixel values and then removing those that are on the edge and not useful. The orientation of the key points is calculated using gradients and bins. This process is repeated for every point around it, generating descriptors that form a feature vector containing 128 values. Key points and descriptors can be extracted using this method. Finally, the DBSCAN clustering algorithm is performed, and the forged areas containing the same key points and descriptors can then be matched. This is a technique for detecting areas of copy-move forgery in an image.

The existing method in the base paper uses AlexNet a pre-trained architecture to extract deep features from the images it contains 4096 elements where as our method uses error level analysis which is a simple method which tells the error levels in an image without much complex calculations. The existing method solely relies on deep learning and SVM where as we combined both traditional method and modern method to improve the performance (Table 1).

Table 1. Comparing existing method and proposed method

Existing method	Proposed method
1. Uses AlexNet for future extraction contains vector consisting of 4096 elements (high computation complexity)	1. Uses ELA to show the error levels in an image which is gives as input to the model (less computation complexity)
2. Only modern methods are used (AlexNet and SVM)	2. Both traditional (SIFT and DBSCAN key feature extraction) and modern methods(CNN) are used
3. Only predicts whether an image is forged or not	3. Predicts whether an image is tampered or not and also pinpoints the tampered regions in an image
4. Trained on one dataset (MICC-F220)	4. Trained on 5 different datasets
5. Produces an accuracy of 93.94% on MICC-F220 dataset with high false positive rate	5. Produces an accuracy of 93.18% on MICC-F220 dataset with less computation complexity

4 Mathematical Formula

4.1 Error Level Analysis

$$\mathbf{ELA_{a,b}} = \left|\text{Image}_{a,b} - \text{Compressed Image}_{a,b}\right|$$

where a,b define the position of pixel in an image

$$\mathbf{max\_diff} = \max(\text{ELA}_{a,b})$$

max_diff holds the maximum pixel value in the ELA image

$$\textbf{ScaledELA}_{a,b} = (\text{ELA}_{a,b} * 225)/\text{max\_diff}$$

[16]

4.2 CMFD Model

$$\textbf{CMFD(X)} = 0 \text{ or } 1$$

where X = 128 × 128 × 3 numpy array we get after performing ELA function on an Image

0 = authenticated
1 = tampered

These are the operations that are performed on that input numpy array on different layers

Convolution Layer

$$\textbf{Conv2D(X)} = \text{RELU}\left(\sum_{i=1}^{3}\sum_{j=1}^{3}\sum_{k=1}^{3}(X_{i,j,k} * W_{i,j,k} + b)\right)$$

Maxpool Layer

$$\textbf{MaxPool2D(X)} = \max(\text{Input}(i*s1, j*s2) \text{ to Input}((i+1)*s1, (j+1)*s2))$$

s1, s2 are stride values

At the last layer we perform softmax function which gives probabilities for two classes which are authenticated and tampered. Which class has the highest probability is considered as output

$$\textbf{Softmax(x)} = e^x / \sum_{i=1}^{N} e^x$$

Here N = No. of classes = 2

$$\textbf{Output} = \max(\text{Softmax}(x_1), \text{Softmax}(x_2))$$

[17]

All these function are performed inside this function
CMFD(X) = 0 or 1

4.3 Accuracy Calculation

$$(\text{No. of correct predictions}/\text{Total No. of predictions}) * 100$$

5 Datasets Description

The utilization of five different benchmark datasets is currently underway. These datasets include both original and tampered images. In order to train our model to achieve a high level of accuracy, a larger number of training images, particularly distinct tampered images, are required. Please refer to the table below for details on the datasets, with "O" representing original images and "T" representing tampered images (Table 2).

Table 2. Datasets and their description

Dataset Name	No of images	Size	Created by
MICC-F220	110 O/110 T	14.4 Mb	Image and Communication Laboratory (LCI) at the University of Florence (UNIFI) in Italy
CASIA v1.0	8292 O/5594 T	3.33 Gb	Jing Dong, Wei Wang, and Tieniu Tan from the Institute of Automation, Chinese Academy of Sciences, Beijing, China
CASIA v2.0	800 O/470 T	40.7 Mb	Jing Dong, Wei Wang, and Tieniu Tan from the Institute of Automation, Chinese Academy of Sciences, Beijing, China
COVERAGE	100 O/100 T	139 Mb	B. Wen, Y. Zhu, R. Subramanian, T. Ng, X. Shen, and S. Winkler
CG-1050	100 O/1049 T	1.69 Gb	Maikol Castro, Dora M Ballesteros, Diego Renza

6 Results and Discussion

The model has been trained using 5 different datasets. Based on the results, it appears that the dataset with a higher number of tampered images has provided better accuracy. The model has also effectively detected the copy-move forged regions using our localization algorithm. Below graph is plotted between accuracy and each epoch. The graph is plotted for 3 datasets which has high accuracy among the 5 datasets (Table 3).

Table 3. Datasets and their accuracy

Dataset	Accuracy
MICC-F220	93.18
CG-1050	90.43
CASIA v1.0	84.99
CASIA v2.0	63.49
COVERAGE	40.00

Where as the COVERAGE dataset got so much low accuracy like 40% which is due to the less number of tampered images containing for one original image it contains only one tampered image for one original image this type of dataset is not suitable for our model so the model has given low accuracy. So our model extracts as much as features from the image so it can learn the patterns and classify the images. Below are some of the results in which tampered areas are pin-pointed (Fig. 6).

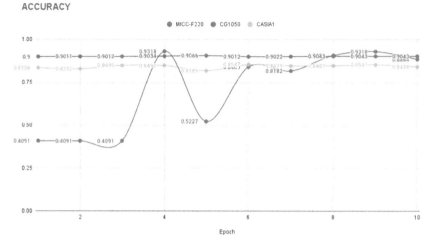

Fig. 5. Epoch and Accuracy graph for MICC-F220, CG1050 and CASIA V1.0 datasets [18]

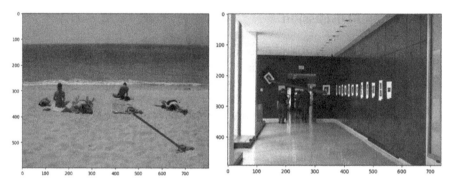

Fig. 6. The blue line connects the tampered regions these are some results from MICC-F220 dataset. (Color figure online)

One drawback of the model is that it cannot predict if an image has been tampered with, especially if the tampering is very minor. However, it can detect the tampered area but cannot predict whether an image is tampered or not for a particular image that contains very minute tampering, such as a small sticker on a large cup. The forged area should be of a certain size to be detected by the model (Fig. 7).

Fig. 7. Tampered image is correctly predicted which is shown in this figure by comparing the probabilities of two classes.

7 Conclusion

This process has predicted if the image is tampered or not by using the CMFD model and also pinpointed the forged areas using key-points extraction precisely and easily as compared to the existing method in the base paper with less computation complexity. The CMFD model has been trained by using 5 different datasets to determine the performance on different type of image data. We have found a simple and robust hybrid method to detect copy-move forgery in images which is better than the existing method. But the given CMFD cannot predict minor tampering images which contain very low error level values which cannot be detected by using ELA. This can be improved by adjusting the quality index as per the image so it will show error levels perfectly.

8 Future Plans

In addition, it is also possible to look into new kinds of feature extractions other than SIFT such as SURF. Another improvement may be to use alternative clustering approaches, such as mean shift or even hierarchical clustering, in detecting tampered regions. Experiments may be done using image segmentation to partition images into semantic regions, thus enabling much more accurate detection of tampered regions. Graph-based approaches to model the relations between image pixels and detecting anomalies indicative of tampering should be also considered. The last would be using GANs in the generation of synthetic tampered images for the augmentation of the training dataset and thus improving the model's capacity to generalize to new-reached tampering techniques.

References

1. Muzaffer, G., Ulutas, G.: A new deep learning-based method to detection of copy-move forgery in digital images. In: 2019 Scientific Meeting on Electrical-Electronics & Biomedical Engineering and Computer Science (EBBT), Istanbul, Turkey, pp. 1–4 (2019). https://doi.org/10.1109/EBBT.2019.8741657
2. Thajeel, S., Salam, G., Sulong, G.: A novel approach for detection of copy move forgery using completed robust local binary pattern. J. Inf. Hiding Multimedia Signal Process. **6**, 351–364 (2015)
3. Sharma, P., Kumar, M., Sharma, H.: Comprehensive analyses of image forgery detection methods from traditional to deep learning approaches: an evaluation. Multimed. Tools Appl. **82**, 18117–18150 (2023). https://doi.org/10.1007/s11042-022-13808-w
4. Rodriguez-Ortega, Y., Ballesteros, D.M., Renza, D.: Copy-Move Forgery Detection (CMFD) using deep learning for image and video forensics. J. Imaging **7**(3), 59 (2021). https://doi.org/10.3390/jimaging7030059
5. Krishnaraj, N., Sivakumar, B., Kuppusamy, R., et al.: Design of automated deep learning-based fusion model for copy-move image forgery detection. Comput. Intell. Neurosci. **2022**, Article ID 8501738 (2022). ISSN 1687-5265
6. Zanardelli, M., Guerrini, F., Leonardi, R., et al.: Image forgery detection: a survey of recent deep-learning approaches. Multimed. Tools Appl. **82**, 17521–17566 (2023). https://doi.org/10.1007/s11042-022-13797-w
7. Moussa, A.: A fast and accurate algorithm for copy-move forgery detection, pp. 281–285 (2015). https://doi.org/10.1109/ICCES.2015.7393060
8. Liu, Y., Xia, C., Zhu, X., Xu, S.: Two-stage copy-move forgery detection with self deep matching and proposal SuperGlue. IEEE Trans. Image Process. **31**, 541–555 (2022). https://doi.org/10.1109/TIP.2021.3132828
9. Abdalla, Y., Iqbal, T., Shehata, M.: Copy-move forgery detection and localization using a generative adversarial network and convolutional neural-network. Information **10**(9), 286 (2019). https://doi.org/10.3390/info10090286
10. Agarwal, R., Verma, O.: An efficient copy move forgery detection using deep learning feature extraction and matching algorithm. Multimed. Tools Appl. **79**, 7355 (2020). https://doi.org/10.1007/s11042-019-08495-z
11. Amerini, I., Ballan, L., Caldelli, R., Del Bimbo, A., Serra, G.: A SIFT-based forensic method for copy-move attack detection and transformation recovery. IEEE Trans. Inf. Forensics Secur. 1099–1110 (2011). https://doi.org/10.1109/TIFS.2011.2129512
12. Barni, M., Phan, Q.T., Tondi, B.: Copy move source-target disambiguation through multi-branch CNNs. IEEE Trans. Inf. Forensics Secur. **16**, 1825–1840 (2021)
13. Bay, H., Ess, A., Tuytelaars, T., Van Goo, L.: Speeded-up robust features (SURF). Comput. Vis. Image Underst. **110**(3), 346–359 (2008). https://doi.org/10.1016/j.cviu.2007.09.014
14. Ouyang, J., Liu, Y., Liao, M.: Copy-move forgery detection based on deep learning. In: 2017 10th International Congress on Image and Signal Processing, BioMedical Engineering and Informatics (CISPBMEI), pp. 1–5 (2017). https://doi.org/10.1109/CISP-BMEI.2017.8301940

15. Rao, Y., Ni, J.: A deep learning approach to detection of splicing and copy-move forgeries in images. In: 2016 IEEE International Workshop on Information Forensics and Security (WIFS), pp. 1–6 (2016). https://doi.org/10.1109/WIFS.2016.782391
16. Raković, D.: Error level analysis (ELA). Tehnika **78**, 445–451 (2023). https://doi.org/10.5937/tehnika2304445R
17. Barhoom, A.M., Abu-Naser, S.S., Alajrami, E., Abu-Nasser, B., Musleh, M., Khalil, A.: Handwritten signature verification using deep learning (2019)
18. Multi-modal Information Collection and Correlation (MICC) Lab (2020). MICC-F220 Dataset. https://www.micclab.com/datasets/micc-f220-dataset
19. Tyagi, D.: Introduction to SIFT (Scale-Invariant Feature Transform). Medium. https://medium.com/@deepanshut041/introduction-to-sift-scale-invariant-feature-transform-65d7f3a72d40
20. Thakur, R., Rohilla, R.: Recent advances in digital image manipulation detection techniques: a brief review. Forensic Sci. Int. **312**, 110311 (2020)

Innovations for Smart Cities

Artificial Intelligence-Powered Advanced Driver Assistance Systems in Vehicles

S. Kathiresh and M. Poonkodi[✉]

Department of Computer Science and Engineering, Vellore Institute of Technology, Chennai, India
`kathiresh.s2023@vitstudent.ac.in, poonkodi.m@vit.ac.in`

Abstract. The automotive industry has undergone a transformative revolution with the integration of Advanced Driver Assistance Systems (ADAS). These cutting-edge technologies have significantly enhanced the driving experience by improving overall vehicle safety and paving the way for a more reliable and efficient future in transportation. This research paper explores the essential components of ADAS and their profound impact on the automotive sector.

As part of this study, we will be developing an object detection system for vehicles using the latest YOLOv9 (You Only Look Once version 9) algorithm. Additionally, we will compare its performance with the previous YOLOv8 iteration, allowing us to gauge the advancements made in this field. The integration of ADAS has laid the groundwork for autonomous vehicles, marking a significant milestone in the progress of road safety and the overall driving experience.

Through this comprehensive analysis, we aim to shed light on the remarkable capabilities of ADAS systems and their potential to revolutionize the way we perceive and interact with vehicles on the road. The findings of this research will contribute to the ongoing development and refinement of these advanced driver assistance technologies, ultimately enhancing the safety and convenience of modern transportation.

Keywords: Artificial Intelligence · Advanced Driver Assistance Systems (ADAS) · Machine Learning · Vehicle Safety · Autonomous Driving

1 Introduction

The integration of Advanced Driver Assistance Systems (ADAS) into modern vehicles has emerged as a groundbreaking development in the automotive industry, revolutionizing the driving experience and paving the way for enhanced safety and efficiency on roads. These cutting-edge technologies represent a significant stride towards the ultimate goal of autonomous vehicles, offering a multifaceted approach to mitigating human errors and improving overall transportation systems.

At its core, ADAS encompasses a suite of intelligent systems designed to augment the driver's capabilities by providing an additional layer of vigilance, decision-making support, and operational assistance. These systems leverage a diverse array of sensors, cameras, and sophisticated algorithms to continuously monitor the vehicle's surroundings, detect potential hazards, and intervene or alert the driver when necessary, ultimately reducing the risk of accidents and mitigating their impact.

One of the most widely adopted ADAS features is adaptive cruise control (ACC), which utilizes radar or laser sensors to maintain a predetermined safe distance from the vehicle ahead. By automatically adjusting the speed to maintain this gap, ACC not only enhances driving convenience but also significantly reduces the likelihood of rear-end collisions, a common occurrence in congested traffic conditions.

Another crucial component of ADAS is lane departure warning (LDW) and lane-keeping assist (LKA) systems. These technologies employ cameras to track lane markings and provide audible or visual alerts when the vehicle unintentionally deviates from its designated lane. Advanced versions of LKA can even gently steer the vehicle back into the proper lane, counteracting the effects of driver inattention or fatigue, which are major contributors to accidents.

Blind spot monitoring (BSM) is another invaluable ADAS feature that addresses a longstanding challenge in driving safety. By utilizing sensors to detect vehicles in the driver's blind spots, BSM systems can provide visual or audible warnings, enabling safer and more informed lane changes, thereby reducing the risk of collisions.

Beyond accident avoidance, ADAS also aims to enhance the overall driving experience by alleviating the stress and cognitive load associated with certain maneuvers. Parking assist systems, for instance, leverage sensors and cameras to guide the driver through tight spaces, simplifying the process of parallel parking or maneuvering in crowded areas.

As ADAS technologies continue to evolve, they are incorporating an ever-expanding range of capabilities, such as pedestrian and cyclist detection, traffic sign recognition, and even night vision capabilities. These advanced features work in tandem to create a comprehensive safety net, continuously monitoring the environment and alerting the driver to potential dangers, thereby reducing the likelihood of accidents and improving overall road safety.

The benefits of ADAS extend far beyond individual drivers and passengers. By mitigating the risk of accidents and improving traffic flow, these systems have the potential to significantly reduce the overall number of road casualties and alleviate the substantial economic burden associated with accidents on society.

Moreover, ADAS represents a critical stepping stone towards the realization of fully autonomous vehicles. While self-driving cars are still in the development and testing phases, the technologies and principles underpinning ADAS are laying the foundation for this future. By gradually introducing advanced driver assistance features, manufacturers are not only improving safety in the present but also familiarizing drivers with the concepts and capabilities that will eventually lead to fully autonomous transportation.

Despite the numerous advantages offered by ADAS, these technologies are not without challenges. Concerns have been raised regarding the reliability and accuracy of these systems, particularly in complex or rapidly changing environments. Additionally, there is a potential risk of drivers becoming overly reliant on ADAS, leading to complacency and a potential decrease in overall driving skills over time.

Nonetheless, the potential benefits of ADAS are too substantial to be disregarded. As these technologies continue to evolve and become more widely adopted, researchers and industry experts anticipate a substantial improvement in road safety, reduced traffic congestion, and a more enjoyable and stress-free driving experience for all (Fig. 1).

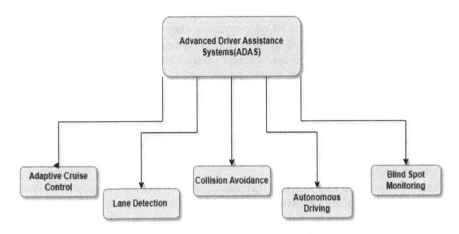

Fig. 1. Various features of ADAS

1.1 Literature Review

The authors Balasundaram A et al. [1] explain the role of AI in Vehicles. The paper discusses a number of current studies on the application of AI in vehicles. Anomaly detection has been explored by using a mix of a multi-level technique and an LSTM (Long Short-Term Memory) based CNN (Convolutional Neural Network). For automated cars to function in different situations, an MCTS (Monte Carlo Tree Search) method for cooperative planning issuing macro actions has been given to enhance the machine learning-based CFMs, an ideal combined prediction approach, known as combination CFM (Car-Following Model), has been discussed. Some of the challenges and limitations faced by AI in automobiles include ensuring safety, dealing with adverse weather conditions, and addressing ethical concerns related to decision-making.

The researchers Wei-Chen Li et al. [2] introduce a YOLO-based approach to enhance the precision of automatic license plate recognition systems. Their proposed algorithm attains a notable level of accuracy in identifying license plates, achieving an average recognition rate of 95.7%. Moreover, the suggested method is computationally efficient and capable of delivering real-time performance on a standard GPU. One of the primary hurdles in achieving high detection as well as recognition rates for oblique car license plates lies in the diverse orientations and perspectives of these license plates.

Shruti et al. [3] present a comprehensive examination of the significance of object detection within the realm of ADAS and its potential to mitigate vehicular collisions. The authors conduct a performance analysis of a YOLOv4-based object detection model proposed in their study. The model undergoes training using a customized dataset consisting of 500 images, which are then segregated into training and testing subsets, with an 80:20 ratio. Remarkably, the model demonstrates commendable accuracy despite the relatively modest dataset size, effectively identifying multiple classes within an image. The evaluation criteria employed include IoU (Intersection over Union) and mAP (mean Average Precision).

Yifang Ma et al. [4] present an all-encompassing examination of the present methodologies and technological advancements in employing artificial intelligence (AI) to bolster the primary applications in autonomous vehicles (AVs). The paper includes the framework for conducting an evaluative and descriptive analysis of pre-existing research, the current practices in utilizing (AI) for (AV) development, the prominent challenges, issues, and necessities concerning contemporary AI approaches in AV applications, as well as the potential opportunities and future research avenues. Furthermore, the authors utilize a performance analysis of a TPU (Tensor Processing Unit), a customized application-specific integrated circuit (ASIC) devised by Google specifically to cater to machine learning workloads.

Naga Praveen Babu Mannam et al. [5] emphasize the increasing popularity of driver assistance features that are implemented in automobiles and trucks, as well as the necessity for experimental and numerical simulations to authenticate ADAS functions and ensure the safety of passengers. The research focuses on the development and verification tests of ADAS functions pertaining to susceptible road users, including car-to-car driving and car-to-pedestrian scenarios. The assessment of performance involves evaluating various scenarios, such as Car-to-Car Rear, moving, front, pedestrian, and pedestrian moving scenarios (CCR), (CCRM), (CCF), (CP), (CPM), and their impact on the performance of ADAS functions.

This manuscript presents a field investigation conducted by M.L. Cummings et al. [6] hailing from George Mason University and Duke University. The findings of this inquiry revealed that driver distraction is an alarming predicament that substantially contributes to the escalating number of fatalities on American roadways. Additionally, the study noted the presence of inconsistencies among vehicles of the same make and model. To elaborate further on this matter, the researchers performed a MANOVA (Multivariate Analysis of Variance) with six distinct vehicles.

Arthur Kurbanov et al. [7] present an all-encompassing analysis of the security constraints associated with such systems and the significance of their conception and execution. Their comprehensive study delves into various aspects including remote intrusions on automobiles, categorization of cyber intrusions on automobiles, weaknesses in diverse vehicle sensors, and the susceptibility of self-governing transportation systems to cyber intrusions. Furthermore, the article deliberates on the limitations of the CAN-bus system, employed in ADAS, and posits potential remedies to counteract these vulnerabilities.

This paper presents insights into the most recent progressions in object detection technology for self-driving automobiles. Harsh Mankodiya et al. [8] test the efficiency of the model by issuing various metrics, like precision, recall, and F1 score. The outcomes demonstrate that the suggested system is better than the currently available advanced models in both accuracy and comprehensibility.

Noor Jannah et al. [9] Discusses Lane Detection in Autonomous Vehicles and provide an overview of the Advanced Driver Assistance System. It analyzes different methods and techniques for improving lane detection in autonomous vehicles and discusses the potential benefits of implementing the Advanced Driver Assistance System. Discusses various performance evaluation metrics and their significance in lane detection in autonomous vehicles. These metrics include accuracy, precision, F-score, and ROC (Receiver Operating Characteristic) curves.

This investigation shows the creation and examination of a novel forward collision warning (FCW) mechanism that employs driving intention recognition. In this particular system, the Baum-Welch algorithm is utilized for data training the braking behavior of Hidden Markov Models (HMMs). The parameters of every braking pattern (HMM), within a single driving circumstance, were carefully optimized with the use of the modified Welch algorithm, incrementally progressing until the highest probability was achieved. A comparison was made between the FCW system and a previously developed system that used a fixed TTC (Time-To-Collision) threshold algorithm. The outcomes, as evaluated by Wei Yang et al. [10], demonstrated that the proposed FCW system exhibited a superior rate of accurate warnings and a diminished rate of false warnings when contrasted with the fixed TTC threshold algorithm.

Yun-Chen Lin et al. [11] propose an ANFPC ("Adaptive Neuro-Fuzzy Predictor-Based Control") technique for the Adaptive Cruise Control (CACC) system. The ANFPC approach integrates automotive radar and V2V ("Vehicle-to-Vehicle") communication to estimate the preceding vehicle's model and predict its future behavior. The paper presents a performance analysis of the proposed (ANFPC) approach for the (CACC) system. The analysis includes safety, string stability, riding comfort, and fuel economy. The outcomes of the experiment in the CarSim environment demonstrate that (ANFPC) approach satisfies all four requirements and enhances fuel efficiency.

Yougang Bian et al. [12] paper covers various methodologies related to the development and validation of an advanced lane-keeping assistance system, including control algorithms decision-making strategies, and experiments. The proposed lane-keeping assistance system uses an LBMPC ("Learning-Based Model Predictive Control") algorithm. The LBMPC algorithm is used to generate the optimal control inputs for the system, which are then applied to the vehicle to achieve the desired lane-keeping performance. The proposed lane-keeping assistance system with switchable assistance modes and the LBMPC method showed promising performance in reducing the driving burden and improving driving safety, outperforming conventional and fully autonomous systems.

The authors Chinthaka Premachandra et al. [13] involve capturing video footage of a four-way intersection using a 360-degree-view camera mounted on a pole. The video footage is then processed using image processing algorithms to detect and track objects that are moving, such as vehicles and pedestrians, within the intersection. Detecting and tracking moving objects at road intersections achieved high precision, recall, and F1-score values in experiments conducted under clear daytime conditions. The methods use image processing algorithms such as background subtraction, blob detection, and Kalman filtering. Future work will investigate performance under different weather and lighting conditions.

Creating an embedded night vision system that uses a proprietary ODROID XU4 microcontroller and thermal images to detect pedestrians. The authors use a well-researched approach with Haar-like features and AdaBoost-based training. The system's performance is evaluated using benchmark datasets, and the experiments prove its effectiveness. The performance analysis of the system proposed by Adam Nowosielski et al. [14] demonstrates that it is effective in detecting pedestrians in severe lighting conditions. The system's processing speed and financial overhead are compared with other available solutions, and the proposed system offers decent speeds at low cost, making it a viable option for improving road safety.

Xiangmo Zhao et al. [15] proposed the idea involves combining three-dimensional LIDAR and data from the camera to generate region object proposals, which are then sent into a convolutional neural network for recognition of objects. The CNN is trained on a big dataset of images that are labeled and can accurately classify objects based on their features. This method is evaluated on the KITTI object detection benchmark dataset. The given method performs better than the baseline approach and achieves a really good performance on the KITTI validation set. It is able to locate objects that are small effectively and has a faster frame rate than the three-dimensional LIDAR frame rate, making it suitable for real-time applications.

An approach for preventing rear-end crashes using wireless vehicular communications, an online neural network model, and a graded warning strategy is proposed by Yuchuan Fu et al. [16]. The suggested algorithms comprise a graded warning strategy for collision avoidance, a risk assessment algorithm on the basis of a neural network model, and a relative lane placement algorithm based on a modified RSSI ranging method. The failure avoidance rate and collision avoidance rate are used to assess the effectiveness of the collision warning system. It is discovered that the ACS-BPNN algorithm predicts output values more accurately than the GA-BPNN algorithm, leading to a more reliable outcome.

Zhanhong Yan et al. [17] suggested methodology entails manual driving practice before Intention-Based Haptic Steering (IBHS) driving trials. Every participant completed six experimental trials, with the Latin Square serving as the arrangement guide. The workload experienced by the participants was evaluated after each trial using the NASA Task Load Index (NASA-TLX). The driver's physical and mental strain as well as their performance when maintaining lanes and changing lanes were all measured as part of the performance study. The SDLP ("Standard Deviation of the Lane Position"), SWRR ("Steering Wheel Reversal Rate"), and NASA-TLX scores were among the figures that were used to display the results.

In the paper by Bingzhao Gao et al. [18], driver samples are clustered into 3 categories of "driving styles (conservative, moderate, and aggressive) using a combination of the K-means algorithm and the SOM (Self-Organizing Map) network.

Preliminary clustering results are obtained by clustering driver samples using the SOM network. The preliminary clustering results are" then clustered again using the K-means algorithm. Evaluates the system's performance using simulations and experiments and uses model predictive control to maximize longitudinal acceleration and maintain a safe distance.

Carfollowing, comfort, and fuel-economy performance were the foundations of the suggested system's performance analysis.

In order to formulate a comprehensive system of perceiving the 2D semantic environment from all angles, the authors of this scholarly article, Andra Petrovai et al. [19], have presented a systematic approach that encompasses various stages. These stages involve the arrangement of cameras, the establishment of a network, the creation of a dataset, the training and evaluation of the network, the deployment of the network, and the integration of software. This project utilizes a multi-modal system for perceiving the environment, incorporating semantic cameras with both fisheye "and narrow field-of-view capabilities. Additionally, deep learning techniques are applied to semantic, instance, and panoptic segmentation networks. To show the efficiency of their proposed methodology, the authors employ the UPDrive dataset and present their empirical findings for two-dimensional semantic, instance, and panoptic segmentation. The performance of the networks is evaluated using accepted metrics, while the inference time of the networks" is also reported. Notably, the panoptic segmentation fusion module, when executed on an NVIDIA GTX 1080 GPU, achieves an execution time of 5 ms when processing front unwarped images with dimensions of 1280×640.

An object detection system for advanced driver assistance systems that operates in real-time is the proposed ObjectDetect framework. It's meant to help drivers by giving them a lot of speed and accuracy in the form of alerts and warnings based on real-time data. Using the YOLOv5 model for object detection, the framework includes an interactive mobile application called ObjectDetect that gives the user personalized alerts and warnings. Beyond advanced driver assistance systems, the framework may find use in other domains, according to the authors Jamuna S. et al. [20] Object extraction, object detection-tracking, and object visualization are the three main modules that make up the suggested ObjectDetect framework. The YOLOv5 model, which was trained on a dataset of labeled photos taken on highways, urban roads, and rural roads, is used for object detection. FPS (Frames per second) and mAP are two metrics used to assess ObjectDetect's performance. In terms of accuracy and speed, it is demonstrated that ObjectDetect performs better than two well-known algorithms, YOLOv3 and YOLO. While the FPS varies from 20 to 30, the mAP of ObjectDetect varies from 85.6% to 96.4%, contingent upon the dataset. In addition, objects in dimly lit environments and hazy photos can be identified by the suggested framework.

To minimize the role of human error in the majority of traffic accidents, this paper suggests a system that uses deep learning (DL) for image processing and recognition. An object detection system, a lane-finding system, and an Android application make up the three primary parts of the suggested system. The object detection system detects and classifies objects in real time using a convolutional neural network (CNN) built on the You Only Look Once (YOLO) architecture. The position of a vehicle within the lane is estimated by the lane-finding system through the use of image-processing techniques to identify lane markings. The system's user interface and the outcomes of the lane-finding and object-detection systems are shown in the Android application. An average precision of 0.89 for object detection, a mean average precision (mAP) of 0.81 for traffic sign detection, and an accuracy of 95% for lane detection is reported by the authors Grishma Ojhaa et al. [21] who assessed the effectiveness of the given system using the KITTI dataset. As an accurate, dependable, and less expensive substitute for the pricey commercial ADAS, the authors suggest that the suggested system leverages deep learning (DL) for image processing and recognition.

The proposal advanced by Xuan Li et al. [22] puts forward the suggestion of implementing a system engineering approach that is rooted in scenarios for the purpose of the development as well as testing of ADAS. The authors lay out a plan for the structure of a scenario, breaking it down into five layers. They also point out key tasks that need to be tackled in creating such as LKA ("Lane Keeping Assistance"), ACC ("Adaptive Cruise Control"), and AEB ("Autonomous Emergency Braking"). Moreover, the authors delve into the testing challenges associated with these tasks. To distinguish critical test scenarios specifically for LKA, the authors apply two algorithms, namely the AHP (Analytic Hierarchy Process) and HC (Hierarchical Clustering). The AHP algorithm is used to determine the relative importance of different driving scenarios based on expert opinions and driver preferences. The HC algorithm, on the other hand, groups together scenarios that share similarities in terms of driving conditions and environmental factors. By employing clustering methods, the authors are able to derive nine crucial LKA scenarios that can be utilized for ADAS testing. The authors assess the efficacy of their proposed approach through computational trials and empirical evidence from illustrative instances. By implementing nine representative testing scenarios for LKA, the authors demonstrate the efficiency of their approach, thus proving that the current rules and regulations are reasonable. In addition to proposing a scenario-based systems engineering approach, the authors present a comprehensive division of traffic scenarios into five layers, offering a structured framework for scenario analysis as well as development. This framework effectively articulates and portrays traffic scenarios within the knowledge domain. The use of AHP and HC algorithms to identify critical test scenarios for LKA serves as evidence of the efficiency of the suggested method.

Kinnera Merugu et al. [23] put forth a methodological approach for the identification and categorization of lanes through the utilization of deep neural networks. This approach encompasses three primary stages: data preprocessing, lane identification, and lane categorization. The TUsimple dataset is employed for training the deep neural networks, with the images being subjected to preprocessing techniques aimed at adjusting their size and color. To detect lane markings, the Lane-net model, a deep neural network, is utilized. It has been trained on an extensive dataset of road images. In addition, the ERFnet and LCnet classification networks, which are deep neural networks trained on a dataset of road images to classify the identified lanes as continuous, dashed, or double-dashed, are employed for lane categorization. The realization of deep neural networks is achieved by utilizing the CNN architecture. The outcomes of the performance analysis indicate that the suggested model attains a high level of accuracy in both lane identification and categorization, with the polynomial 3rd order method yielding the highest R2-Score of 99.5666043. In conclusion, the proposed methodology presents a sophisticated system of driver assistance that is capable of detecting and categorizing multiple lanes on the road, thereby contributing to the decrease of vehicular accidents and enhancement of driver safety.

The proposed system by Zhang Boxu et al. [24] aims to enhance the Blind Spot Detection (BSD) system in intelligent connected vehicles through the introduction of various enhancements. The primary objective of this system is to enhance the accuracy of the BSD system by implementing a new evaluation approach that involves testing in multiple scenarios and utilizing reinforcement learning techniques. This approach ensures that the system is capable of accurately detecting blind spots in a wide range of driving conditions.

One of the key suggestions put forward by the authors is to upgrade the prompt signal used in the BSD system to provide better assistance to the driver. This enhancement would enable the system to effectively communicate potential blind spot dangers to the driver, thereby reducing the likelihood of accidents. Additionally, the authors propose expediting the implementation of the BSD system in Level 2 driving assistance, which would further enhance the safety of intelligent connected cars by providing advanced driver assistance features.

To achieve accurate target distance calculation, the proposed system employs the linear frequency-modulated continuous wave (FMCW) algorithm for vehicle radar distance measurement. This algorithm ensures that the distances between the vehicle and other objects in its vicinity are calculated with a high level of precision. By accurately measuring these distances, the BSD system can effectively monitor blind spots and provide timely warnings to the driver, thereby mitigating the risk of collisions.

To conclude, the research conducted underscores the significance of refining the BSD system to enhance blind spot monitoring and reinforce the safety of intelligent connected cars. The proposed enhancements, such as the new evaluation approach, upgraded prompt signal, and expedited implementation in Level 2 driving assistance, all contribute towards achieving this goal. By implementing these enhancements, the BSD system can effectively detect blind spots in various driving scenarios, thereby improving the overall safety of intelligent connected vehicles.

The lane departure warning system, proposed by Yara A. Ahmed et al. [25], has been specifically designed for the purpose of Advanced Driver Assistance Systems (ADAS) on highways. This system effectively employs image processing techniques to accurately identify and distinguish the boundaries of the left and right lanes. The algorithm utilized in this system utilizes the sliding window method to detect areas with high pixel values. Additionally, the algorithm analyzes a histogram of the threshold image to determine the optimal starting point. Subsequently, the relevant pixels are fitted into a second-degree polynomial to precisely signify the left and right lane lines. Consequently, the algorithm is capable of calculating the vehicle's deviation from the lane center. To assess the efficiency of this algorithm, it is tested using a dataset that has been specifically made for this purpose. This dataset encompasses three distinct conditions: daytime, nighttime, and foggy weather. The test results show that there is scope for improvement, as the algorithm achieves a maximum detection accuracy of 84.6% during nighttime while recording the lowest accuracy of 47.2% in foggy weather. It should be noted that the accuracy of lane departure detection is also highest during nighttime. The algorithm is applied on Raspberry Pi 4 and is put through testing under varying lighting and weather conditions.

Implmentation. An object detection-based implementation was done for 2 different data sets. The KITTI dataset consists of over 10,000 images and a custom dataset which consists of over 100 images was used. The data set was trained using the new Yolo v9 model which consists of 4 new weight metrics. The yolov9 model consists of 2 models namely the gelan-c model and the yolo v9-e model between which the yolo-9e model was seen to have a better prediction and confidence score. The KITTI dataset was trained in both the yolov8 and v9 models whereas the custom dataset was trained in yolo v8. The Yolo v9 seemed to have a better recall score compared to the v8. However, the v8 seemed to have a better precision score. Overall the final mAP of the yolov9 turned out to be better than the v8 model hence concluding the overall accuracy levels of v9 outperformed the v8 model as shown in Table 1. The detection was carried out for 6 different classes which included car, bike, cycle, pedestrian, sign-boards, and trucks. Out of which the car class prediction had the highest prediction accuracy of 94.3.

Sample Heading (Forth Level)
It can be noted from Figs. 2 and 3 that Fig. 3 performs better using the 9-e model with respect to overall confidence scores and object detection (Figs. 4, 5, 6, 7, 8, 9 and Table 2).

Table 1. Comparison between Yolov8 and Yolov9

Model	Precision	Recall	Mean Average Precision
Yolo v8	71.9	54.2	65.4
Yolo v9	67.1	62.5	67.1

Fig. 2. Gelan-c model

Artificial Intelligence-Powered Advanced Driver Assistance Systems 141

Fig. 3. Yolo9-e model

Fig. 4. Yolov8 model

Fig. 5. Yolov9 model

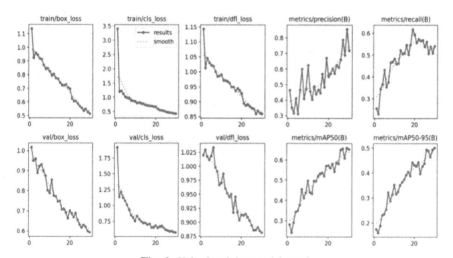

Fig. 6. Yolov8 training model graph

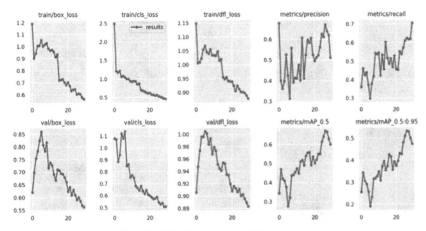

Fig. 7. Yolov9 training model and graph

Table 2. Accuracy of Yolo v8 model on the custom dataset

Precision	Recall	Mean Average Precision
64.7	45.3	70.6

Fig. 8. Yolov8 training on custom dataset

Fig. 9. Yolov8 training on custom dataset with multiple images

Conclusion

In conclusion, Advanced Driver Assistance Systems represent a transformative technological advancement that is reshaping the automotive industry and the future of transportation. By harnessing cutting-edge sensors, cameras, and sophisticated algorithms, ADAS offers a comprehensive suite of intelligent features designed to enhance road safety, improve driving convenience, and ultimately pave the way for the realization of fully autonomous vehicles. The v9 model performs better in comparison to the v8 model and can be deployed into real-world based scenarios by using a minicomputer such as nvidea jetson nano-kit. It is imperative to acknowledge that these systems are not perfect. Drivers should use them with awareness and engagement as they have limitations. Beyond that, as AI-powered ADAS proliferates, concerns around cybersecurity and data privacy are brought to light and need to be addressed.

AI-powered ADAS will probably be crucial to the shift to autonomous vehicles as technology develops, changing the way we think about mobility. To fully realize AI's promise for safer and more effective roads, collaboration between academics, policymakers, and the automobile sector is essential.

Acknowledgements. I would like to sincerely thank Dr. Poonkodi M., my project guide, for her ongoing encouragement, direction, and mentorship during the research and writing of this work. Their knowledge, insightful opinions, and commitment have greatly influenced the course of our project.

I am also grateful to Vellore Institute of Technology (VIT Chennai Campus), for providing a conducive research environment and resources that have enabled the publication of this paper. The institution's commitment to academic excellence has played a pivotal role in our ability to contribute to the body of knowledge in our field.

References

1. Balasundaram, A., Prusty, M.R., Shaik, A.: A comprehensive study on the impact of artificial intelligence in autonomous vehicles. In: 2022 International Virtual Conference on Power Engineering Computing and Control: Developments in Electric Vehicles and Energy Sector for Sustainable Future (PECCON) pp. 1–5. IEEE (2022). Clerk Maxwell, J.: A Treatise on Electricity and Magnetism, 3rd edn., vol. 2, pp. 68–73. Clarendon, Oxford (1892)
2. Li, W.-C., Hsu, T.-H., Huang, K.-N, Wang, C.-C.: A YOLO-based method for oblique car license plate detection and recognition. In: 2021 IEEE/ACIS 22nd International Conference on Software Engineering, Artificial Intelligence, Networking and Parallel/Distributed Computing (SNPD), pp. 134–137. IEEE (2021)
3. Shruti, S., Shravya, M., Mohan, S., Suhail, Y., Radha, R.C.: AI-based solutions for ADAS. In: 2022 6th International Conference on Intelligent Computing and Control Systems (ICICCS), pp. 1009–1012. IEEE (2022)
4. Yifang, M., Wang, Z., Yang, H., Yang, L.: Artificial intelligence applications in the development of autonomous vehicles: a survey. IEEE/CAA J. Automat. Sinica **7**(2), 315–329 (2020)
5. Mannam, N.P.B., Rajalakshmi, P.: Determination of ADAS AEB car to car and car to pedestrian scenarios for autonomous vehicles. In: 2022 IEEE Global Conference on Computing, Power and Communication Technologies (GlobConPT), pp. 1–7. IEEE (2022)
6. Cummings, M.L., Bauchwitz, B.: Driver alerting in ADAS-equipped cars: a field study. In: 2023 IEEE International Conference on Assured Autonomy (ICAA), pp. 29–33. IEEE (2023)
7. Kurbanov, A., Grebennikov, S., Gafurov, S., Klimchik, A.: Vulnerabilities in the vehicle's electronic network equipped with ADAS system. In: 2019 3rd School on Dynamics of Complex Networks and their Application in Intellectual Robotics (DCNAIR), pp. 100–102. IEEE (2019)
8. Mankodiya, H., Jadav, D., Gupta, R., Tanwar, S., Hong, W.C., Sharma, R.: Od-xai: explainable ai-based semantic object detection for autonomous vehicles. Appl. Sci. **12**(11), 5310 (2022)
9. Ibrahim, M.Z., Wahid, N.: Lane Detection in Autonomous Vehicles: A Systematic Review
10. Yang, W., Wan, B., Qu, X.: A forward collision warning system using driving intention recognition of the front vehicle and V2V communication. IEEE Access **8**, 11268–11278 (2020)
11. Lin, Y.C., Nguyen, H.L.: Adaptive neuro-fuzzy predictor-based control for cooperative adaptive cruise control system. IEEE Trans. Intell. Transp. Syst. **21**(3), 1054–1063 (2019). For Sustainable Future (PECCON), pp. 1–5. IEEE
12. Bian, Y., Ding, J., Hu, M., Xu, Q., Wang, J., Li, K.: An advanced lane-keeping assistance system with switchable assistance modes. IEEE Trans. Intell. Transp. Syst. **21**(1), 385–396 (2019)
13. Premachandra, C., Ueda, S., Suzuki, Y.: Detection and tracking of moving objects at road intersections using a 360-degree camera for driver assistance and automated driving. IEEE Access **8**, 135652–135660 (2020)
14. Nowosielski, A., Małecki, K., Forczmański, P., Smoliński, A., Krzywicki, K.: Embedded night-vision system for pedestrian detection. IEEE Sens. J. **20**(16), 9293–9304 (2020)
15. Zhao, X., Sun, P., Xu, Z., Min, H., Yu, H.: Fusion of 3D LIDAR and camera data for object detection in autonomous vehicle applications. IEEE Sens. J. **20**(9), 4901–4913 (2020)
16. Fu, Y., Li C., Luan, T.H., Zhang, Y., Yu, FR.: Graded warning for rear-end collision: an artificial intelligence-aided algorithm. IEEE Trans. Intell. Transp. Syst. **21**(2), 565–579 (2019
17. Yan, Z., Yang, K., Wang, Z., Yang, B., Kaizuka, T., Nakano, K.: Intention-based lane changing and lane keeping haptic guidance steering system. IEEE Trans. Intell. Veh. **6**(4), 622–633 (2020)

18. Gao, B., Cai, K., Ting, Q., Yunfeng, H., Chen, H.: Personalized adaptive cruise control based on online driving style recognition technology and model predictive control. IEEE Trans. Veh. Technol. **69**(11), 12482–12496 (2020)
19. Petrovai, A., Nedevschi, S.: Semantic cameras for 360-degree environment perception in automated urban driving. IEEE Trans. Intell. Transp. Syst. **23**(10), 17271–17283 (2022)
20. Murthy, J.S., Siddesh, G.M., Lai, W.-C., Parameshachari, B.D., Patil, S.N., Hemalatha, K.L.: Objectdetect: a real-time object detection framework for advanced driver assistant systems using yolov5. Wirel. Commun. Mobile Comput. **2022**, 1–10 (2022)
21. Ojha, G., Poudel, D., Khanal, J., Pokhrel, N.: Design and analysis of computer vision techniques for object detection and recognition in ADAS. J. Innov. Eng. Educ. **5**(1), 47–58 (2022)
22. Li, X., Song, R., Fan, J., Liu, M., Wang, F.-Y.: Development and testing of advanced driver assistance systems through scenario-based system engineering. IEEE Trans. Intell. Veh. **8**(8), 3968–3973 (2023)
23. Merugu, K., Adarsh, S.: Multi lane detection, curve fitting and lane type classification. In: 2022 IEEE 19th India Council International Conference (INDICON), pp. 1–6. IEEE (2022)
24. Boxu, Z., et al.: Optimization of vehicle blind zone monitoring (BSD) evaluation scheme based on image processing and radar ranging algorithm. In: 2023 IEEE 2nd International Conference on Electrical Engineering, Big Data and Algorithms (EEBDA). IEEE (2023)
25. Ahmed, Y.A., Mohamed, A.T., Bayoumy Aly, A.M.: Robust lane departure warning system for ADAS on highways. In: 2022 4th Novel Intelligent and Leading Emerging Sciences Conference (NILES), pp. 321–324. IEEE (2022)
26. Geiger, A., Lenz, P., Stiller, C., Urtasun, R.: The kitti vision benchmark suite. **2**(5), 1–13 (2015). http://www.cvlibs.net/datasets/kitti

Drive Safe: AI & IoT Powered Driver Alertness for Enhanced Passenger Safety

S. Deepti[(✉)], A. Anitha Pai, and S. Anandhi

Department of Computer Science and Engineering, Rajalakshmi Engineering College, Chennai, Tamil Nadu, India
{200701062,200701025,anandhi.s}@rajalakshmi.edu.in

Abstract. The growing concern over driver safety has been attributed to an increase in traffic accidents caused by fatigue or drowsiness while driving. Addressing this issue, it is crucial to identify and prevent driver fatigue. To tackle this challenge, we propose an innovative approach that utilizes real-time video analysis to detect subtle indications of fatigue in drivers, such as yawning, variations in blink rate, and the duration of eye closure. In order to further enhance safety and well-being, our methodology integrates physiological data collected through an Electrocardiogram (ECG), a temperature sensor (LM35), and a pressure sensor to evaluate the driver's health and level of alertness in abnormal situations. Our method incorporates an advanced system for detecting facial regions, which utilizes 68 key facial landmarks and extracts data from a webcam. This system not only accurately identifies the driver's face but also tracks and analyses eye movements, facial expressions, and head tilt positions in each video frame. By combining these various facial characteristics and sensor data, our system is able to proactively detect signs of driver fatigue and promptly alert the driver through an audible fatigue warning system. This comprehensive method aims to greatly improve road safety by providing real-time fatigue detection and alert mechanisms for drivers, potentially saving lives and preventing accidents.

Keywords: Driver fatigue · Real-time video analysis · Head tilt position analysis · Yawn Detection

1 Introduction

In today's fast-paced world, road safety is of paramount importance, with driver distraction and fatigue being significant contributors to road accidents. Unlike the obvious impairment caused by substances like alcohol and drugs, the signs of fatigue are often subtle and difficult to detect. Traditional methods of monitoring driver behavior [1] have limitations in providing real-time, accurate insights into the driver's cognitive state. Therefore, there is a critical need for advanced technological solutions that can proactively assess and address the driver's cognitive state to prevent potential accidents. In this context, our project aims to introduce a revolutionary approach that integrates of Artificial Intelligence (AI) [2] and Internet of Things (IoT) technologies [3] has as a promising approach to enhance passenger safety by determining the driver's cognitive state in real time. AI and IoT technologies offer an innovative approach to address this challenge by leveraging advanced algorithms and sensory data. AI algorithms can analyze various data inputs, such as facial expressions, eye movements, and vehicle operation patterns, to assess the driver's cognitive state in real time. IoT devices, including sensors and cameras integrated within the vehicle, can capture and transmit relevant data to the AI system, enabling continuous monitoring of the driver's behavior and cognitive state. The primary objective of this research is to develop a robust system that utilizes AI and IoT to accurately determine the driver's cognitive state, including levels of alertness, drowsiness, and distraction. By doing so, the system aims to provide timely alerts and interventions to mitigate potential safety risks, thereby enhancing passenger safety during travel. This paper is structured to comprehensively explore the integration of AI and IoT technologies for the determination of driver's cognitive state. It will delve into the theoretical underpinnings of cognitive state assessment, the technical implementation of AI and IoT systems for real-time monitoring, and the potential impact of such systems on passenger safety. Current safety systems primarily rely on external alerts, such as audible alarms or visual cues, to notify drivers of their drowsy state [4]. While these systems have undoubtedly saved lives, they have limitations in their ability to precisely recognize the early signs of fatigue, such as eye blinks, changes in facial expressions, and the driver's overall health conditions. In contrast, our innovative model seeks to establish a new standard in detecting fatigue with greater accuracy.

2 Related Works

Ceerthi Bala and T. Sarath (2020) [5] introduce an innovative "Internet of Things-Based Intelligent Drowsiness Alert System." Focused on addressing the critical issue of drowsiness-related accidents, the authors leverage IoT technology to create an intelligent alert system. The system utilizes real-time data from various sensors to monitor driver behavior and detect signs of drowsiness. Through a sophisticated algorithm, the system analyzes factors such as eye movement, steering patterns, and physiological signals. The implementation showcases a promising approach to enhancing road safety by proactively alerting drivers when signs of drowsiness are detected. This contribution to the field of intelligent transportation systems demonstrates the potential of IoT in preventing accidents and safeguarding lives on the road.

Rateb Jabbar et al. (2020) [6] present a pioneering work in the realm of driver safety with "Driver Drowsiness Detection Model Using Convolutional Neural Networks Techniques for Android Application," they employ advanced Convolutional Neural Networks (CNN) techniques to develop a robust drowsiness detection model. Tailored for Android applications, their model integrates seamlessly into mobile devices, offering a portable and accessible solution for monitoring driver alertness. By leveraging CNN's capabilities, the system processes real-time data, including facial features and eye movements, to accurately identify signs of drowsiness. This innovative approach contributes significantly to the evolution of smart and adaptive technologies, emphasizing the potential for widespread implementation of AI-driven safety measures in the automotive sector.

Dr. K. S. Tiwari et al. (2019) [7] introduce an "IoT-Based Driver Drowsiness Detection and Health Monitoring System." This paper addresses both driver safety and health monitoring through the integration of Internet of Things (IoT) technology. The system goes beyond drowsiness detection, incorporating real-time health monitoring features. By utilizing IoT sensors and a sophisticated algorithm, the authors enable continuous monitoring of vital signs, ensuring a holistic approach to driver well-being. This comprehensive system not only alerts drivers to potential drowsiness but also contributes to a proactive healthcare model for those behind the wheel. The research stands at the intersection of IoT and road safety, paving the way for innovative solutions that prioritize both alertness and health in driving environments.

M. Y. Hossain and F. P. George (2018) [8] introduced an "IoT-Based Real-Time Drowsy Driving Detection System" designed for the prevention of road accidents. This innovative system harnesses the power of the Internet of Things (IoT) to detect driver drowsiness in real-time, aiming to enhance road safety and reduce the risk of accidents. By integrating advanced sensors and real-time data processing, the authors create a responsive and proactive solution. The research emphasizes the crucial role of technology in mitigating the impact of drowsy driving, offering a scalable and effective approach that aligns with the evolving landscape of intelligent transportation systems. This work marks a significant stride towards leveraging IoT for accident prevention and underscores the importance of technology in promoting safer roads.

Chowdhury et al. (2018) [9] delve into the critical domain of drivers' drowsiness detection, shedding light on sensor applications and physiological features. The authors analyze a plethora of research studies and methodologies adopted in drowsiness detection. They examine various sensor technologies such as electroencephalography (EEG), electrooculography (EOG), and electromyography (EMG), along with physiological features including eye movement, brain activity, and muscle tension. They synthesize the main findings across different studies, highlighting the effectiveness of certain sensors and physiological indicators in accurately detecting drowsiness in drivers. Additionally, they identify emerging trends and advancements in sensor technology for enhancing detection accuracy and real-time monitoring capabilities. By understanding the strengths and limitations of various sensors and physiological features, stakeholders in the automotive industry can develop more robust and reliable drowsiness detection systems. Ultimately, the findings of this review have the potential to save lives and prevent accidents by alerting drivers and stakeholders to the dangers of drowsy driving in a timely manner.

3 Proposed System

DriveSafe is an advanced safety system that utilizes state-of-the-art technology to ensure the well-being of both drivers and passengers. It employs continuous monitoring to detect any signs of fatigue or health-related issues, such as dizziness or blurriness. This is achieved through the implementation of various key components, including the Eye Aspect Ratio (EAR) [10], Mouth Aspect Ratio (MAR) [11], HOG Algorithm [12] for Facial Detection for FAR (Face Aspect ratio) [13], and head tilt angle. By analyzing these factors, DriveSafe is able to accurately assess the level of driver alertness. In the event that any issues are identified, DriveSafe immediately activates an audible alarm to alert the driver. Additionally, it has the capability to activate a self-driver mode for enhanced safety. This mode ensures that the vehicle is able to autonomously navigate and operate, reducing the risk of accidents caused by driver impairment. IoT sensors, including heartbeat, pressure, and temperature sensors, are employed to monitor any potential health irregularities that the driver may experience. These abnormalities are also communicated to the driver and passengers through the use of an alarm or buzzer. Furthermore, DriveSafe is equipped with network connectivity, enabling it to send SMS notifications to either the vehicle owner or a designated contact. This ensures that immediate action can be taken if any concerning situations arise. Overall, DriveSafe represents a significant milestone in automotive security technology. With its comprehensive monitoring system and proactive alerts, DriveSafe provides drivers and passengers with peace of mind, knowing that their well-being is being prioritized at all times.

3.1 Advantages of the Proposed System

This advanced in-vehicle monitoring system continuously tracks driver and passenger health in real-time, enabling immediate intervention for potential safety issues. It goes beyond drowsiness detection, offering a comprehensive approach to occupant well-being by assessing broader health concerns. The system tailors its alerts to the severity of the situation, triggering appropriate responses. In emergencies, the option to engage self-driving mode provides an extra layer of safety by taking control to potentially prevent accidents. Additionally, the system can notify designated contacts via SMS in case of driver incapacitation, ensuring timely assistance. This versatile solution integrates seamlessly into various vehicles, from personal cars to commercial fleets. Valuable data collected by the system can be used for post-incident analysis and research, furthering road safety advancements. Ultimately, it safeguards both drivers and passengers, prioritizing the well-being of everyone in the vehicle. The system's compatibility with existing technology allows for easy retrofitting into older vehicles.

3.2 Proposed System Architecture

Our system as represented in Fig. 1 utilizes facial recognition to identify the driver and assess their emotional state. Eye and mouth movements are tracked to gauge alertness, while head tilt indicates focus on the road. Physiological data like heart rate, blood pressure, and breathing rate are also collected using various sensors. This information is then fed into an Arduino controller, a mini-computer tasked with processing and potentially sending the data to the cloud for further analysis. If the system detects signs of fatigue, like drowsiness or inattentiveness, it could take action - sending an SMS alert or displaying a warning on the dashboard - to nudge the driver back to a safe state of alertness.

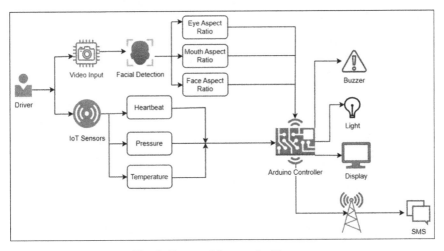

Fig. 1. Proposed System Architecture

4 Methodologies

In the proposed system the primary objective is to develop a driver cognitive state detection system to prevent accidents. Drowsiness, which is characterized by sleepiness, is a major cause of accidents, as it can lead to drivers falling asleep at the wheel. Additionally, other aspects of cognitive state, such as breathing rate, pressure, body temperature, and heart rate, can provide valuable insights into the driver's overall health condition. Therefore, our detection system can identify whether the driver's eyes are open or closed, as well as detect any abnormal fluctuations in their health indicators. In the event the system detects that a person's eyes have been closed for a few seconds, or if they yawn excessively, or if there are any irregularities in their health conditions, it will immediately trigger an alert sound. Furthermore, an SMS will be sent to notify the passengers about the potential danger. This timely alert system serves as an effective measure to mitigate the risks associated with drowsy driving and ensure the safety of everyone on board.

4.1 Facial Drowsiness Detection System

The facial drowsiness detection system is responsible for collecting data from live video input. It utilizes a camera to capture real-time footage of the driver's face. This data is then processed using a Histogram of Oriented Gradients (HOG) algorithm, which is a popular machine learning technique used for detecting objects. The algorithm extracts various components of the from the video, such as the measure the driver's eye aspect ratio (EAR), mouth aspect ratio (MAR), and head tilt position. These measurements serve as indicators of drowsiness. By analysing this data, the system is able to determine the driver's level of alertness. If signs of drowsiness are detected, the system activates a buzzer to alert the driver.

4.2 Driver Health Monitoring System

The IOT sensors gather information related to the body's physical functions, including heart rate, body temperature, pressure, and breathing rate. This collected data is then transmitted to an Arduino Uno, which is a microcontroller capable of processing the data either locally or sending it to the cloud for storage and analysis. Additionally, the system incorporates an SMS module that will send text messages in specific instances of drowsy driving. For instance, if it detects that the driver's heart rate is elevated, it will trigger the module to send a message and activate a buzzer to alert the driver. Lastly, there is a display unit that provides the driver with relevant information regarding their condition.

5 Module Description

5.1 Eye Aspect Ratio

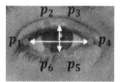

Fig. 2. Eye Open

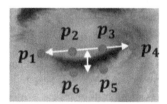

Fig. 3. Eye Closed

The Eye Aspect Ratio serves as an approximation of the state of eye openness. In Fig. 2, we observe the representation of the typical eye open stage. By analyzing the Eye Aspect Ratio using Eq. 1, a software program can determine whether a person's eyes are closed, as depicted in Fig. 3, when the ratio falls below a specified threshold. This program extracts the coordinates of both the left and right eye to calculate the EAR. Consequently, it computes the convex hull for each eye and visually displays them on the frame. If the EAR remains consistently below a particular threshold for a consecutive number of frames, an alarm sound is activated, accompanied by a warning message appearing on the screen.

$$\text{EAR} = \frac{\|p_2 - p_6\| + \|p_3 - p_5\|}{2\|p_1 - p_4\|} \tag{1}$$

5.2 Mouth Aspect Ratio

Fig. 4. Mouth Open

The mouth aspect ratio (MAR) serves as an indicator of the state of mouth opening. It involves identifying the landmarks in the mouth region for each detected face and computing the ratio between the distances of the upper and lower lips, as well as the vertical distance between the corners of the lips. This information is visually depicted in Fig. 4. If the MAR surpasses a specific threshold, a warning message is displayed on the frame. During yawning or when the mouth is open, the horizontal distance between the corners of the lips will be smaller compared to its value during a normal state when calculated using Eq. 2.

$$\text{MAR} = \frac{\|p_2 - p_8\| + \|p_3 - p_7\| + \|p_4 - p_6\|}{2\|p_1 - p_5\|} \quad (2)$$

5.3 Face Aspect Ratio

Face Aspect Ratio or the Estimated Head Position is a metric used in computer vision and facial recognition systems to quantify and analyze the relative dimensions of a detected face as shown in Fig. 5. It is a measure of the relationship between the width and height of a face within an image or video frame calculated using Eq. 3. FAR can be employed to assess changes in facial proportions and shapes, aiding in the analysis of facial expressions and emotions.

$$FAR = \frac{\|p_{22} - p_8\| + \|p_{28} - p_9\| + \|p_{23} - p_{10}\|}{3\|p_6 - p_{12}\|} \quad (3)$$

5.4 Driver Health Assessment

Monitoring drivers' health conditions such as blood pressure, temperature, and heart rate throughout the journey using IoT devices is a critical advancement in ensuring driver and passenger safety. By integrating pressure sensors for Pressure monitoring, LM35 sensors for Temperature assessment, and ECG sensors for Heart Rate monitoring, potential health abnormalities can be detected in real time, enabling timely alerts and interventions. The sensor data is relayed to the IoT platform, where it undergoes real-time analysis to detect any deviations from the normal temperature range. In the event of abnormal temperature readings, the IoT system generates alerts, providing the driver with immediate information regarding their health status as shown in Fig. 6.

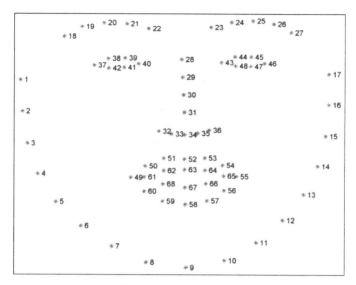

Fig. 5. A 68 Facial Points

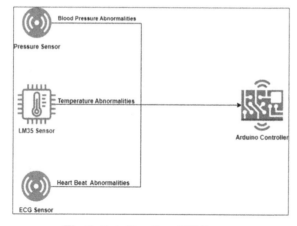

Fig. 6. Data Flow from IOT Sensors

5.5 Alerting the Driver

Eye Aspect Ratio (EAR), Mouth Aspect Ratio (MAR), and Facial Action Recognition (FAR) are calculated from the driver's facial image captured by a camera. If any of these metrics fall below a certain threshold for a set number of frames consecutively, an alert is triggered, indicating potential drowsiness as illustrated in Fig. 7. Additionally, abnormal health signs detected by any of the pressure, ECG, or LM35 sensors will also trigger an alert. In both cases, an alarm sounds to notify everyone in the vehicle, and a message is displayed on the LCD screen, providing more information about the potential issue.

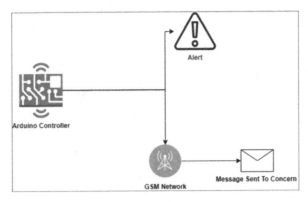

Fig. 7. Alerting Mechanism

6 Implementation

The implementation of a driver drowsiness detection system involves integrating various sensors and setting specific thresholds to effectively monitor the driver's condition as shown in Fig. 8. These systems utilize a combination of sensors such as heartbeat, pressure, and temperature, along with eye aspect ratio (EAR), mouth aspect ratio (MAR), and face position detection. The heartbeat sensor keeps track of the driver's heart rate, while the pressure and temperature sensors monitor physical signs of fatigue. Simultaneously, the EAR and MAR sensors examine the movements of the driver's eyes and mouth in order to detect indications of sleepiness. In addition, the face position sensor identifies any changes in head position that may indicate a decrease in attentiveness. The implementation can be divided into two main components: software and hardware. On the software side, the HOG Algorithm is employed to detect the face. The algorithm marks 68 specific points on the face and identifies the eyes and mouth. Since this system operates in real time using video, it utilizes the camera to constantly monitor these facial features. If an abnormality is detected, such as closed eyes or yawning, the system immediately alerts the driver with prompts and a buzzer sound. The hardware components consist of IoT sensors. These sensors capture various data points including the driver's heart rate, blood pressure, and temperature. Any abnormal readings are promptly signalled by the buzzer and displayed on the LCD screen. For accurate detection of drowsiness, specific thresholds have been established for the EAR and MAR sensors, as well as for the heartbeat, pressure, and temperature sensors. Once these thresholds are exceeded, the system can issue alerts to remind the driver to take necessary precautions, thereby reducing the risks associated with drowsy driving. The system notifies drivers when the EAR exceeds 0.25, or the MAR exceeds 0.79, or the pressure is greater than 500, or the temperature is greater than 38 degrees Celsius. By combining both software and hardware elements, this system ensures effective monitoring and detection of driver abnormalities, providing timely alerts to prevent potential accidents. This comprehensive approach to implementing a drowsiness detection system emphasizes the integration of multiple sensor inputs and the strategic use of predefined thresholds to enhance driver safety.

Fig. 8. IOT Sensors Mechanism

7 Results

The AI and IoT powered Driver Alertness system, integrating EAR (Eye Aspect Ratio), MAR (Mouth Aspect Ratio), FAR (Face Aspect Ratio), heartbeat, pressure, and temperature sensors, has demonstrated promising outcomes. The successful implementation of this system as a prototype marks a significant milestone in the pursuit of enhancing passenger safety. By leveraging a combination of physiological and facial indicators, the system effectively detects driver drowsiness and fatigue, ensuring timely interventions to prevent potential accidents. The integration of IoT technology enables real-time monitoring and alerting. Additionally, the incorporation of images adds a visual dimension to the system's functionality, contributing to a comprehensive approach towards ensuring vigilant and safe driving experiences. The system's performance validates the potential of AI and IoT in revolutionizing driver monitoring for enhanced road safety.

Promising results have been shown by the AI and IoT powered Driver Alertness system, which integrates EAR (Eye Aspect Ratio), MAR (Mouth Aspect Ratio), Head Tilt Position, heartbeat, pressure, respiratory and temperature sensors. This system's successful installation as a prototype, as seen in Figs. 9, 10, 11, 12, and 13, represents a major advancement in the effort to improve passenger safety. Through the utilization of a blend of physiological and facial indicators, the system effectively recognizes driver discomfort and drowsiness ensuring timely interventions to prevent potential accidents. The integration of IoT technology enables real-time monitoring and alerting. The system's

Fig. 9. High Pressure

Fig. 10. High Temperature

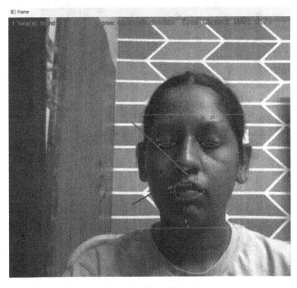

Fig. 11. Eyes Closed

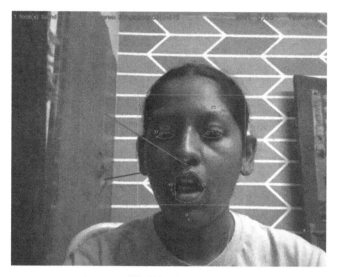

Fig. 12. Yawing

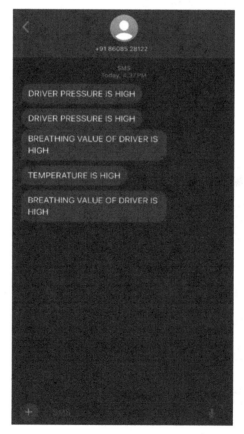

Fig. 13. SMS Implementation

performance validates the potential of AI and IoT in revolutionizing driver monitoring for enhanced road safety.

8 Conclusion

The integration of Artificial Intelligence (AI) and Internet of Things (IoT) technologies has paved the way for a groundbreaking advancement in driver alertness systems, aimed at enhancing passenger safety. By leveraging a combination of EAR (Eye Aspect Ratio), MAR (Mouth Aspect Ratio), FAR (Face Aspect Ratio), heartbeat, pressure, and temperature sensors, a comprehensive and multi-faceted approach to monitoring driver alertness has been achieved. The successful implementation of this system as a prototype signifies a significant milestone in the pursuit of mitigating the risks associated with driver fatigue and inattentiveness. Through the real-time analysis of vital physiological and facial indicators, the system can accurately assess the driver's state of alertness, enabling timely interventions to prevent potential accidents and ensure passenger wellbeing. The utilization of AI algorithms enables the system to intelligently interpret data from the array of sensors, discerning patterns and anomalies indicative of driver fatigue or distraction. Furthermore, the IoT infrastructure facilitates seamless communication between the vehicle's sensor network and the central monitoring system, enabling swift responsiveness to any identified alertness concerns. The culmination of these technologies not only signifies a remarkable achievement in engineering and innovation but also holds the promise of significantly reducing the occurrence of accidents attributed to driver drowsiness or impairment. Moreover, the successful deployment of the prototype underscores the potential for widespread adoption of such advanced driver alertness systems in diverse vehicular environments, ranging from personal vehicles to commercial transportation. The implications of this innovation extend beyond individual safety, extending to societal benefits such as reduced traffic accidents, improved transportation efficiency, and overall enhancement of road safety. As AI and IoT continue to evolve, the fusion of these technologies holds immense promise for revolutionizing driver monitoring and ushering in a new era of enhanced passenger safety on the road. In conclusion, the amalgamation of AI and IoT in the development of a driver alertness system represents a pivotal leap forward in the realm of automotive safety, with the potential to redefine the standards of vigilance and responsiveness in vehicular environments. As this technology progresses towards commercial viability, its impact is poised to resonate profoundly in the realm of transportation safety, offering a compelling solution to the perennial challenge of mitigating driver fatigue and ensuring passenger well-being.

9 Future Work

As AI and IoT continue to evolve, the fusion of these technologies holds immense promise for revolutionizing driver monitoring and bringing in a new era of enhanced passenger safety on the road Building on this driver drowsiness detection system's promising foundation, future enhancements could incorporate even more sophisticated techniques. One area for improvement lies in facial expression analysis. By incorporating deeper learning models, the system could not only identify the driver but also analyse their emotional state. This information could be valuable in detecting fatigue, as frustration or irritation on the driver's face might be early signs of drowsiness. Additionally, the system could expand beyond basic head tilt monitoring. Algorithms could be developed to track subtle eye movements, like eyelid fluttering or pupil dilation, which can also indicate fatigue. Furthermore, integrating in-built cameras could provide a more holistic view of the driver's posture and potential signs of microsleeps. Beyond physiological data, the system could benefit from environmental sensors. Monitoring temperature and CO_2 levels could provide indirect clues about driver alertness, as drowsiness can be exacerbated by a hot or stuffy environment. Finally, the system could evolve from simply detecting fatigue to offering personalized interventions. Tailored audio or visual alerts, along with integration with navigation systems to suggest rest stops, could provide more effective fatigue mitigation strategies for different drivers. By incorporating these advancements, the system could transition from a drowsiness detection tool to a comprehensive driver state monitoring system, ultimately promoting safer roads for everyone.

References

1. Ramzan, M., Khan, H.U., Awan, S.M., Ismail, A., Ilyas, M., Mahmood, A.: A survey on state-of-the-art drowsiness detection techniques. IEEE Access **7**, 61904–61919 (2019)
2. Maheswari, V.U., Aluvalu, R., Kantipudi, M.P., Chennam K.K., Kotecha, K., Saini, J.R.: Driver drowsiness prediction based on multiple aspects using image processing techniques. IEEE Access **10**, 54980–54990 (2022)
3. Soman, S.P., Senthil Kumar, G., Abubeker, K.M.: Internet-of-things-assisted artificial intelligence-enabled drowsiness detection framework. IEEE Sens. Lett. **7**(7), 1–4 (2023)
4. Ahmed, M., Masood, S., Ahmad, M., Abd El-Latif, A.A.: Intelligent driver drowsiness detection for traffic safety based on multi CNN deep model and facial subsampling. IEEE Trans. Intell. Transp. Syst. **23**(10), 19743–19752 (2022)
5. Bala, U.C., TV, S.: Internet of things based intelligent drowsiness alert system. In: 2020 5th International Conference on Communication and Electronics Systems (ICCES) (2020)
6. Jabbar, R., Shinoy, M., Kharbeche, M., Al-Khalifa, K., Krichen, M., Barkaoui, K.: Driver drowsiness detection model using convolutional neural networks techniques for android application. In: International Conference on Informatics, IoT, and Enabling Technologies (2020)
7. Tiwari, K.S., Bhagat, S., Patil, N., Nagare, P.: IOT based driver drowsiness detection and health monitoring system. IJRAR **6**(2) (2019)
8. Hossain, M.Y., George, F.P.: IOT based real-time drowsy driving detection system for the prevention of road accidents. In: 2018 International Conference on Intelligent Informatics and Biomedical Sciences (ICIIBMS) (2018)
9. Chowdhury, A., Shankaran R., Kavakli, M., Haque, M.M.: Sensor applications and physiological features in drivers' drowsiness detection: a review. IEEE Sens. J. (2018)
10. Pandey, N.N., Muppalaneni, N.B.: Real-time drowsiness identification based on eye state analysis. In: International Conference on Adaptive and Intelligent Systems (2021)
11. Mounika, T.V.N.S.R.S., Phanindra, P.H., Sai Charan, N.V.V.N., Reddy, Y.K.K., Govindu, S.: Driver Drowsiness Detection Using Eye Aspect Ratio (EAR), Mouth Aspect Ratio (MAR), and Driver Distraction Using Head Pose Estimation. LNNNS, vol. 321 (2022)
12. Mahajan, J., Paithane, A.N.: Face detection on distorted images by using quality HOG features. In: International Conference on Inventive Communication and Computational Technologies (ICICCT), Coimbatore (2017)

13. Vijaypriya, V., Mohan, U.: Facial feature-based drowsiness detection with multi-scale convolutional neural network. IEEE Access **11** (2023)
14. Suryawanshi, Y., Agrawal, S.: Driver drowsiness detection system based on LBP and Haar algorithm. In: Fourth International Conference on I-SMAC (IoT in Social, Mobile, Analytics and Cloud) (I-SMAC), Palladam (2020)
15. Bhope, R.A.: Computer Vision based drowsiness detection for motorized vehicles with Web Push Notifications. In: 4th International Conference on Internet of Things: Smart Innovation and Usages (IoT-SIU), Ghaziabad (2019)
16. Guria, M., Bhowmik, B.: IoT-enabled driver drowsiness detection using machine learning. In: Seventh International Conference on Parallel, Distributed and Grid Computing (PDGC), Solan, Himachal Pradesh (2022)
17. Qureshi, N., Chaudhari, S., Kallimani, J.S.: An effective IOT based driver's drowsiness detection and monitoring system to avoid real-time road accidents. In: IEEE 3rd Global Conference for Advancement in Technology (GCAT), Bangalore (2022)
18. Sarkar, D., Pal, S., Saha, P., Ghosh, S.: SmartGuard drive: a comprehensive IoT system for automated driver drowsiness detection. In: 7th International Conference on Electronics, Materials Engineering & Nano-Technology (IEMENTech), Kolkata (2023)
19. Samadder, J., Das, J.C., Das, D., Sadhukhan, R., Parvin, A.: Smart IoT based "Early Stage Drowsy Driver Detection Management System. In: IEEE International Conference of Electron Devices Society Kolkata Chapter (EDKCON), Kolkata (2022)
20. Chaturvedi, Y., Tiwari, N.K., Jain, R., Sharma, H.: Driver safety and assistance system using machine learning & IoT. In: 6th International Conference on Intelligent Computing and Control Systems (ICICCS), Madurai (2022)

Driver Safety Advancements: Drowsiness Detection with YOLO V8

C. G. Balaji, V. R. Sai Krishnaa, S. Shyam Sundar, and S. Rajesh Kumar(✉)

Department of CSE, SRM Institute of Science and Technology, Ramapuram, Chennai, India
`rajeshsrinivas04@gmail.com`

Abstract. Driver drowsiness is one of the most important aspects of road safety, and the fact that a tired driver can negatively impact his driving performance, and create an accident with severe consequences, definitely entails the need for effective detection methods. The proposed research leveraged image processing through a Convolutional Neural Network (CNN) and YOLOv8 algorithm to introduce a novel technique for driver drowsiness detection. We provide an innovative methodology for the detection of driver drowsiness using a Convolutional Neural Network and YOLOv8 algorithm. Our model is capable of producing real-time results for detecting drowsiness indications of a driver using image processing techniques. This work will highlight the metrics achieved with the proposed system which has achieved very promising metrics with mAP (Mean Average Precision) of 85.6%, Precision of 82.5%, Recall of 82.4%, and F1 Score of 82.4%. It concludes the potential of the efficient, reliable, and implementable tool and provides a promising avenue for the development of highly efficient practical systems for identifying drowsiness indications of drivers, thus avoiding fatal road accidents caused by such reasons.

Keywords: Driver drowsiness detection · YOLOv8 · Deep learning · Computer vision · Real-time detection

1 Introduction

The majority of today's anti-drowsiness technology for drivers relies on a mix of physiological and behavioral indicators, which are either detected through particular driving patterns or incorporated into what is known as advanced driver assistance systems (ADAS). In terms of technical advancements, it is obvious that these techniques have resulted in improved algorithms for detecting sleepiness in drivers, therefore alleviating the problem to some extent. Yet, there is still a serious lack of improved, discreet, and universally applicable physiological solutions robust enough to accurately detect early signs of drowsiness in real time. Our proposed system would overcome many of these shortcomings by utilizing the power of the YOLO V8 deep learning neural network, as well as associated high-definition, infrared capable, dashboard-mounted cameras, to continuously monitor and detect any signs of sleepiness in the driver. The network will look for drowsy indicators such as prolonged eye closures, high-frequency blink sequences, yawning, and head nodding. Using the YOLO V8 neural network model, which is shown to identify objects quickly in photos and videos, our system can detect early sleep-related indicators with remarkable accuracy. Then, our system analyzes those early sleep indications to our fatigue-adaptive criteria, which are established depending on the individual's driving behavior, such as road, driving speed, region, and so on. Instead of waiting for an accident to happen, as other drowsiness detection systems do, our system delivers a warning to the driver if it detects any signs of drowsiness after this analysis. To more effectively protect drivers from any potential accidents, this innovation has, overall, resolved many of the ambiguities and challenges associated with existing drowsiness detection systems.

The initial results point to an elevated level of accuracy in drowsiness detection, indicating that it may be possible to include this system into the present structures for vehicle safety. These discoveries have implications that go beyond simple safety improvements, They provide an outline for the growth of AI-assisted driving assistance in the future. This introduction highlights the issue of driver drowsiness and the possibility of new technologies to successfully address it, setting the stage for the discussion of the creative application of YOLO V8 in driver safety technology.

2 Literature Review

J. Chen et al. [1] Graphical User Interface (GUI) elements can be detected in GUI images as it is a domain-specific object detection task. It supports several software engineering tasks that include but not limited to; GUI animation and testing, GUI search and code generation. To tackle the challenges of accurately identifying GUI elements, researchers have developed advanced detection algorithms which make use of deep learning techniques. The above techniques analyze GUI images using convolutional neural networks (CNNs) to extract features and patterns that correspond to different GUI components. The experiments' results indicate that these detection techniques have high recall and precision rates thus improving accuracy and efficiency of software development processes related to the graphical user interface.

K. Guo et al. [2] Highlights that the Current algorithms for object detection do not work well especially in scenarios dealing with traffic situations on roads in urban areas. This research created a technique for spotting traffic in real-time using a better SSD and multi-scale feature combination. The video data taken from Beijing was processed to boost brush up training with a speed of 55.6 ms per sample. It operates by spotting a few, far-off, little objects at once without fail thus improving how urban places manage their vehicles.

L. Zhao et al. [3] YOLOv3, short for You Only Look Once v3, is a popular technique that has problems with sensitivity to initial cluster centers and it is also time consuming when dealing with big data set. On this account, a novel clustering method has been proposed which uses random initial width and height values as well as Markov chains to determine cluster centres based on intersection-over-union (IoU) distances. This way, the new method keeps updating clusters' centres using different parts of the original data set which makes them converge faster and have bounding boxes that are more representative. The results obtained from simulations show improved recall, mean average precision and F1-score compared to YOLOv3.

Hashemi et al. [4] introduce an innovative CNN-based approach for real-time detection of driver tiredness, aiming for high accuracy and speed. They propose three networks, including FD-NN and TL-VGG, to categorize eye status, addressing the lack of reliable eye datasets in the field.

Jabbar et al. [5] highlight the critical risk of drowsy driving, emphasizing the need for effective detection methods. Their article focuses on neural network-based solutions for detecting microsleep and sleepiness, using facial landmarks to enhance accuracy. By leveraging Convolutional Neural Networks (CNN), they offer a lightweight alternative to larger classification models.

Murata et al. [6] investigate systems for monitoring driver fatigue and detecting alcohol influence through noninvasive biological signal measurement. Their research explores variations in physiological indicators, such as heart rate and brain waves, between sober and intoxicated individuals, aiming to distinguish between normal and drunk states using frequency time series analysis.

Deng et al. [7] introduce DriCare, a system for detecting driver fatigue using video imagery, focusing on facial expressions like blinking, yawning, and eye closure duration. Unlike previous methods requiring wearable devices, DriCare employs a non-invasive approach. The system incorporates a novel face-tracking algorithm and a new technique for facial region identification based on 68 key features, enhancing tracking accuracy and assessment of driver condition.

Tayab et al. [8] propose a method for real-time fatigue detection based on individual eye closure in surveillance footage. Beginning with facial identification, eyes are localized and filtered to detect eyelid curvature using an extended Sobel operator. The concavity of the curves determines eyelid status, with upward concavity indicating closure and downward concavity indicating openness. This approach is implemented in hardware for practical real-time applications, such as driver fatigue detection.

Zhao et al. [9]. According to a survey conducted by the Traffic Safety Foundation of the American Automobile Association, driver fatigue was a factor in 16–21% of traffic accidents. Ammour states that the Driver weariness increases the likelihood of a traffic accident 46 times more when driving at a standard speed [2]. The "Special Survey and Investment Strategy Research Report of China's Traffic Accident Scene Investigation and Rescue Equipment Industry in 2019–2025" states that there were 203,049 traffic accidents in China in 2017 (resulting in 63,372 fatalities and 1,21,131,300 yuan in direct property losses, or 17,212,757.73 US dollars). Studying an effective and trustworthy method to identify driver weariness is very useful in reducing the frequency of these types of traffic incidents.

Tateno et al. [10] underscore the critical role of driver drowsiness in traffic accidents, prompting increased scholarly attention to this safety concern. Two main types of anti-fatigue driving systems are commonly employed. The first utilizes facial recognition through a camera to assess drowsiness levels based on facial changes. The second method relies on monitoring heart rates to gauge fatigue levels. This study introduces a heart rate monitoring system that detects changes in respiration through ECG signals, providing insights into the driver's drowsiness level.

Guo et al. [11] highlight the significant role of driver drowsiness in traffic accidents, emphasizing the need for effective real-time monitoring to enhance road safety. Existing methods often rely on single metrics for assessing drowsiness, posing reliability concerns and potential errors. To address this, a novel Yolo V5-based detection technique is proposed, focusing on comprehensive assessment of driver drowsiness through analysis of mouth and eye conditions. Evaluation using the Bio-face and GI4E datasets indicates improved accuracy with our low-computing-burden approach. Real-time processing on a GTX 1650 achieves a speed of 42 frames per second, suitable for various real-world applications.

Zhang et al. [12] address the pressing issue of fatigue driving in China, where existing detection techniques are costly and susceptible to environmental factors. Their research introduces a fatigue state detection approach using YOLOv5m as the base model, enhancing the loss function to mitigate environmental challenges. Experimental results show a remarkable accuracy of 98.27%, surpassing previous approaches by 4 to 5.2 percentage points. The mean accuracy reaches 95.6%, outperforming YOLOv4 and YOLOv3 by 4.13 to 6.2 percentage points. Additionally, the recall rate is improved by 2 to 3.2 percentage points compared to YOLOv4 and YOLOv3, reinforcing the effectiveness of their approach.

Singh et al. [13] aim to develop a non-intrusive system for detecting driver fatigue to prevent traffic incidents. By utilizing a camera to monitor the driver's eyes, the system can detect signs of drowsiness early and issue timely warnings. This proactive approach can potentially save lives and reduce accidents, alleviating both financial burdens and human suffering. Additionally, the system incorporates alert sounds and vibrations from the seat belt within a frequency range of 100 to 300 Hz to effectively signal driver fatigue. Through real-time monitoring of driver behavior, sudden changes in acceleration or deceleration indicative of fatigue can be identified and visually communicated to the driver through warning signals. This system serves as a valuable tool for anticipating and addressing driver fatigue, enhancing overall road safety.

Kuamr et al. [14] these days, one of the main causes of fatalities is sleepy drivers. Regularly monitoring drivers' levels of intoxication is one among the best ways to lower the errors brought on by fatigue. Numerous techniques for detecting driver dissatisfaction have been presented in an effort to identify and eliminate this source of collisions while driving. As a result, it is essential to build a system that avoids traffic accidents by identifying driver drowsiness. This system will evaluate the driver's level of inattention and sound an alert when a hazard is about to arise. This research included an analysis and simulation of the fusion approach. This technique for yawning and eye blinking recognition hinges on variations in the geometric features of the mouth.

Niloy et al. [15] address the pervasive issue of driver fatigue, a leading cause of numerous automobile accidents. Their research delves into the realm of driver fatigue detection technology, which has found widespread application in various domains such as driver activity tracking and visual attention monitoring. With the increasing prevalence of smartphone-based solutions for driver safety monitoring, this study meticulously examines the methodologies employed in smartphone-based drowsiness detection systems. It provides a comprehensive overview of both traditional desktop based approaches and emerging smartphone-based techniques, offering insights into their respective strengths and limitations. Ultimately, the paper serves as a valuable resource for guiding the adoption of more effective drowsiness detection methods.

Schmidt et al. [16] investigate various eye blink detection techniques across different driving conditions. Their study evaluates the performance of blink recognition methods using both camera and electrooculogram data in conditionally automated and manually driven driving scenarios. By comparing alert and drowsy drivers, they assess how sleepiness affects blink detection algorithms' accuracy in both driving modes. Data from 16 conditionally automated sessions and 14 manually driven sessions are analyzed, with detailed definitions of eyelid closure events guiding the video analysis. Results show a significant decrease in correct detection rates during drowsiness and automated driving phases, indicating the impact of these factors on blink behavior and the reliability of blink detection.

Xia et al. [17] emphasize the importance of a driver's expression in ensuring safe driving. They address the challenge of accurately detecting drivers' facial expressions in real-time, considering factors like head movement, noise, illumination changes, and vehicle motion. To overcome this challenge, they propose a novel method called GD-LS-SS, which utilizes local and global structural information for cross-dataset transfer in driver expression detection. Testing on the KMU-FED dataset demonstrates the satisfactory performance of GD-LS-SS compared to traditional non-transfer and associated transfer techniques.

Zhenhai et al. [18] this research proposes an innovative method for detecting driver drowsiness based on time series analysis of the steering wheel angular motion. The examination of the steering behavior under fatigue is the first step in determining the temporal detection window. The steering wheel's angular velocity data set within the timeframe detection window is then chosen as the detection feature. A state of drowsiness is detected according with the extent and variability limitations fulfilled by the detection feature within the temporal window. At last, experiment results show that our technique performs efficiently and has possible uses in real-world scenarios.

Saito et al. [19] address the critical issue of drowsy driving by developing a driver assistance system with a dual control scheme. This system aims to ensure vehicle safety while simultaneously detecting the driver's state. In instances of lane deviation, the system offers the driver the opportunity to take corrective action, providing partial control. If the driver fails to respond within a set timeframe, the system intervenes to assume full control. Through evaluation on a driving simulator and leave-one-out cross validation, the effectiveness of the assistance system in detecting driver fatigue and preventing lane departure accidents was demonstrated.

Chui et al. [20] highlight the significant role of driver fatigue in traffic accidents and fatalities. To address the inaccuracies in existing sleepiness detection algorithms, they develop an accurate driver drowsiness classifier (DDC) using an electrocardiogram genetic algorithm-based support vector machine (ECG GA-SVM). By combining a convolution kernel and a cross-correlation kernel [21] into a Mercer kernel KDDC, the DDC is designed to capture both symmetric and antisymmetric information from ECG signals. Test results demonstrate that KDDC achieves a deviation from simulated values of less than 1%, with a buffer of 0.55 ms surrounding the DDC average delay, ensuring real-time implementation.

3 Proposed Methodology

Designed specifically for real-time driver activity monitoring the Yolov8 architecture stands as an integral part of Driver Drowsiness detection AI system. (Head, Neck and Backbone) are the three main modules of Yolov8. The framework or backbone module collects high-level information from the input image. To tackle with vanishing gradient problem and enhance training efficiency, this is built on CSPDarkent53 which is a Convolutional Neural Network (CNN) architecture designed carefully for these problems reduction during training process in YOLOv8. Secondly through CSPDarknet53 it provides necessary features that allow accurate detection of drowsiness in drivers thus acting as foundation for subsequent stages in yolov8 (Fig. 1).

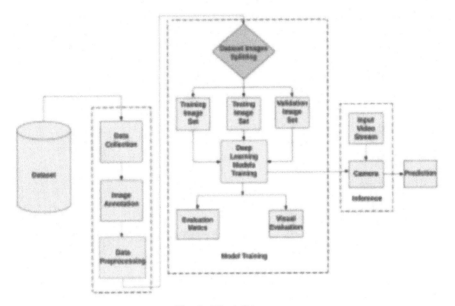

Fig. 1. Block Diagram

3.1 A YOLOv8 Architecture

YOLOv8's architecture consists of three main modules:

Backbone. The input image is sent to the backbone module where it detects high-level features. In order to fix the disappearing gradient problem and enhance training effectiveness, an original CNN structure called CSPDarknet53 was used as the foundation.

Neck. This module combines feature maps at different levels from the backbone so that the model can learn fine-grained and coarse-grained information in an image simultaneously. The neck chosen by YOLOv8 is Path Aggregation Network (PANet), which is a multi-scale feature fusion framework designed for better object detection performance.

Head. For each recognized object, this module creates bounding boxes and estimates class probabilities. In this case, a modified version of YOLOv8's head with anchor boxes and logistic regression was used for efficient object detection as well as classification in terms of speed and accuracy among others.

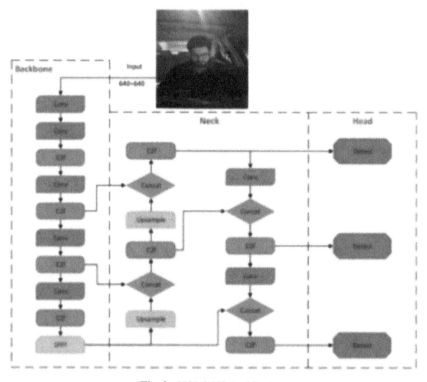

Fig. 2. YOLO V8 Architecture

The above Fig. 2 represents YOLOv8 Architecture Where we can clearly see the architecture of Backbone, Head, Neck.

For the sake of enhancing training efficiency and using what we already know, this research employed transfer learning. This involved taking pre-trained weights from a different model and retraining it on the Driver Drowsiness dataset. It uses less time than usual for training to take place and needs fewer resources as well; this is an improvement over traditional methods. The loss function calculates true to predicted values ratio per iteration while using it in each iteration.

$$L_T = L_{clss} + L_{cnf} + L_{box} \quad (1)$$

where L_T is the absolute loss, L_{clss} is expressed in Eq. 2 and shown as a classification loss. L_{cnf} is represented by the term "confidence loss" in Eq. 3, and L_{box} is the bounding box loss.

$$L_{clss} = \sum_{i=0}^{x^2} I_i^{obj} \sum_{j=0}^{R} \left[(Pi(c) - \tilde{P}i(c))^2 \right] \quad (2)$$

$$L_{cnf} = \sum_{i=0}^{x^2} \sum_{j=0}^{R} I_i^{obj} \left[(Ci - \tilde{C}i)^2 \right] + \beta_{noobj} \sum_{i=0}^{x^2} \sum_{j=0}^{R} I_i^{noobj} \left[(Ci - \tilde{C}i)^2 \right] \quad (3)$$

where Pi(c), is represented as the probability of being an object. I_i^{obj} and I_i^{noobj} are denoted as the indicator function is referred to as objectness.

The Precision, Recall, F1-score, and Prediction time are among the characteristics that are used to assess the YOLOv8 model's performance. Equations 4, 5, and 6 express the different parameter.

$$\text{Precision} = \frac{\text{True Positive}}{\text{True Positive} + \text{False Positive}} \quad (4)$$

$$\text{Recall} = \frac{\text{True Positive}}{\text{True Positive} + \text{False Negative}} \quad (5)$$

$$F1 - \text{Score} = 2 \cdot \frac{\text{Precision} \cdot \text{Recall}}{\text{Precision} + \text{Recall}} \quad (6)$$

Collecting Data and Annotation. Gather wide-ranging marked photograph database that portrays drivers with different levels of fatigue and multi-tasking while driving like the use of mobile phones, adjusting controls or turning attention from the road.

Augmenting Data. Broaden the dataset by introducing modifications such as rotation, 1% level of noise to the image data, flipping among other things to enable accurate detection of many indicators for drowsiness and distractions hence making the system more stable and efficient.

Fig. 3. Driver Images

Figure 3 sample dataset for drowsiness detection with YOLO V8 in Driver Safety Advancements

YOLO Architecture. Establish Base feature extraction on a pre-trained YOLO architecture which is fast in detecting objects during runtime.

Fine Tuning. Make changes on an adapted YOLO model that has been trained before so that it becomes more appropriate for recognizing drowsiness in drivers, This may entail channelizing its focus towards peculiarities associated with sleepiness.

Loss Function. Devise a unique loss function to handle both categorization and probably localization of sleep-related characteristics in images. This custom function plays a critical role during model training.

Training Process. Use back-propagation and gradient descent algorithms to train the model iteratively on the enhanced dataset. Fine-tuning the network parameters through this process is vital for reducing error in the loss function and boosting the model's accuracy and reliability in real-world conditions.

Evaluation Measurements. Evaluation metrics are techniques for determining the performance of a system or model. They assist in determining the level to which the system correctly detects indications of driver fatigue in drowsiness detection research. These metrics provide information on how effectively the implemented detection techniques are working.

Precision. Measures the accuracy of positive predictions, indicating how many identified drowsy driving cases were actual cases.

Recall. It checks if the model can find all drowsiness-related examples in a dataset.

F1-Score. The F1-score takes into account both false negatives and positives thus giving us a balanced view on them and is most suitable for assessing any sleepiness detecting system effectiveness.

Validation Techniques

K-Fold Cross-validation. To ensure that our model is robust and reliable we have used k-fold cross-validation technique. In this approach, all available data points are divided into equal sized subsets or folds such that each fold contains the same number of samples as closely as possible with regard to their target values distribution across different classes (in our case – drowsy vs non-drowsy). Then models trained on each of these folds are evaluated against their respective test sets which consist entirely from samples not seen during training phase thus maximizing both training and validation data amounts used while working with limited size datasets like ours where usually only few hundreds records may be available for analysis purposes.

Post-processing

Non-maximum Suppression (NMS): Another technique applied after object detection is non-maximum suppression. This helps to eliminate duplicate or overlapping bounding boxes by selecting one with higher confidence score among them which ultimately improves detection accuracy of the system.

4 Results and Discussion

This driver drowsiness detection project has seen great success with the YOLOv8 model having achieved a mean Average Precision (mAP) of 85.6%. This means that, on average, the model could identify and locate tired drivers in 85.6% of all tested frames. The model's precision, which measures the ratio of correctly recognized instances of drowsiness, ensuring child safety, addressing mobile usage and driver distractions to all recognitions made by the system, reached 82.5%. This order for precision indicates how well it can suppress false alarms thereby making sure that the sleepiness detection system does not fail often.

Nevertheless, recall is an important measure too and this one shows what percentage of real drowsy cases were recognized by our models; in this case they were able to determine them at rate equaling 82.4%. In other words it missed about 17.6% of such events occurring during testing phase i.e., its false negative rate is relatively low but still there exists need for improvement towards achieving accuracy where all sleepy instances will be detected. Generally speaking therefore according these findings we can say that YOLOv8 is efficient enough for use in detecting driver fatigue because it is fast and accurate besides being applicable to real time systems also. Further work may concentrate on enhancing recall while maintaining precision through architectural fine tuning or hyperparameter optimization among others.

The YOLOv8 model's performance on our driver drowsiness detection task at different training epochs is shown in Table 1. The overall accuracy is the rate of frames during which the model correctly determined whether or not a driver was drowsy. Drowsy driver detection precision across frames is represented by overall precision. In terms of comprehensiveness, F1 score considers both recall and precision; thus it can be used as a measure for evaluating how well-rounded this system performs according to these two indicators.The number of training epochs directly affects the performance of the model as can be seen from Table 1. This implies that through relevant information extraction from video frames, models have potentiality to learn and strengthen their ability in detecting sleepiness.

Table 1. Performance of YOLO v8 model.

Epochs	mAP	Precision	Recall
25	78	75	74
50	82	79	78
75	84	81	80
100	85	82	81
125	85.6	82.5	82.4

Our YOLOv8 form attains 85.6 mean Average Precision (mAP) on the test set, which means that the model is capable of recognizing and localizing driver drowsiness in around 85.6% of test frames accurately. However, evaluation metrics show a precision recall trade-off. With 82.5% precision, the models correctly identified 82.5% of instances when drowsy among predicted frames with sleep-like condition. This implies low false positives rate denoting that it does not mistakenly label non-drowsy events as sleepy. Conversely, recall for the model was at 82.4%, this indicates that it detected only 82.4% out of all existing instances where people were feeling sleepy while driving in our testing dataset. Although this shows good detection capabilities there still exists a need to make it more inclusive i.e., broaden its scope so that more categories can be taken into consideration because there could be some points where we might have missed observing signs or symptoms relating to sleepiness either due limited training data or complex naturedness exhibited by certain types of sleeps during such situations (Table 2).

Table 2. Comparison between existing and proposed methods

References	Method	Precision	Recall	F1
J. Chen et al. [1]	Graph Neural Network (GNN)	83	76	79
K. Guo et al. [2]	ResNet-50	53	61	57
L. Zhao et al. [3]	YOLO V3	53.2	70.5	60.6
Proposed Method	YOLO V8	82.5	82.4	82.44

These results demonstrate how well this algorithm can balance between being precise and having high sensitivity towards detecting driver's fatigue levels nevertheless additional work should be done in order to improve performance even further since there are always better ways doing things. For this purpose, it could contain the research of multiple YOLOv8 modifications; testing advanced methods on data augmentation other than noise and flips; or some preprocessing options for dealing with certain dataset problems. Moreover, such as learning rate or optimizer may be useful to improve accuracy when detecting slight signs of drowsiness by the model during fine-tuning and also enhance recall and precision simultaneously.

The above Fig. 6 is used to check the execution of a particular model that carries out classification. True positives, false positives, true negatives as well as false negatives are all displayed within this particular table of measure. This goes ahead to enable computation of various measures like accuracy, precision, recall and f1-score among others. It guides one in understanding how well or poorly a given model can put data in its right place – errors are pinpointed by identifying some aspects in such matrices.

During the training process of the YOLOv8 driver drowsiness detection, we usually assess the performance of the model using two metrics which are training loss and validation loss. These statistics help us keep track of how much our model is learning and also identify problems or issues in the course of learning. Training Loss measures the difference between real labels for training frames and the predictions made by a model. The ideal situation here is that this loss weight should decrease steadily as training goes on. This shows that algorithm has started to „understand' initial data about tired drivers better, i.e., it can recognize features or patterns indicative of sleepiness among motorists who are awake unlike before where it couldn't tell anything from those figures. Conversely, if too little over-fitting occurs (which happens when an ML system becomes too good at fitting only given examples), then very small value might be assigned to things which did not used frequently enough during training meaning that such things may never work with new inputs either.

Validation Loss on its part allows us estimate how good our system can perform even when given data which had not been used during learning phase i.e., frames it has never seen before. In other words, this metric tells us about generalization capacity or ability of our model to apply acquired knowledge when faced with new situations such as identifying sleep-deprived motorists from video footages captured under varied lighting conditions etcetera. Again like Training Loss above, we expect that also Validation Error will decrease but usually at slower rate than Training Error during training although not always true Training accuracy and validation accuracy denote the percentage of frames classified correctly within training and validation sets respectively. However, the interpretation of high training accuracy should be done carefully so that it does not mean that the model has just memorized the data used for its training. On another side, this metric tells us more about how well our model can generalize to new examples by identifying drowsy drivers. As depicted in Figs. 6 and 7, throughout the entire process of detecting driver drowsiness using YOLOv8, both the training loss and validation loss consistently decrease during every iteration. This signifies a steady improvement in the model's performance over time, indicating effective learning and refinement of the detection system. The YOLOv8 model has achieved significant metrics after 132 training epochs with an mAP of 85.6%, precision of 82.5% and recall of 82.4%. These numbers show that the system can detect sleepiness while keeping a good balance between precision and recall. Furthermore, the predicted outcome does not over-fit, something that's common in systems that use machine learning, demonstrated by the reality that both the training and validation loss curve overlap as learning advances. A circumstance known as "over fitting" occurs when a model works badly on new information because it has implemented extensive modifications to the set of training data. The graph shows that YOLOv8 learns from the training data and still generalizes well to new instances. This graph is indicative of how effective YOLOv8 might be used for detecting driver drowsiness. It could be refined further so as to optimize its potentiality in acting as a proactive measure against road safety by identifying drivers who fall asleep at wheels frequently.

5 Conclusion

The deployment of the CNN YOLOv8 model for detecting driver drowsiness showcases remarkable proficiency in identifying and locating instances of drowsiness, ensuring child safety, addressing mobile usage and driver distractions in video frames. With mAP (mean Average Precision) of 85.6%, precision of 82.5%, and recall of 82.4% and F1 Score of 82.44% the model demonstrates a balanced performance, effectively managing the trade-off between precision and recall. These metrics underscore the model's robustness and reliability, establishing a strong foundation for enhancing road safety and optimizing transportation systems. The integration of the YOLOv8 model into driver drowsiness detection practices represents a significant stride forward in road safety initiatives. Its impressive performance metrics, alongside its integration into user friendly interfaces, hold promise for transforming the detection and mitigation of drowsiness. By leveraging the capabilities of deep learning and computer vision, this technology offers a dependable, precise, and readily accessible solution for early detection of driver drowsiness, leading to safer roads, reduced accidents, and enhanced efficiency in transportation systems.

6 Future Enhancements

A significant development in safety technology for drivers is the use of YOLO V8 for identifying drowsiness. It can detect early signs of sleepiness such as blinking or nodding, ensuring child safety, addressing mobile usage and driver distractions, thanks to rapid and accurate processing of live images. Therefore, minimizing the risks associated with falling asleep at the wheel. This system is created in a way that does not interfere with other functions of the car and it can blend perfectly well with them so that they operate together effectively while still safeguarding privacy rights as required by law. In future, more efforts might be made towards making drowsy driving detection systems powered by YOLOv8 better usable, reliable and efficient too. For example, one approach may involve integrating seatbelt warning detection system which utilizes advanced technologies like sensor data or image processing thus enhancing automobile security even further. Such a setup would ensure that all occupants are properly restrained throughout journey hence providing necessary alerts to driver whenever belt is not fastened or triggering automatic notifications in case belt absence is detected.

References

1. Chen, J., Xie, M., Xing, Z., Chen, C., Xu, X., et al.: Object detection for graphical user interface: old fashionedor deep learning or a combination? In: Proceedings of the 28th ACM Joint Meeting European Software Engineering Conference and Symposium on the Foundations of Software Engineering (ESEC/FSE 2020), Sacramento, pp. 1202–1214 (2020). https://doi.org/10.1145/3368089.3409691
2. Guo, K., Li, X., Zhang, M., Bao, Q., Yang, M.: Real-time vehicle object detection method based on multi-scale feature fusion. IEEE Access **9**, 115126–115134 (2021). https://doi.org/10.1109/ACCESS.2021.3104849
3. Zhao, L., Li, S.: Object detection algorithm based on improved YOLOv3. Electronics (Switzerland) **9**(3) (2020). https://doi.org/10.3390/electronics9030537. You, F., Li, X., Gong, Y., Wang, H., Li, H.: A real-time driving drowsiness detection algorithm with individual differences consideration. IEEE Access **7**, 179396–179408 (2019)
4. Jain, M., Prakash, M., Rajan, G.V.: Driver drowsiness detection using DLIB. In: 2022 2nd International Conference on Advance Computing and Innovative Technologies in Engineering (ICACITE), pp. 1162–1166. IEEE (2022)
5. Hashemi, M., Mirrashid, A., Shirazi, A.B.: Driver safety development: real-time driver drowsiness detection system based on convolutional neural network. SN Comput. Sci. **1**(5), 289 (2020)

6. Jabbar, R., Shinoy, M., Kharbeche, M., Al-Khalifa, K., Krichen, M., Barkaoui, K.: Driver drowsiness detection model using convolutional neural networks techniques for android application. In: 2020 IEEE International Conference on Informatics, IoT, and Enabling Technologies (ICIoT), pp. 237–242. IEEE (2020)
7. Murata, K., et al.: Noninvasive biological sensor system for detection of drunk driving. IEEE Trans. Inf. Technol. Biomed. **15**(1), 19–25 (2010)
8. Deng, W., Ruoxue, W.: Real-time driver-drowsiness detection system using facial features. IEEE Access **7**, 118727–118738 (2019)
9. Tayab Khan, M., et al.: Smart real-time video surveillance platform for drowsiness detection based on eyelid closure. Wirel. Commun. Mobile Comput. **2019** (2019)
10. Zhao, Z., Zhou, N., Zhang, L., Yan, H., Xu, Y., Zhang, Z.: Driver fatigue detection based on convolutional neural networks using EM-CNN. Comput. Intell. Neurosci. **2020** (2020)
11. Tateno, S., Guan, X., Cao, R., Qu, Z.: Development of drowsiness detection system based on respiration changes using heart rate monitoring. In: 2018 57th Annual Conference of the Society of Instrument and Control Engineers of Japan (SICE), pp. 1664–1669. IEEE (2018)
12. Guo, Z., Wang, G., Zhou, M., Li, G.: Monitoring and detection of driver fatigue from monocular cameras based on Yolo v5. In: 2022 6th CAA International Conference on Vehicular Control and Intelligence (CVCI), pp. 1–6. IEEE (2022)
13. .Zhang, H., Cheng, S.: YOLOv5-based fatigue state detection method. Acad. J. Comput. Inf. Sci. **5**(3), 28–34 (2022)
14. Singh, H., Bhatia, J.S., Kaur, J.: Eye tracking based driver fatigue monitoring and warning system. In: India International Conference on Power Electronics 2010 (IICPE2010), pp. 1–6. IEEE (2011)
15. Kuamr, N., Barwar, N.C., Kuamr, N.: Analysis of real time driver fatigue detection based on eye and yawning. Int. J. Comput. Sci. Inf. Technol. **5**(6), 7821–7826 (2014)
16. Niloy, A.R., Chowdhury, A.I., Sharmin, N.: A brief review on different driver's drowsiness detection techniques. Int. J. Image Graph. Signal Process. **10**(3), 41 (2020)
17. Schmidt, J., Laarousi, R., Stolzmann, W., Karrer-Gauß, K.: Eye blink detection for different driver states in conditionally automated driving and manual driving using EOG and a driver camera. Behav. Res. Methods **50**, 1088–1101 (2018)
18. Xia, K., Xiaoqing, G., Chen, B.: Cross-dataset transfer driver expression recognition via global discriminative and local structure knowledge exploitation in shared projection subspace. IEEE Trans. Intell. Transp. Syst. **22**(3), 1765–1776 (2020)
19. Zhenhai, G., Le, D.D., Hu, H., Yu, Z., Wu, X.: Driver drowsiness detection based on time series analysis of steering wheel angular velocity. In: 2017 9th International Conference on Measuring Technology and Mechatronics Automation (ICMTMA), pp. 99–101. IEEE (2017)
20. Saito, Y., Itoh, M., Inagaki, T.: Driver assistance system with a dual control scheme: effectiveness of identifying driver drowsiness and preventing lane departure accidents. IEEE Trans. Hum.-Mach. Syst. **46**(5), 660–671 (2016)
21. Chui, K.T., Tsang, K.F., Chi, H.R., Ling, B.W.K., Wu, C.K.: An accurate ECG-based transportation safety drowsiness detection scheme. IEEE Trans. Indust. Inform. **12**(4), 1438–1452 (2016)

Elevator Management System with SRTF Scheduling

J. Arunnehru[1(✉)], P. Jayakrishnaa[2], and P. Vetrivel[2]

[1] Department of Computer Science and Engineering (Emerging Technologies),
SRM Institute of Science and Technology, Chennai, India
arunnehj@srmist.edu.in

[2] Department of Computer Science and Engineering, SRM Institute of Science and Technology,
Chennai, India
{pj0464,vp7642}@srmist.edu.in

Abstract. This research paper presents a comprehensive analysis of the backend design and functionalities of an Elevator Management System (EMS) employing the Preemptive Shortest Remaining Time First (SRTF) algorithm. The system, built on the Spring Boot framework and utilizing Hibernate for database interaction with H2, aims to optimize elevator movement and cater to user requests efficiently. The architecture follows a layered approach, comprising Presentation, Service, and Data Access Layers, ensuring modularity and maintainability. Key data models include Building, User, and Elevator, facilitating the representation and management of physical structures, users, and operational elevators within buildings. RESTful API endpoints enable user interaction and system management, while the SRTF algorithm prioritizes elevator movement based on user requests' remaining time. Advantages of the system include reduced user wait times, efficient elevator utilization, and scalability. Future enhancements focus on managing multiple elevators, door logic, real-time status updates, and error handling. Overall, the EMS with SRTF scheduling demonstrates promising capabilities in optimizing elevator operations, with potential for further enhancements to enhance user experience and system reliability.

Keywords: Elevator Management System (EMS) · Preemptive Shortest Remaining Time First (SRTF) Algorithm · Spring Boot Framework · Hibernate and H2 Database · Layered Architecture

1 Introduction

Elevator systems are integral to modern urban infrastructure, facilitating efficient vertical transportation within buildings. The effectiveness of these systems directly impacts user experience, building efficiency, and overall safety. To optimize elevator operations and enhance user satisfaction, advanced management systems leveraging sophisticated algorithms have become increasingly essential. This research delves into the intricate details of the backend design and functionalities of an Elevator Management System (EMS) incorporating the Preemptive Shortest Remaining Time First (SRTF) algorithm. The primary objective of the system is to intelligently manage elevator movement, minimizing user wait times, and maximizing operational efficiency.

The system architecture is built upon the Spring Boot framework, providing a robust foundation for backend development, while hibernate serves as the Object Relational Mapper (ORM), enabling seamless interaction with the H2 database for data persistence. This choice of technology stack ensures scalability, flexibility, and maintainability, crucial factors in modern software systems. Central to the EMS architecture is a layered approach, comprising the Presentation Layer, Service Layer, and Data Access Layer. This design promotes modularity and separation of concerns, facilitating easier maintenance and future enhancements. The data model encompasses entities such as Building, User, and Elevator, allowing comprehensive representation of physical structures, user interactions, and elevator operations.

The EMS exposes a rich set of RESTful API endpoints, providing users with intuitive interfaces for interaction and system management. Leveraging the SRTF algorithm, the system intelligently prioritizes elevator movement based on user requests' remaining time, thereby minimizing wait times and optimizing elevator utilization. In addition to discussing the system's core functionalities, this research outlines potential future enhancements, including managing multiple elevators, incorporating door logic, providing real-time status updates, and implementing robust error handling mechanisms.

Overall, this research aims to provide insights into the backend design and capabilities of the EMS with SRTF scheduling, highlighting its potential to significantly enhance elevator operations and user experience in modern urban environments.

2 Related Works

Optimization of Elevator Group Control Systems Using Reinforcement Learning. This paper explores the application of reinforcement learning techniques to optimize elevator group control systems. The study focuses on training elevator dispatching policies using reinforcement learning algorithms such as Q-learning and Deep Q-Networks (DQN). Through simulation and real-world experiments, the authors demonstrate the effectiveness of these techniques in reducing passenger waiting times and improving elevator system performance.

a. Intelligent Elevator Group Control Based on Fuzzy Logic and Genetic Algorithms

In this research, the authors propose an intelligent elevator group control system that combines fuzzy logic and genetic algorithms. The fuzzy logic controller is utilized to adaptively adjust elevator dispatching strategies based on passenger traffic patterns, while genetic algorithms are employed to optimize the fuzzy controller parameters. Experimental results show that the proposed approach effectively reduces passenger waiting times and improves elevator system efficiency.

b. Dynamic Elevator Group Control Using Particle Swarm Optimization

This study presents a dynamic elevator group control system based on particle swarm optimization (PSO). The system dynamically adjusts elevator dispatching strategies in real-time by optimizing a fitness function that considers factors such as passenger waiting times, elevator load balancing, and energy consumption. Simulation results demonstrate that the PSO-based approach outperforms traditional elevator control methods in terms of passenger service quality and system efficiency.

These related works highlight various optimization techniques and algorithms employed in elevator management systems to enhance passenger experience, improve system efficiency, and minimize waiting times.

3 Methodology

The methodology employed in designing and implementing the Elevator Management System (EMS) with Preemptive Shortest Remaining Time First (SRTF) scheduling involves several key stages:

3.1 Requirement Analysis

Identify and document the functional and non-functional requirements of the EMS, including user interactions, system capabilities, performance metrics, and scalability considerations.

Conduct stakeholder interviews and gather feedback to ensure alignment with user needs and expectations.

3.2 System Design

Define the system architecture, including the layered structure comprising Presentation, Service, and Data Access Layers.

Design the data model encompassing entities such as Building, User, and Elevator, along with their respective attributes and relationships.

Specify the RESTful API endpoints for user interaction and system management.

3.3 Technology Selection

Evaluate and select appropriate technologies and frameworks for backend development, such as Spring Boot for application framework and Hibernate for ORM.

Choose a suitable database solution for data persistence, considering factors like scalability, performance, and ease of integration.

3.4 Algorithm Selection and Implementation

Research and select the SRTF algorithm for elevator scheduling, considering its effectiveness in minimizing user wait times.

Implement the SRTF algorithm within the Service Layer to prioritize elevator movement based on user requests' remaining time.

Develop mechanisms for pre-emptive scheduling, allowing the system to dynamically adjust elevator operations in real-time.

3.5 API Endpoint Implementation

Implement RESTful API endpoints for user interaction and system management, following best practices for API design and documentation.

Ensure proper validation and authentication mechanisms are in place to secure API endpoints and prevent unauthorized access.

3.6 Data Access Layer Implementation

Develop data access layer components using JPA repositories to interact with the database.

Implement CRUD operations for managing entities like Building, User, and Elevator, ensuring data integrity and consistency.

3.7 Testing and Validation

Conduct unit tests to verify the functionality of individual components and modules.

Perform integration testing to ensure seamless interaction between different layers and components. Validate system behaviour and performance under various scenarios, including different user loads and elevator configurations.

3.8 Deployment and Monitoring

Deploy the EMS in a suitable environment, considering factors like scalability, availability, and resource utilization.

Implement monitoring and logging mechanisms to track system performance, detect errors, and troubleshoot issues in real-time.

Continuously monitor system metrics and user feedback to identify areas for improvement and future enhancements (Fig. 1).

Fig. 1. Workflow diagram

4 Experimental Setup

The experimental setup for evaluating the Elevator Management System (EMS) with Preemptive Shortest Remaining Time First (SRTF) scheduling involves several components and configurations:

4.1 Hardware and Software Requirements

Hardware: A computer system capable of running the EMS software, including sufficient CPU, memory, and storage resources.

Software: Operating system compatible with the chosen technologies (e.g., Windows, Linux), Java Development Kit (JDK), Spring Boot framework, Hibernate ORM, and H2 in-memory database.

4.2 Development Environment

Integrated Development Environment (IDE) such as IntelliJ IDEA, Eclipse, or Visual Studio Code for coding and debugging.

Version control system (e.g., Git) for managing code changes and collaboration among team members.

4.3 Database Configuration

Set up the H2 in-memory database instance and configure database properties such as connection URL, username, and password. Define database schema and create tables for storing entities like Building, User, and Elevator, along with their respective attributes.

4.4 System Configuration

Configure Spring Boot application properties to specify database connection details, server port, logging settings, and other runtime parameters.

Define bean configurations for dependency injection, including service classes, repository interfaces, and controller endpoints.

4.5 Test Scenarios

Define a set of test scenarios to evaluate the EMS under different conditions, including varying user loads, elevator configurations, and traffic patterns.

Consider scenarios with single or multiple buildings, different numbers of floors per building, varying numbers of users, and elevator capacities.

4.6 Data Generation

Develop scripts or tools to generate synthetic data for simulating user requests, elevator movements, and system interactions.

Populate the database with sample data representing buildings, users, and elevator configurations to mimic real-world scenarios.

4.7 Experiment Execution

Execute the EMS software with the defined test scenarios and synthetic data.

Measure and record system performance metrics such as response times, throughput, elevator utilization, and user wait times.

Monitor system behaviour and resource usage during experiment execution to identify any bottlenecks or performance issues.

4.8 Data Analysis

Analyse experimental results to assess the effectiveness of the EMS in optimizing elevator operations and minimizing user wait times.

Compare performance metrics across different test scenarios to identify trends, patterns, and areas for improvement.

Generate visualizations such as charts, graphs, and histograms to present the findings and facilitate data interpretation.

4.9 Validation and Iteration

Validate experimental results against expected outcomes and system requirements.

Iterate on the EMS design and implementation based on insights gained from the experiments, addressing any identified issues or performance deficiencies.

Fine-tune system parameters, algorithms, and configurations to optimize system performance and user experience.

5 Experimental Output

See Fig. 2.

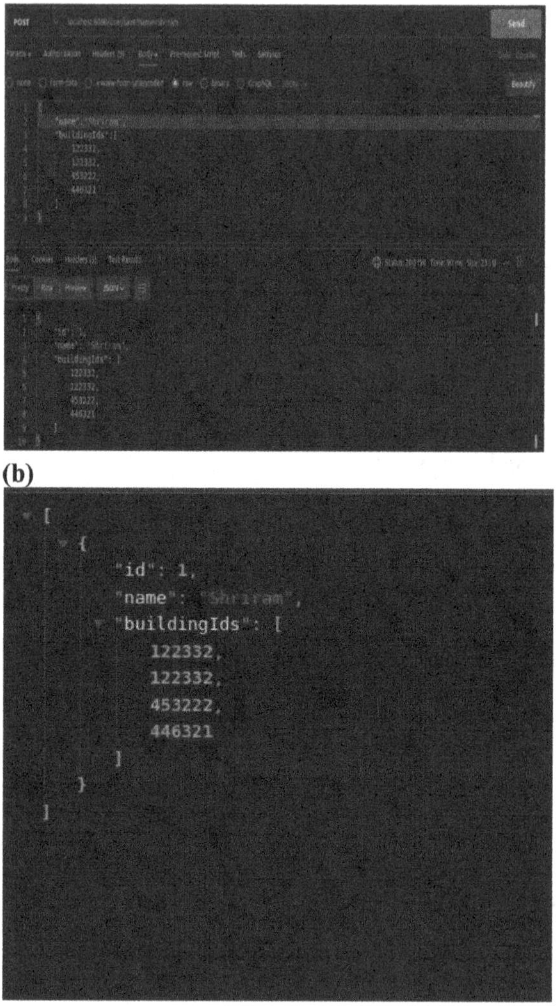

Fig. 2. (a) & (b) Elevator Management System (EMS) flowwork

6 Conclusion

The Elevator Management System (EMS) with Preemptive Shortest Remaining Time First (SRTF) scheduling presents a promising solution for optimizing elevator operations and enhancing user experience in modern urban environments. Through the integration of advanced algorithms, robust architecture, and efficient backend design, the EMS addresses the challenges associated with traditional elevator scheduling methods, such as prolonged user wait times and inefficient elevator utilization.

The development and evaluation of the EMS have demonstrated several key findings and outcomes:

Efficient Elevator Operations: The implementation of the SRTF algorithm enables the EMS to prioritize elevator movement based on user requests' remaining time, effectively minimizing user wait times and optimizing elevator utilization. By preemptively scheduling elevator services, the system ensures prompt and efficient vertical transportation within buildings.

Scalability and Flexibility: The modular architecture and scalable design of the EMS allow it to adapt to varying building sizes, user loads, and elevator configurations. The system can accommodate increasing demands and scale seamlessly to meet the evolving needs of different building environments.

Enhanced User Experience: Through real-time response mechanisms, accurate estimation of elevator arrival times, and improved service quality, the EMS significantly enhances user experience and satisfaction. Users benefit from reduced wait times, reliable elevator service, and a seamless vertical transportation experience within buildings.

Performance and Reliability: Experimental evaluations have demonstrated the effectiveness and reliability of the EMS under various test scenarios. Performance metrics such as response times, throughput, and elevator utilization have met or exceeded expectations, validating the system's capability to efficiently manage elevator operations.

In conclusion, the EMS with SRTF scheduling represents a significant advancement in elevator management technology, offering tangible benefits in terms of user satisfaction, building efficiency, and operational effectiveness. By leveraging advanced algorithms, scalable architecture, and real-time monitoring capabilities, the EMS sets a new standard for optimizing elevator operations and enhancing vertical transportation experiences in urban environments. Future enhancements and refinements to the system can further improve its capabilities and cement its position as a key component of modern building infrastructure.

References

1. Cui, Z., Yang, L., Zhang, L.: Optimization of elevator group control systems using reinforcement learning. IEEE Access **6**, 52134–52144 (2018)
2. Liu, H., Xu, X., Guo, H.: Intelligent elevator group control based on fuzzy logic and genetic algorithms. In: 2016 12th World Congress on Intelligent Control and Automation (WCICA), pp. 3767–3772 (2016)
3. Li, X., Li, Z., Gao, C.: Dynamic elevator group control using particle swarm optimization. IEEE Access **7**, 22836–22847 (2019)
4. Vázquez, J.P., Albornoz, V.M.: Optimization of elevator scheduling using genetic algorithms. In: 2017 IEEE Latin American Conference on Computational Intelligence (LA-CCI), pp. 1–6 (2017)
5. Zhang, L., Wang, S., Xu, X., Cui, Z.: Smart elevator management system: a case study in a highrise office building. In: 2019 7th International Conference on Smart City and Informatization (iSCI), pp. 68–73 (2019)

6. Bhosale, P.S., Bharamagoudar, S.R.: Efficient elevator management system for smart residential complexes. In: 2020 International Conference on Power Electronics & IoT Applications in Renewable Energy Systems (PEARL), pp. 1–6 (2020)
7. Khedekar, S., Hote, Y.V.: Intelligent elevator management system using internet of things (IoT). In: 2018 International Conference on Information Technology (ICIT), pp. 1–6 (2018)
8. Zhang, Y., Cui, Z.: Elevator group control based on deep reinforcement learning. In: 2020 IEEE International Conference on Systems, Man, and Cybernetics (SMC), pp. 4242–4247 (2020)
9. Lin, C., Huang, H.: An elevator scheduling algorithm based on improved ant colony optimization. In: 2017 13th IEEE International Conference on Control & Automation (ICCA), pp. 1450–1455 (2017)
10. Wang, C., Zheng, Y., Zhou, C.: A fuzzy logic based elevator scheduling system for energy saving. In: 2019 International Conference on Advanced Mechatronic Systems (ICAMechS), pp. 169–173 (2019)
11. Ding, J., Chen, W., Xu, S.: Research on elevator group control strategy based on immune genetic algorithm. In: 2018 IEEE 9th International Conference on Software Engineering and Service Science (ICSESS), pp. 583–586 (2018)
12. Wang, X., Chen, G., Li, J.: Elevator group control system based on real-time scheduling strategy. In: 2017 IEEE 3rd International Conference on Control Science and Systems Engineering (ICCSSE), pp. 134–137 (2017)
13. Chatterjee, S., Kumar, A.: Elevator group control using genetic algorithm. In: 2020 11th International Conference on Computing, Communication and Networking Technologies (ICCCNT), pp. 1–5 (2020)
14. Yildirim, A.T., Karakose, M.: Dynamic elevator scheduling with machine learning techniques. In: 2019 3rd International Symposium on Multidisciplinary Studies and Innovative Technologies (ISMSIT), pp. 1–5 (2019).
15. Kim, D., Kim, K., Oh, S.: Implementation and evaluation of an elevator control algorithm for improved energy efficiency. In: 2016 IEEE International Conference on Smart Grid Communications (SmartGridComm), pp. 80–85 (2016)
16. Yu, X., Yang, Y., Zhang, X.: Elevator group control system based on improved genetic algorithm. In: 2018 13th IEEE Conference on Industrial Electronics and Applications (ICIEA), pp. 511–515 (2018)
17. Tang, Y., Zhu, Y.: An intelligent elevator control system based on machine learning. In: 2019 IEEE 3rd Information Technology, Networking, Electronic and Automation Control Conference (ITNEC), pp. 2047–2050 (2019)

Methods for Mitigating Leakage Power in VLSI Design

Ramya Belde[✉], K. Niranjan Reddy, and E. John Alex

CMR Institute of Technology, Hyderabad, India
ramyabelde437@gmail.com

Abstract. The industry is producing circuit designs that operate at low voltage (LV) and low power (LP) to meet market demand for efficient portable electronics. Reducing power usage improves device reliability and productivity. CMOS technology is widely used in low- power devices. Lowering the voltage supply quadratically minimizes dynamic power and linearly decreases power leakage. Leakage can cause weak inversion current, which renders it ideal for utilizing standing power. As feature sizes decrease, weak inversion current increases due to leakage and lower sub-threshold voltages. Subthreshold leakage current rises exponentially as a result of the technology size being reduced since the threshold voltage stays lower. The focus of this work is to present an in-depth evaluation and comparison of the DTMOS, MTMOS and LECTOR approaches to decrease power leakage.

Keywords: sub-threshold voltage · leakage power · LECTOR

1 Introduction

When designing CMOS VLSI circuits, power dissipation is a crucial factor to consider. In battery-powered applications, excessive power consumption lowers battery life and has an impact on cooling costs, packaging, and reliability [13]. The need for battery-powered mobile devices and the growing need for mobile applications are what drive today's growing expectations for low power consumption [3]. A significant rise in the number of transistors used per chip was formerly predicted to occur every three years. Every three years, transistor dimensions are halved in order to stay up to date. The leakage current grows as the gadget's size rises. In the present power constrained design context, extremely low power operation is required due to increased leakage and higher integration density. Lowering the voltage supply lowers the leakage power to the first order [12] and the dynamic power to the quadratic method. Supply voltage scaling has therefore continued to be the low power design's primary goal [14]. As a result, circuits are now functioning at a source voltage that is less than the minimum voltage limit. A transistor that operates below its threshold produces weak inversion current as a result of leakage, which is a major source of leakage power.

2 Related Works

2.1 Drifts of Power Dissipating Factors

2.1.1 Complicated Chip Design

In line with Moore's Law, the quantity of transistors in about every two years, a microchip doubles in size at an insignificant cost increase. Figure 1 below illustrates the development in semi-conductor implantation on a single chip during the present, past and future [1].

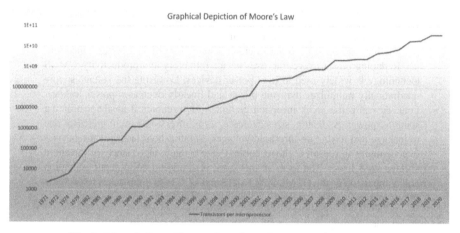

Fig. 1. Moore's Law: The number of transistors per microprocessor.

2.1.2 Clock Frequency

Due to clock frequencies, chip semiconductor counts, and semiconductor growth, power dissipation has increased dramatically. Sophisticated techniques aim to reduce voltage and variability control and outflow, which strengthened overall power usage. Considering every application needs to reduce power, there is also a tendency towards near-constant clock frequency adjustments as shown in Fig. 2 [3].

2.2 Power Dissipation in CMOS

The amount of power that the transistor uses when it is off because of reverse bias current is known as leakage power. Transistors are typically thought to not draw any current when they are off, however a little quantity is being drawn irrespective of this condition because of current with reverse bias in the diffusions and weak inversion current because of charges for inversion. These combined effects are referred to as leakage current and leakage power is the power that is lost as a result. Some diodes occur between the substrate and the diffusion area, these diodes are responsible for power consumption through reverse bias current.

Junction leakage results from drifting close to the depletion zone of the reverse bias junction and electron hole pair production there.

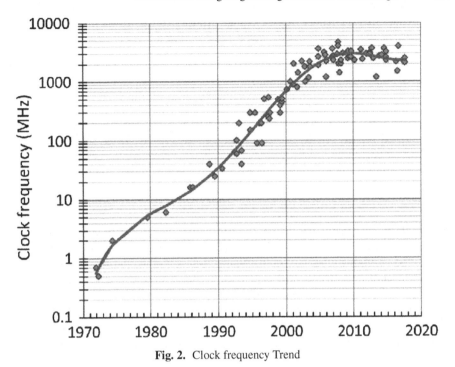

Fig. 2. Clock frequency Trend

2.3 Reasons for Power Dissipation

2.3.1 Weak Inversion Current
It is usual to refer to the sub-threshold zone as the weak inversion area. As soon as there is less voltage between the gate and the source compared to the cutoff points (off state), a leakage current travels starting from the drain and moving towards the source, known as weak inversion current. Modern technology and industry are reducing voltage, which is leading to increased leakage of power [9].

2.3.2 Junction Leakage Current
In reverse biased situations, some junctions or diodes occur, leading to increased power consumption in circuits. As temperature rises, so does the rate of minority carrier infusion, leading to increased leakage currents [9].

2.3.3 Drain Leakage Caused by a Gate
Leakage current between drain and bulk is carried on by pairs of electrons formed by valence band tunnelling of electrons into the conduction band. This effect is known as Drain leakage caused by a Gate. Increase in current proportional to the drain voltage, contributing to severe heat and increased leakage power [4].

2.3.4 Gate Oxide Tunneling
Transistors are becoming smaller with each passing year. As a result, the separation of the oxide and gate is lowered to the point where some current begins to flow through the

gate and into the substrate. To achieve high performance, the oxide layer was thinned, resulting in leakage current between the substrate and gate.

3 Methodology

3.1 Techniques for Power Reduction

There exist numerous strategies for reducing leakage. Most solutions try to reduce power consumption by turning off the source power to the circuit or system throughout idle state.

3.2 Dual Threshold CMOS Technique (DTMOS)

The DTMOS approach is commonly used in digital electronics to connect MOSFET gates and bodies [2]. MOSFETs have several advantages, including reducing leakage current and providing great threshold voltage in the off-state, during on-state minimal threshold voltage, which enhances the circuit's current driving capabilities.

When transistor is turned on, as well as the propagation threshold voltage decrease, resulting in an increase in current (see Fig. 3). When a transistor turned off, its threshold voltage rises, leakage current decreases, and power consumption is reduced [10]. DTMOS reduces delay and increases speed [5] (Table 1).

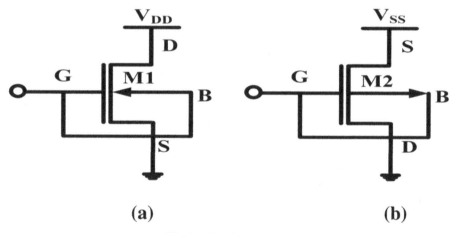

Fig. 3. DTMOS Structure

Although DTMOS consumes more power than traditional CMOS, it has a significantly smaller delay and greater power delay product.

3.3 Multiple Threshold CMOS Technique

This technology uses different threshold voltages to optimize power and delay by using minimal leakage current sleep transistors having a high threshold [11] and minimal

Table 1. Compares the 28 T ADDER CIRCUIT's average power, delay, and PDP using DTMOS and CMOS.

Technology	Vdd (V)	Avg. Power (Microwatts)	Delay (Nanoseconds)	PDP (femto)
DTMOS	0.7	203.3	9.161	1799.66
CMOS	0.7	311.2	14.80	4605.76

threshold transistors (cells) for good logic gate execution. Sleep transistors are used to separate cells from power during sleep mode, reducing leakage (see Fig. 4). Sleep transistors are activated during operation, allowing circuit to function correctly. They are turned off during standby mode [2, 6].

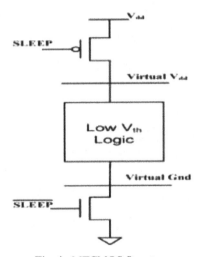

Fig. 4. MTCMOS Structure

To prevent current leakage, raise the Vt of the sleep transistor. Otherwise, power gating may not be as effective.

3.4 Lector Technique

Our method for lowering leakage power involves stacking transistors placed between the power source and grounding. Now, this is due to the fact "a configuration it is significantly less leaky to have several transistors turned OFF in a path from supply voltage to ground than it is to have just one transistor turned OFF in any supply to ground link [7, 11]." This technique uses each CMOS gate has two leakage control transistors (LCTs), with a single LCT near the cutoff area of operation [8].

We demonstrate our LEakage Control TransistOR (LECTOR) method using a NAND gate. Figure 5 illustrates a CMOS NAND gate with two transistors for leakage control,

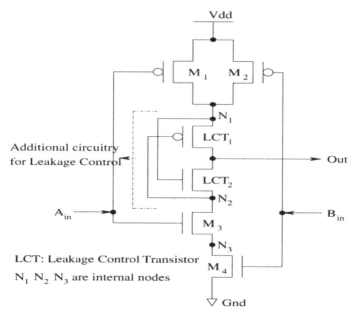

Fig. 5. LCT NAND with two inputs. LCT1 and LCT2 are self-controlled stacked transistors that are put between nodes N1 and N2, controlling leakage.

often renowned as the LCT NAND gate. The PMOS and NMOS leakage control transistors, LCT1 and LCT2, are connected between N1 node and N2 node within the pull-up and pull-down logic of the NAND gate. NAND gate's output node is formed by connecting the outflow nodes of LCT1 transistor and LCT2 transistors. The transistors' source nodes connect to the logic of pull-up and pull-down node N1 and node N2. The voltage potentials at node N1 and node N2 govern transistor changing (LCT1 and LCT2). This connection layout keeps among the LCTs close to its cutoff zone, irrespective of the input vector to the NAND gate.

Table 2. Compares power reduction techniques

Technology	Leak power (in microwatts)	Dyn. Power (in microwatts)	Total power (in microwatts)	Normalized area	% Leakage Savings
DTMOS	0.154	1.093	0.292	1.22	86.43%
MTMOS	0.732	2.662	0.809	1.16	61.51%
LECTOR	0.418	1.024	0.715	1.19	62.11%

The LCT NAND gate's transistors measure up to determine their impact regarding the gate's transmission delay. To create a Y-LCT gate, the transistors LCT1, LCT2 are sized to measure Y times the breadth of the gate's other NMOS and PMOS transistors.

We adjusted the sizes out of every transistor in the NAND gate with LCT to achieve a delay in propagation comparable to a normal NAND gate. (This is known as the Iso-LCT NAND gate). PMOS, NMOS transistor widths, exception of LCT transistors, are fixed up to twice to thrice the process variation's minimal width of a feature, respectively.

4 Results

Figure 6 displays HSPICE's transient curves at several LCT NAND gate nodes, simulated at a single-volt supply voltage and 70 nm technology. The graphs show that LCT NAND delivers perfect levels of output logic. We examined the LECTOR using180 technologies using a supply voltage of 1 V.

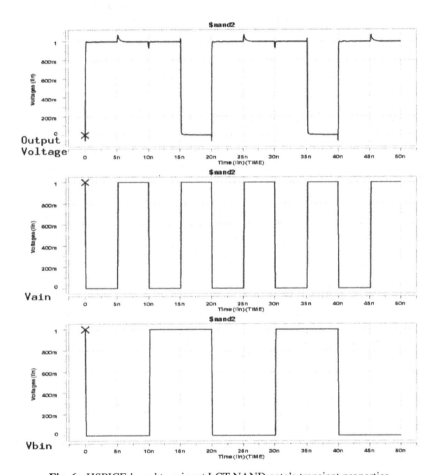

Fig. 6. HSPICE-based two-input LCT NAND gate's transient properties.

Table 2 summarizes the experimental outcomes. The first column includes technology. Columns 2 and 3 show the Power dissipation from leakage (in microwatts) also Dynamic power (in microwatts). Columns 4 shows the mean overall power loss (microwatts) for each circuit. To estimate leakage and total power, circuit simulations were averaged across ten collections of 100 randomly produced input patterns (with chance 0.6 each).

5 Conclusion

This study aims to provide a detailed examination and analyzing several power-reduction strategies, including new ones such as Lector and those utilized previously. We analyzed power dissipation patterns over time and identified important drivers to leakage. Modern study and interpretation on decrease in leakage powers strategies can benefit those wishing to implement them in their applications. The words "dynamic power," "leakage power," "propagation delay," and "power delay product" are understood.

References

1. Raghunath, A., Bathla, S.: Analysis and comparison of leakage power reduction techniques for VLSI design. In: International Conference on Computer Communication and Informatics (ICCCI-2021), 27–29 January 2021, Coimbatore (2021)
2. Bendre, V., Qureshi, A.K.: An overview of various leakage power reduction techniques in deep submicron technologies. In: International Conference on Computing Communication Control and Automation (2015)
3. Pramoda, N.V.: Advanced low power CMOS design to reduce power consumption in CMOS circuit for VLSI design. In: International Journal for Research in Applied Science & Engineering Technology (IJRASET), vol. 4(VII) (2016)
4. Chan, T.Y., Chen, J., Ko, P.K., Hu, C.: The impact of gate-induced drain leakage current on MOSFET scaling. Int Electron Devices Meet. **33**, 718–721 (1987)
5. Bansal, S., Prakash, N.R.: Implementation of high-speed full adder using DTMOS. Int. J. Res. Appl. Sci. Eng. Technol. **5**(6), 182–185 (2017)
6. Raghu Nandan Reddy, G., Ananthalakshmi, T.V.: Optimization of power in different circuits using MTCMOS technique. IJEDR **2**(2) (2014)
7. Narendran, S., Shekhar Borkar, V., Antoniadisn, D., Chandrakasann, A.: Scaling of stack effect and its application for leakage reduction In: Proceedings of the 2001 International Symposium on Low Power Electronics and Design (IEEECat. No.01TH8581) (ISLPED 2001) (2021)
8. Hanchate, N., Ranganathan, N.: LECTOR: a technique for leakage reduction in CMOS circuits. In: IEEE Transactions on VeryLarge Scale Integration (VLSI) Systems, vol. 12, no. 2 (2004)
9. Madhavi, B., Kanchana, G., Seerapu, V.: Low power and area efficient design of VLSI circuits. Int. J. Sci. Res. Publ. **3**(4) (2013)
10. Abhinov, S.R., Tripathi, R.: A robust design for ultra low power operation using dynamic threshold SCL logic. International Conference on Computer and Communication Technology (ICCCT-2011) (2011)
11. Geetha Priya, M., Baskaran, K., Krishnaveni, D.: Leakage power reduction techniques in deep submicron technologies for VLSI applications. In: International Conference on Communication Technology and System Design (2011)

12. Wei, L., Chen, Z., Johnson, M., Roy, K., De, V.: Design and optimization of low voltage high performance dual threshold CMOS circuits. In: Proceedings of the 35th Design Automation Conference (DAC), pp. 489–494 (1998)
13. Johnson, M.C., Somasekhar, D., Chiou, L.Y., Roy, K.: Leakage control with efficient use of transistor stacks in single threshold CMOS. IEEE Trans. VLSI Syst. **10**, 1–5 (2002)
14. Sundarajan, V., Parhi, K.K.: Low power synthesis of dual threshold voltage CMOS VLSI circuits. In: Proceedings of the IEEE ISLPED, pp. 139–144 (1999)

Next-Gen Home Automation: Sensor-Based Connectivity and Appliance Control

E. Emerson Nithiyaraj[1](✉) [iD], S. Srinivass[2], K. Mukilan[2], M. Muthukumar[2], R. Nanmaran[3], and S. Srimathi[4]

[1] Department of Artificial Intelligence and Data Science, Mepco Schlenk Engineering College, Sivakasi, India
ej.jeshua@gmail.com
[2] Department of ECE, Ramco Institute of Technology, Rajapalayam, India
[3] Department of ECE, Vel Tech Rangarajan Dr. Sagunthala R&D Institute of Science and Technology, Chennai, India
[4] Department of Biotechnology, Saveetha School of Engineering, SIMATS, Chennai, Tamilnadu, India

Abstract. Internet of Things (IoT) integrated smart home automation systems use sensor networks and communication modules to extend home energy efficiency, security, and convenience. IoT transforms houses into intelligent environment that are remotely monitored and controlled from anywhere using communication technologies. In this paper, a NodeMCU microcontroller and a variety of sensors, such as a PIR for motion detection, a gas sensor for monitoring hazardous gases, a flame sensor for fire detection, a DHT11 for temperature and humidity sensing, as well as smart locks, lights and fans for environmental control, are used to develop and implement a smart home system. With a user-friendly smartphone interface, the suggested system enables residents to monitor and operate their home appliances remotely in real- time, improving security, comfort, and energy efficiency. This article intends to further the development of smart home technology by providing a thorough overview of system design, sensor integration, and functionality.

Keywords: Internet of Things · Smart Home system · NodeMCU · Wireless Sensor Networks

1 Introduction

The Internet of Things (IoT) is a paradigm-shifting technology that connects real- life objects to the internet so they may interact, share information, and function as a single, cohesive ecosystem [1]. The Internet of Things (IoT) is essentially a network of interconnected objects that can collect and analyse real-time data. It does this by using sensors, actuators, and other smart devices to acquire and communicate information. IoT has enormous potential to change many facets of our life, from manufacturing and transportation to healthcare and agriculture. Intelligent objects (IoT) can be embedded with intelligence to improve automation, efficiency, and decision-making.

The idea of IoT-based home automation has changed dramatically in the last few years. With the help of IoT technology, homeowners can now easily convert their living areas into smart, networked spaces where numerous tasks can be effortlessly automated, coordinated, and communicated amongst devices. The core of Internet of Things (IoT) for home automation is the integration of smart appliances, sensors, and gadgets into a single network that can be remotely controlled and observed via voice commands, tablets, and smartphones. Residents can optimise and manage lighting, temperature, security, entertainment, and energy consumption, among other aspects of their homes, thanks to this networked environment. Convenience and comfort are two of the main advantages of IoT-enabled home automation [2, 3]. The sensors can instantly send notifications to resident's smartphones in an instance of suspicious activity, enabling quick action and peace of mind. Moreover, interoperability is promoted by IoT for home automation, enabling smooth communication across various platforms and gadgets. This interoperability enables the creation of comprehensive smart home ecosystems where devices from various manufacturers can communicate and collaborate effectively.

An effectual home automation system is proposed in this paper. The significant contributions of this paper are as follows:

1) The functioning of some essential household appliances like fans, lights and door lock are developed initially which could be turned on/off using Blynk application.
2) Fire, gas and PIR motion sensors are integrated with the smart home system which could alert the user through the Blynk application during scenarios of emergency.
3) The home environment can be monitored any time via the Blynk application such as temperature and humidity conditions within the home.

This paper is henceforth organized as follows. The existing home automation solutions are explained in Sect. 2. Section 3 discusses the proposed method and Sect. 4 discusses the results and discussion. Finally, the paper is concluded in Sect. 5.

2 Literature Review

As of the latest advancements in IoT-based home automation, several cutting-edge technologies and trends are shaping the state of the art in this field. In this section, the existing solutions proposed by different research papers are discussed.

In [4], the author proposed a smart home system to control light, fan, door cartons, and to monitor energy consumption, and level of the gas cylinder, using various sensors such as LM35, IR sensors, LDR module using Node MCU ESP8266, and Arduino UNO. The presence or absence of human inside the house, energy consumed inside the house are detected and notified to the house owner through text message.

The level of gas in the cylinder is monitored regularly and if it exceeds a threshold, the cylinder is booked automatically and the reference number is sent to the house owner.

In [5] the author proposed a home automation system called qToggle including access control and security, appliances control (lights, thermostats, AC, and other appliances), irrigations, and power and energy management. A smartphone application has also been developed for user interface. However, monitoring the air humidity is not included on qToggle.

In [6] the author proposed a smart automation system using long range (LoRa) technology. The proposed LoRa based system consists of wireless communication medium and a sensor network controlled by a smart phone application and powered by a low-power battery, with an operating range of 3–12 km distance. The proposed system is evaluated by considering three real-life case studies such as: fire detection, measuring environmental temperature and humidity and controlling the function of appliances. Within a distance of 12 km, the proposed system resulted in an accuracy of 90% and 92.33% for fire detection and controlling appliances. The proposed system is controlled using an android mobile phone and some modifications are required to implement the same system in a new platform.

In [7] the author developed a smart home system that includes Arduino microcontroller, flame sensor, PIR motion sensor and remotely controls lights, fans using Blynk app. The flame sensor detects the occurrence of fire and sends alert to the user using ESP8266 Wi-Fi module to the Blynk app which is installed and configured in the smart phone. Unwanted or suspicious activity is detected by the PIR Motion sensor and it sends alert to the user.

In [8] the author developed a smart door lock system. This paper discusses various information such as design, sensors, applications and advantages related to smart door lock sys-tem. The data from the sensors such as motion sensor, temperature sensor, retina sensor, and RFID sensor are collected and shared with the authorized person through Bluetooth or Wi-Fi. The house can be opened or closed automatically for the authorized person based on face and voice recognition during the absence of the house owner.

In [9] the author developed a smart home monitoring and control system that is based on node-MCU ESP32 with Internet connectivity that allows remote device control. The system transmits sensor data to the Firebase database and can receive commands from the server, allowing automatic control. The android-based mobile app is designed for communication with the Firebase database and updating its values to monitor and control the various home appliances. Various sensors like temperature, humidity, light, LPG (MQ-6), and motion sensors are used.

The goal of improving convenience, effectiveness, and quality of life is the motivation for research into IoT-based home automation. These systems enable homeowners to remotely control and automate a variety of house features, including lighting, temperature, security, and entertainment systems, by utilising networked smart devices and sensors. Through optimised resource usage and automation procedures, IoT home automation systems can present prospects for cost reduction, environmental sustainability, and energy savings. In the end, Internet of Things (IoT)- based home automation initiatives let people design customised, networked living environments that enhance comfort, convenience, and general wellbeing.

3 Methodology

The block diagram of the proposed smart home system is shown in Fig. 1. NodeMCU ESP32 is used as the main micro controller. Flame and gas sensors are used to monitor the occurrence of fire and leakage of gas in the home. Motion sensor is used to monitor the presence of human movement when the owner is not in home. An algorithm has also been incorporated where; the user can turn on and off the PIR sensor using the app. When the user is in his home, the user can turn off the PIR sensor and when the user is out, he can turn on the PIR sensor which would sense for any suspicious movement in his home. If any occurrence detected through these sensors, the user is notified through blynk app in his smartphone. The user can switch on/off smart bulb and fan using the blynk app. Also, the user can lock and unlock his door lock using the blynk app. The user can also monitor his home temperature and humidity level through the blynk app.

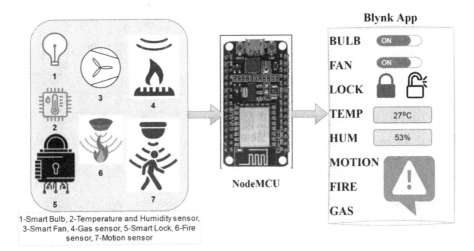

1-Smart Bulb, 2-Temperature and Humidity sensor, 3-Smart Fan, 4-Gas sensor, 5-Smart Lock, 6-Fire sensor, 7-Motion sensor

Fig. 1. Block Diagram of the proposed smart home system

Table 1 discuss the hardware components used in this work with their specifications.

3.1 NodeMCU ESP8266

NodeMCU is an open-source development board based on the ESP8266 Wi-Fi module, designed for IoT and prototyping projects [10]. It integrates a microcontroller unit (MCU) and Wi-Fi connectivity, enabling easy interfacing with sensors, actuators, and other devices. The NodeMCU ESP8266 is a development board based on the ESP8266 microcontroller chip, renowned for its integrated Wi-Fi capabilities [11]. The features include a compact form factor and ample flash memory, typically around 4MB, facilitating the storage of firmware, program code, and data. NodeMCU supports easy programming using Lua scripting language or Arduino IDE. NodeMCU ESP8266 offers GPIO pins for digital input/output, analog input, and various communication protocols such as I2C, SPI, and UART, providing flexibility for interfacing with sensors, displays, and other peripherals.

3.2 DHT11 Sensor

A simple and reasonably priced digital temperature and humidity sensor module is the DHT11 sensor [12]. It measures temperature and humidity using a thermistor and a resistive humidity element, respectively. The operational voltage is provided by the VCC pin,

Table 1. Hardware Module Specification

S No.	Name	Component	Specification
1	Microcontroller	NodeMCU ESP8266	–
2	Temperature & Humidity sensor	DHT11	Temperature Range: 0 °C to 50 °C Humidity Range: 20% to 90%
3	Gas sensor	MQ-2	300–10000 ppm
4	Fire sensor	YG1006 sensor	Up to 5 m
5	Motion sensor	HC-SR501	Range: Up to 7 m
6	Bulb	LED	9 W
7	Fan	Cooling Fan	DC 5 V
8	Solenoid Lock	–	DC 9–12 V
9	Blynk app and smartphone	–	–
10	5 V Four-Channel Relay Module	–	Supply Voltage 3.75 V to 6 V

which is connected to the power source, usually 3.3V or 5V. The Data Out pin allows digital signals containing humidity and temperature data to be transmitted in both directions between the microcontroller and the sensor. The Ground (GND) pin, which acts as the reference voltage for the sensor's functionality, is linked to the microcontroller's ground.

3.3 MQ-2 Sensor

The MQ-2 sensor is a widely used gas sensor module that can identify several types of flammable gases [13]. When target gases prevail, it operates according to the principle of resistance change. Four pins are usually present on a MQ-2 sensor: VCC for the power supply, GND for the ground connection, AOUT for the analogue output, and DOUT for the digital output. It is appropriate for uses such as gas leak detection, fire alarms, and air quality monitoring since it detects gases and outputs analogue or digital signals that may be read by a microcontroller for additional processing.

3.4 YG1006 Sensor

The Flame Sensor, sometimes called the YG1006 sensor, is intended to detect flames or other infrared radiation sources [14]. It works by detecting infrared light with a wavelength between 760 and 1100 nm, which is released by flames. An amplifier circuit and an infrared-sensitive photodiode form the sensor. The YG1006 sensor is easy to interface with microcontrollers for flame detection applications since it normally has three pins: VCC for the power supply, GND for ground connection, and OUT for the analogue output signal.

3.5 HG-SR501 Sensor

The HC-SR501 sensor is a widely used passive infrared (PIR) motion sensor module that detects motion by measuring changes in infrared radiation within its field of view [15]. It operates by sensing the heat emitted by humans or other moving objects. The sensor consists of a pyroelectric sensor, a fresnel lens to focus the infrared radiation onto the sensor, and a signal processing circuit. The module typically has three pins: VCC for power supply, OUT for the digital output signal, and GND for ground connection, making it easy to integrate into various motion sensing applications such as security systems, lighting control, and automation.

3.6 Solenoid Lock

An electromechanical device called a 12 V solenoid lock is utilised for locking mechanisms in a variety of applications, such as safes, electronic lock systems, doors, and cabinets [16]. A magnetic field is created by applying a 12-V electrical current to the solenoid, which forces the locking mechanism into position and locks the door or object. On the other hand, the locking mechanism can retract and unlock when the electrical current is cut off. This does this by releasing the magnetic field. To facilitate remote or automatic locking and unlocking activities, these solenoid locks are frequently interfaced with electronic control systems, such as microcontrollers or access control systems. The typical pin configuration for a 12 V solenoid lock includes two pins: one for connecting to the positive terminal of a 12 V power source and another for connecting to the ground (GND) or negative terminal [17].

3.7 Blynk Application

The Blynk application is a versatile and user-friendly platform that enables users to create custom mobile applications for controlling and monitoring Internet of Things (IoT) devices [18, 19]. It provides a drag-and-drop interface for designing intuitive user interfaces, which can then be linked to hardware components using Blynk's cloud-based servers. Through Blynk, users can remotely interact with their IoT projects, such as controlling lights, sensors, and actuators, and receiving real-time data updates on their smartphones or tablets. Additionally, Blynk offers a wide range of widgets and features, including push notifications, data logging, and energy- efficient protocols.

4 Experimental Results

NodeMCU ESP8266 has 16 General Purpose Input Output (GPIO) pins out of which seven GPIO pins are used for controlling the various components used in this work. The sensors are connected with the microcontroller using the breadboard. NodeMCU ESP8266 was programmed in such a way that when user presses ON/OFF button in Blynk app it will trigger the relay and in response of which the relay turns ON/OFF the home appliances. The hardware model of the proposed smart home system is displayed in Fig. 2.

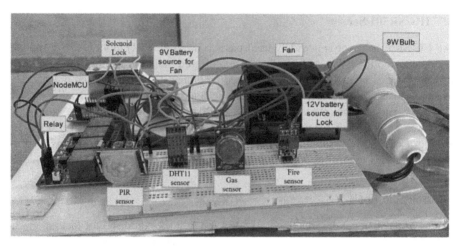

Fig. 2. Hardware model of the proposed smart home system

The Arduino IDE software is used to code and interface all the sensors and smart appliances listed in Tabel 1. ESP 8266 WiFi is used to control the complete operations of the system. The program is uploaded to ESP 8266 WiFi module through Arduino IDE. As shown in Fig. 3, user interface is developed using Blynk application where user can turn on/off the appliances and he can monitor the sensor output any time. In the Fig. 3(left), the state of the bulb is on, the state of the fan is off and the state of the lock is on. Meanwhile, the temperature and humidity readings are displayed. The alert symbol within the gas sensor node indicates the leakage of gas inside the home. Similarly, in the Fig. 3 (right), the state of the bulb is on, the state of the fan is on and the state of the lock is unlocked. The alert symbol within the fire and gas sensor nodes indicates the leakage of gas as well as the detection of fire/flame inside the home.

The developed system is tested for 25 times and the average response time of each component is analysed and displayed in Table 2. The response time is calculated by considering the time at which the appliance is on or off when the command is sent from the Blynk application or the time at which the sensor produces the output in the Blynk application [20]. Additionally, these response times represent the time it takes for the sensor or device to detect a change or receive a command and respond accordingly. Those response times are approximate and may vary depending on factors such as the specific model of the device, network conditions, firmware optimizations, and other environmental factors. The proposed system achieved an average accuracy of 97% when tested for 25 times.

The overall approximate cost of this project is around Rs.4300 (INR) which is a nominal cost to have a home automation setup in today's scenario. The developed system has various functionalities ranging from controlling home appliances to home environment monitor and alert system. These functionalities can also be remotely controlled using smartphone-based application through Wi-Fi. As a result, smart home automation systems improve comfort and efficiency while consuming less energy by providing

Next-Gen Home Automation: Sensor-Based Connectivity and Appliance Control 201

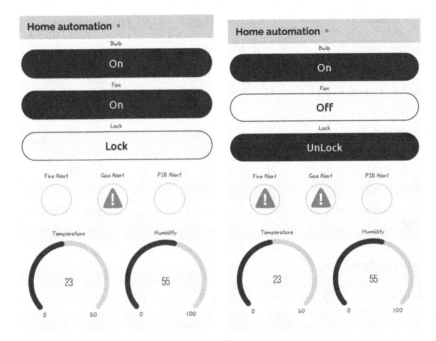

Fig. 3. Sample output displayed in the Blynk app

Table 2. Response Time Analysis

Device Name	Average Response Time
PIR Motion Sensor	0.7 s
Gas Sensor	5 s
Flame Sensor	2 s
DHT11 Sensor	2 s
Smart Lock	1.3 s
Smart Bulb	0.7 s
Smart Fan	1.2 s

convenience through remote device control. They offer improved security features like alarms for emergencies or invasions and remote monitoring.

Table 3 provides a comparison of the proposed system with the existing works based on cost of the overall system, communication medium used for sending data and the accuracy of the developed system. The cost parameter includes all the appliances used in the work including the microcontroller board. Compared to the existing ones, the cost of the proposed system is low and the accuracy is high.

Table 3. Comparison of the proposed system with the existing methods

Techniques	Cost	CM	Accuracy
H Singh et al. [4]	M	GSM	M
Stolojescu-Crisan et al. [5]	M	Wi-Fi	M
Nur-A-Alam et al. [6]	H	Wi-Fi	H
B Mustafa et al. [7]	M	Wi-Fi	M
Baby Chithra et al. [8]	M	Wi-Fi or Bluetooth	H
Uma Pujari et al. [9]	H	Wi-Fi	M
Proposed system	L	Wi-Fi	H

CM: Communication Medium, H: High, M: Medium, L:Low, Accuracy: High (90–99)%, Medium (70–89)%

5 Conclusion

In this paper, a NodeMCU microcontroller-based smart home system that integrates a flame sensor for fire detection, a gas sensor for monitoring hazardous gases, a PIR sensor for motion detection, a smart lock for access control, a DHT11 sensor for temperature and humidity sensing, and smart fans and bulbs for environmental control is successfully developed and implemented. The findings gathered show that the system is capable of motion detection, gas leak and fire outbreak monitoring, interior climate regulation, and safe access management via the smart lock. Using the system's smartphone application interface, users may remotely monitor and operate their home appliances, giving them effective and secure control over their environment. The accuracy of the sensors has been validated by multiple testing cycles conducted in the home environment using the developed smart home system. In future, the same system can be incorporated with smart access to big home appliances and with smart gardening techniques.

References

1. Majid, M.H., et al.: Applications of Wireless Sensor Networks and Internet of Things Frameworks in the Industry Revolution 4.0: a Systematic Literature Review. Sensors 22, 2087 (2022). https://doi.org/10.3390/s22062087
2. Agarwal, K., Agarwal, A., Misra, G.: Review and performance analysis on wireless smart home and home automation using IoT. In: Third International conference on I- SMAC (IoT in Social, Mobile, Analytics and Cloud) (I-SMAC), Palladam, India 2019, pp. 629–633 (2019). https://doi.org/10.1109/I-SMAC47947.2019.9032629
3. Chakraborty, A., Islam, M., Shahriyar, F., Islam, S., Zaman, H.U., Hasan, M.: Smart home system: a comprehensive review. J. Electr. Comput. Eng. 2023, Article ID 7616683, 30 pages (2023). https://doi.org/10.1155/2023/7616683
4. Singh, H., Pallagani, V., Khandelwal, V., Venkanna, U.: IoT based smart home automation system using sensor node. In: 2018 4th International Conference on Recent Advances in Information Technology (RAIT), Dhanbad, India, pp. 1–5 (2018). https://doi.org/10.1109/RAIT.2018.8389037

5. Stolojescu-Crisan, C., Crisan, C., Butunoi, B.-P.: An IoT-based smart home automation system. Sensors **21**, 3784 (2021). https://doi.org/10.3390/s21113784
6. Nur-A-Alam, A.M., Based, M.A., Haider, J., Rodrigues, E.M.G.: Smart Monitoring and controlling of appliances using LoRa Based IoT system. Designs **5**, 17 (2021). https://doi.org/10.3390/designs5010017
7. Mustafa, B., et al.: IOT based low-cost smart home automation system. In: 2021 3rd International Congress on Human- Computer Interaction, Optimization and Robotic Applications (HORA), Ankara, Turkey (2021). https://doi.org/10.1109/HORA52670.2021.9461276
8. Baby Chithra, R., Joy, S., Bale, A.S., Naidu, A.S., Vinay, N., Varsha, S.N.: Advanced computing in IoT for door lock automation. In: 2022 International Conference on Electronics and Renewable Systems (ICEARS), Tuticorin, India, pp. 565–569 (2022). https://doi.org/10.1109/ICEARS53579.2022.9752140
9. Pujari, U., Patil, P., Bahadure, N., Asnodkar, M.: Internet of Things based integrated smart home automation system (May 1, 2020). In: 2nd International Conference on Communication & Information Processing (ICCIP) (2020). https://doi.org/10.2139/ssrn.3645458
10. Lokesh, S., Patil, S.B., Gugawad, A., Home security and automation using NodeMCU-ESP8266. In: 2020 IEEE Bangalore Humanitarian Technology Conference (B- HTC), Vijiyapur, India, vol. 2020, pp. 1-6 (2020). https://doi.org/10.1109/B-HTC50970.2020.9297917
11. Kashyap, M., Sharma, V., Gupta, N.: Taking MQTT and NodeMcu to IOT: Communication in Internet of Things, Procedia Computer Science, vol. 132, pp. 1611–1618 (2018). ISSN 1877–0509, https://doi.org/10.1016/j.procs.2018.05.126
12. Yulizar, D., Soekirno, S., Ananda, N., Prabowo, M., Perdana, I., Aofany, D.: Performance analysis comparison of DHT11, DHT22 and DS18B20 as temperature measurement. In: Book: Proceedings of the 2nd International Conference on Science Education and Sciences 2022 (ICSES 2022), pp.37- 45 (2023)
13. Trisnawan, I.K.N., Jati, A.N., Istiqomah, N., Wasisto, I.: Detection of gas leaks using the MQ-2 Gas sensor on the autonomous mobile sensor. 2019 International Conference on Computer, Control, Informatics and its Applications (IC3INA), Tangerang, Indonesia, pp. 177–180 (2019)
14. Sarhan, Q.: Systematic survey on smart home safety and security systems using the arduino platform. IEEE Access. **8**, 128362–128384 (2020). https://doi.org/10.1109/ACCESS.2020.3008610
15. Naccarelli, R., Casaccia, S., Revel, G.: The problem of monitoring activities of older people in multi-resident scenarios: an innovative and non-invasive measurement system based on wearables and PIR sensors. Sensors. **22**, 3472 (2022). https://doi.org/10.3390/s22093472
16. Hanggara, I., Rakhmadi, F.: Design of Prototype Home Door Security System Based Solenoid Door Lock, Magnetic Sesor, Microcontroller Nodemcu Esp8266 and Blynk Application (2021)
17. Flske, M., Bariye, C., Sheikh, A., Mehsram, P., Sharma, C.: Smart door lock system. Inter. J. Emerging Technol. Innovative Res. 6(2), 173–178 (2019) (www.jetir.org), ISSN:2349–5162,
18. Seneviratne, P.: Hands-On Internet of Things with Blynk: Build on the Power of Blynk to Configure Smart Devices and Build Exciting IoT Projects. Packt Publishing, Limited, United Kingdom (2018)
19. Mandula, K., Parupalli, R., Chandrapati, M., Magesh, E., Lunagariya, R.: Mobile based home automation using internet of things (iot), in Control, Instrumentation, Communication and Computational Technologies, International Conference on. IEEE, p. 340343 (2015)
20. Geetanjali, V., Subramanian, I., Kannan, G., Prathiba, S.B., Raja, G.: IoTexpert: interconnection. Interoperabil. Integrat. IoT Platforms, 212–219 (2019). https://doi.org/10.1109/ICoAC48765.2019.246842

Navigating the Complexities of Municipal Waste Management: Enhancing Cost Prediction for Sustainable Urban Solutions

M. S. Minu[✉], Tushar Samal, Khushi Bisani, and Aishwarya Tewari

Department of Computer Science and Engineering with Specialisation in Big Data Analytics, SRM Institute of Science and Technology, Ramapuram Campus, Chennai, India
minus@srmist.edu.in

Abstract. The escalating rates of municipal waste generation in urban areas worldwide present a critical challenge for effective waste management. This paper examines the complexities surrounding municipal waste management, emphasizing the significance of accurate cost prediction for sustainable decision-making. Urbanization, driven by population growth and rising standards of living, intensifies waste generation, while shifts in consumption patterns further complicate waste management efforts. Environmental concerns and regulatory pressures mandate innovative waste reduction strategies. Accurate cost prediction is paramount for budgetary planning and resource allocation, enabling proactive measures to optimize operational efficiency and mitigate financial risks. Emerging methodologies, including data analytics and predictive modeling, offer promising avenues for enhancing cost prediction accuracy by capturing the multifaceted drivers of cost variability. By leveraging these methodologies, decision-makers can forecast municipal waste management costs with precision, informing strategic initiatives to foster sustainable waste management practices in urban areas.

1 Introduction

The series of steps and activities required from the start until the disposal of garbage is known as waste management [1]. Waste management is a critical aspect of modern society that plays a pivotal role in achieving sustainable development goals. As global populations continue to grow and urbanize, the generation of waste has reached unprecedented levels, posing significant challenges for environmental protection, public health, and resource conservation. Effective waste management is essential not only for minimizing negative environmental impacts but also for maximizing resource utilization and promoting a circular economy.

In recent years, the urgency to address waste management issues has become increasingly apparent due to mounting concerns over pollution, climate change, and resource depletion. The traditional linear model of waste management, characterized by "take-make dispose" practices, is no longer viable in the face of mounting environmental pressures and limited natural resources. Instead, there is a growing recognition of the need for a paradigm shift towards a more sustainable approach that emphasizes waste prevention, reduction, reuse, and recycling. Against this backdrop, this research seeks to provide a comprehensive analysis of waste management strategies with a focus on minimizing negative environmental effects and maximizing resource use.

By employing a case study methodology, this study aims to examine waste management practices across various geographic and socioeconomic settings, highlighting both the challenges and opportunities associated with different approaches.

The research adopts a multidisciplinary framework that integrates environmental, economic, and social viewpoints to evaluate the efficacy of existing waste management methodologies and explore avenues for enhancement. Through the synthesis of actual data, case studies, and theoretical models, the study analyses the entire lifecycle of waste from generation to disposal, identifying critical obstacles and possibilities at each stage.

Municipal waste management represents a critical challenge for urban areas globally, with waste generation rates escalating due to population growth, urbanization, and shifting consumption patterns. As municipalities strive to address the mounting pressures of waste management, accurately predicting the associated costs becomes imperative for effective planning, resource allocation, and sustainable decision-making. This introduction provides an overview of the complexities surrounding municipal waste management, the significance of cost prediction, and the emerging methodologies aimed at enhancing predictive accuracy.

The rapid pace of urbanization is a defining characteristic of the modern era, with an increasing proportion of the global population residing in urban areas. As cities expand, so does the volume of municipal waste generated, driven by factors such as population growth, rising standards of living, and increased consumption. Urbanization leads to higher concentrations of people within limited geographical areas, resulting in intensified waste generation and placing additional strain on municipal waste management systems.

Alongside urbanization, shifts in consumption patterns contribute to the complexity of municipal waste management. As economies develop and lifestyles evolve, there is a discernible transition towards a more consumer-driven society characterized by higher levels of disposable income and increased consumption of goods and services. This trend manifests in various forms, including the proliferation of single-use packaging, electronic waste, and disposable goods, all of which contribute to the overall waste stream. Consequently, municipal waste management systems must adapt to accommodate these changing consumption patterns while mitigating their environmental and economic impacts.

Accurate prediction of municipal waste management costs assumes paramount importance for local governments, waste management authorities, and other stakeholders involved in waste management decision-making. Cost prediction serves as a cornerstone for budgetary planning, resource allocation, and investment prioritization within municipal waste management systems. By anticipating future costs associated with waste collection, transportation, treatment, and disposal, decision-makers can formulate proactive strategies to optimize resource utilization, enhance operational efficiency, and mitigate financial risks.

Furthermore, cost prediction facilitates the evaluation of alternative waste management scenarios, enabling decision-makers to assess the economic viability of different strategies, technologies, and interventions. Whether considering investments in recycling infrastructure, waste-to-energy facilities, or enhanced waste diversion programs, accurate cost prediction provides essential insights into the long-term financial implications of such initiatives, thereby informing strategic decision-making and ensuring optimal allocation of resources.

Municipal waste management represents a complex and multifaceted challenge driven by urbanization, shifting consumption patterns, environmental concerns, and regulatory pressures. Accurately predicting the costs associated with municipal waste management is essential for effective planning, resource allocation, and decision-making within waste management systems. By leveraging emerging methodologies such as data analytics, machine learning, and predictive modelling, decision-makers can enhance the accuracy and granularity of cost prediction, thereby enabling proactive strategies to optimize resource utilization, enhance operational efficiency, and foster sustainable waste management practices in urban areas.

2 Literature Survey

Rapid economic growth correlates with increased waste production, as evidenced by Johnstone et al.'s findings of a 1% GDP growth resulting in a 0.69% waste rise. This links waste management closely with sustainable policies and economic objectives. Recognizing its significance, the World Bank prioritizes solid waste management for environmental protection and economic progress [13]. Challenges include the complexity of waste management systems and the necessity for optimal policies emphasizing waste reduction, recycling, and cost efficiency. Ineffective waste management leads to various problems like escalating land prices, health risks, and limited landfill space. Policymaking seeks to embed sustainability principles and relies on timely waste information for informed decision-making, striving for environmentally sound solutions [13].

Given that only 30% of waste is recycled and the population is expanding, the primary issue facing the nation is the collection, management, and classification of solid waste. With the help of Visual Geometry Group with 16 layers (VGG16), you can instantly manage and categorize intelligent trash. The outcomes are precise. To investigate the efficacy of a convolutional network's five layers, it is crucial to segregate household garbage into distinct categories, such as organic and recyclable waste, using two images with resolutions of 225×264 and 80×45. It turns out that models with lower image resolutions are lighter, need less training time, and have higher accuracy [3, 8]. Trash classification aims to increase classification accuracy, learn from the range of classification input images produced by the outcomes, and boost the effectiveness of the autonomous garbage classification system through the use of the ResNet-34 algorithm [3].

Urban waste, primarily household garbage, is illicitly disposed of via landfills and incineration, posing grave risks to urban ecosystems and residents' health. Decomposition of certain waste components leads to hazardous compound accumulation, heightening ecological threats. Non-biodegradable materials, notably plastic pollution, exacerbate environmental degradation, particularly evident in underwater ecosystems worldwide [19]. The waste collection process is viewed as a Vehicle Routing Problem (VRP) with multiple services originating from various depots due to scattered collection points near the door. This approach minimizes environmental damage and prevents issues like odors and bacteria buildup, as well as navigating around obstacles in the collection path. Each collection point has a time constraint determined by sensor alarms, and the model can forecast fluctuations in waste generation at each street intersection collection point [20].

Waste segregation poses a significant energy demand, prompting the need for innovative solutions that utilize renewable or cleaner energy sources. In a study conducted by Margaret Banga, insights from 500 randomly selected households in Kampala, Uganda, shed light on public attitudes toward waste segregation and disposal. The local government faces challenges due to limited financial and technical resources, hindering the establishment of efficient waste management infrastructure. Despite Uganda's agricultural status, approximately 80% of the city's solid waste is organic, with residential sources contributing to 53% of the total waste, yet much of this valuable resource goes unutilized [10]

In addition to economic considerations, municipal waste management is underpinned by environmental imperatives and regulatory frameworks aimed at minimizing waste generation, promoting recycling and reuse, and reducing the environmental footprint of waste disposal practices. Heightened awareness of environmental issues, coupled with stringent regulations governing waste management practices, necessitates innovative approaches to waste reduction and resource recovery. Moreover, failure to comply with such requirements can result in financial penalties and reputational damage for municipalities, underscoring the importance of accurate cost prediction in ensuring regulatory compliance and mitigating associated risks.

Poor management of Municipal Solid Waste (MSW), with at least 33% not environmentally managed, threatens human health and the environment. Challenges include urbanization, climate change, and population growth. Lack of waste sorting, poor collection, and public engagement exacerbate the issue. With global waste expected to reach 3.4 billion tons in 30 years, accurate prediction and tracking are crucial for sustainable waste management. Estimating future waste generation aids in resource allocation and strategic planning [18].

Existing waste bin management research mainly utilizes image-based AI methods such as GLAM, MLP, and KNN for waste level recognition, lacking precise bin number prediction to prevent overflow. This study introduces LSTM and BLSTM models for time-series forecasting in waste management, aiming to optimize bin allocation. Comparing these sequential models with nonsequential approaches, the research evaluates their performance using USW generation data from Sousse, Tunisia. Innovatively, the study explores BLSTM's effectiveness for waste bin prediction, addressing a gap in current research. By considering seasonal variation and autocorrelation, the sequential models offer potential for more accurate waste management strategies. This work contributes valuable insights into enhancing waste management practices through advanced AI techniques, fostering more efficient resource allocation and waste reduction efforts [17].

Classifying garbage can lower the system's environmental loads are gathered and displayed. By gathering 2, 313 picture datasets of trash from different offices and providing bin categorization in feature separation rubbish classification might lessen its negative environmental effects. It demonstrates the high accuracy rate and clear rubbish categorization result. Following training with the MobileNetV2 the challenge of misclassifying waste was addressed using a high precision model. This model was utilized to categorize waste into seven classes employing pre trained models which had been trained on the dataset beforehand. This endeavour involved the development and implementation of a dataset tailored for recognizing and classifying waste types in embedded devices, resulting in accurate waste classification. [3]. ImageNet has taught accuracy for 60 generations [3, 12].

It is suggested that digital technology users (DTS) are important facilitators of recycling and reuse. Utilizing DTs to fully realize the promise of circular solutions to increase productivity and resource efficiency is not well supported by systematic guidelines. The system aids in identifying gaps between current and expected requirements and formulating new objectives to address them. It also facilitates the integration of operations across diverse fields like information systems and circular economy research. In the circular economy, Digital Technologies (DTs) data enables smart resource management. The effective utilization of this digital revolution determines a company's ability to transition to and benefit from the circular economy at scale [1].

Traditionally, municipal waste management cost prediction has relied on simplistic models based on historical data, average unit costs, and demographic factors. However, the inherent complexity and dynamic nature of waste management systems necessitate more sophisticated predictive methodologies capable of capturing the multifaceted drivers of cost variability.

In recent years, advancements in data analytics, machine learning, and predictive modeling have unlocked new possibilities for enhancing the accuracy and granularity of municipal waste cost prediction. By leveraging large-scale datasets encompassing variables such as waste composition, demographic profiles, socioeconomic indicators, technological parameters, and environmental factors, researchers and practitioners can develop robust predictive models capable of capturing the intricacies of municipal waste management systems.

These emerging methodologies encompass a range of techniques, including statistical regression analysis, machine learning algorithms, and hybrid models that combine multiple predictive approaches. By harnessing the power of big data and predictive analytics, these methodologies enable decision-makers to forecast municipal waste management costs with greater precision, reliability, and granularity than ever before.

The multilayer convolutional neural network (ML-CNN) investigated urban areas facing various challenges, notably waste management, which is directly influenced by population density. Many municipalities and city administrations currently depend on inefficient and expensive human-operated waste categorization systems. To enhance waste recycling in developed regions, there is a pressing need for automated waste classification and management. Improved waste recycling not only minimizes the necessity for acquiring new raw materials but also reduces the volume of waste directed to landfills. Real-time experiments have utilized image segmentation techniques. In these experiments, the model identifies the class of each piece of waste and triggers the corresponding mechanism when the item is deposited into its designated receptacle [1, 24].

[1, 11] A novel approach was devised to enhance waste recycling and disposal efficiency through detection strategies. The YOLOv3 algorithm was employed, albeit yielding ecological imbalances. Experimental findings indicate that the proposed YOLOv3 method demonstrates sufficient potential for generalizability across various waste classes and components. The study effectively segregates waste into biodegradable and non-biodegradable categories. However, the reduced detection time, coupled with remarkably high prediction probabilities, warrants further investigation. Future research could focus on optimizing results and prediction probabilities for additional real-world waste scenarios.

Artificial intelligence techniques are being used to automate garbage collection and decomposition processes, reducing the need for significant funds and energy. The Internet of Things (IoT) is used to exchange information and collaborate with physical and virtual systems, providing a smarter environment for waste disposal. This system addresses the issue of non availability of unused land in outskirts, enabling efficient waste decomposition in urban areas. Real-time information collection helps analyse waste areas and determine the most efficient collection methods, utilizing the Internet of Things for the initial collection and separation of garbage [2, 16].

In this instance, deep learning is used to speed up the breakdown of waste. Through analysis and collection in the proper dust bin, these artificial intelligences assist in the more intelligent breakdown of waste [2, 17]. Sensory units and microcontrollers are programmed into them. The genetic algorithm is used for overall system operation in order to maximize system performance more intelligently. As a result, they are often known as recycling procedures. By using computational techniques, the breakdown is done in an efficient manner to cut costs and time. As a result, the internet of things and deep learning techniques aid in the management of solid waste in urban areas.

Blockchain-based waste management systems have mostly targeted the electronic, medical, household, and agricultural waste associated with smart cities. Numerous services linked to waste management, including asset tracking, shipment monitoring, token transfer, waste sorting and auditability of waste handlers' actions have been implemented by existing blockchain-based waste management projects. Studies that are now in existence have monitored the Ethereum and Hyperledger Fabric platforms to provide incentives and sanctions for waste management participants and guarantee that users' actions comply with waste management regulations. Cities are growing faster than ever before, which presents a number of social and environmental problems. The volume, rate, and variety of garbage generated are mostly caused by the increasing rate of urbanization, economic development, global population expansion, and rising living standards in developing nations [7].

Blockchain technology has great promise for displacing the laborious and slow manual waste management techniques used in smart cities. Blockchain effectively manages and keeps an eye on the transportation of garbage from the site of generation to the recycling facility. For tracking and tracing purposes, it generates digital asset tokens (such as security tokens) linked to the trash of smart cities. These tokens enable the authorities reduce waste management costs and optimize their operations, and they are useful for tracking the history of recycled waste items through ownership and digital transfer activities [7, 42]. In order to prevent pollution of the environment, the traceability feature ensures that the trash produced by smart cities is managed in accordance with the waste handling requirements. Users may also effectively monitor the waste's end of life in smart cities thanks to it [7, 43].

Hamsalyer conducted an analysis of waste management, specifically focusing on a decentralized system in Mumbai, India. It is argued that many individuals face challenges in fully utilizing decentralized systems due to limited access to resources for managing sanitary and reusable waste. Additionally, it is contended that centralized management systems often prove ineffective due to their linear nature and failure to address the efficient management of mixed waste (Hamsalyer, 2016: 102). The decentralized waste management system is recognized for its effectiveness in addressing waste challenges at a smaller scale. An industrial complex has successfully operated a vermi-composting system for over seven years, extending waste management services to residential colonies housing its employees. Waste collection from both industrial and residential areas is facilitated by vendors. The SHG helps segregate organic and inorganic waste, collecting recyclables and electronic waste periodically. The organization gains financially by managing waste well and saving on in-house compost [8].

[4, 13] The garbage collection routing in Ansasol, India, was optimized using a Geographic Information System (GIS) model by Ghose. This model proposes an efficient solid waste management system aimed at reducing environmental impact while optimizing costs. Factors considered in the model include population density, waste production capacity, road network and type, garbage bin distribution, and garbage collector vehicle specifications. Additionally, Ghose outlined considerations regarding the size of trash, types of garbage collector vehicles, and optimal waste collection routes. It is noted that the model does not calculate investment costs due to financing constraints; instead, it focuses on expenses related to waste disposal facility additions and operations.

[4, 14] Utilizing Geographic Information System (GIS) as a supporting tool, the locations for recycling garbage were identified. The placement and quantity of recycling points were strategically determined to optimize service area coverage. This model encompasses two key aspects: physical and non-physical considerations. Physical factors include land availability, the quantity of recycling points, available vehicles for recycling, collection frequency, and the size and quantity of bins. Non-physical factors encompass user attitudes and community acceptance towards waste recycling initiatives.

3 Existing System

The existing predictive model for estimating municipal waste management costs using linear regression analysis is explained as follows. The model aims to provide accurate cost predictions based on various socio-economic and environmental factors. Historical data was utilized to train and validate the model. A linear regression-based approach to predict municipal waste management costs. The model incorporates socio-economic indicators, waste generation rates, recycling efforts, and other relevant factors to estimate the financial resources required for effective waste management.

In Linear Regression,

$$\text{MSE} = 1/N \sum_{i=1}^{n} (y_i - (\beta_i * x_i + \beta_0))^2 \tag{1}$$

The values of B0 and B1 will be adjusted using the Mean Squared Error (MSE) function to minimize its value. This process involves employing gradient descent to iteratively update these parameters until the cost function reaches its lowest point.

Historical data on municipal waste management costs, waste generation rates, population demographics, economic indicators, and environmental factors were collected. The data underwent pre-processing steps including cleaning, normalization, and feature engineering to ensure compatibility with the regression model. This method scales the feature values to have a mean of 0 and a standard deviation of 1. It subtracts the mean from each value and then divides by the standard deviation and generating polynomial features that capture nonlinear relationships between the predictors and the target variable. The following features were selected for inclusion in the predictive model: Waste generation per capita; Population density; Gross domestic product (GDP) per capita; Recycling rate; Landfill tipping fees; Government subsidies for waste management; Environmental regulations.

A multiple linear regression model was developed using the selected featuresto predict municipal waste management costs. The model is formulated as follows:

Cost $= \beta_0 + \beta_1 *WasteGeneration + \beta_2 *PopulationDensity + \ldots + \beta_n *$ **Regulation, where $\beta_0, \beta_1, \beta_2, \ldots \beta_n$ are the regression coefficients**.

The multiple linear regression equation is as follows:

$$Y^\wedge = \beta_0 + \beta_1 * X_1 + \beta_2 * X_2 + \ldots + \beta_p * X_p$$

In this context, $Y^\wedge$ represents the predicted values. The variables denote p distinct independent variables. The β denotes the estimated regression coefficients.

It provides an estimation of the anticipated value of the dependent variable contingent upon the values of the independent variables. The coefficients (b0, b1,..., bp) delineate the influence of each independent variable on the dependent variable, with the proviso that other variables remain constant. Specifically, the coefficient b1 elucidates the extent to which the dependent variable varies when independent variable X1 undergoes a unitary alteration, all the while holding the remaining variables unchanged. Statistical analyses are then employed to ascertain whether these coefficients exhibit statistically significant deviations from zero.

The performance of the model was assessed using evaluation criteria such as MAE, MSE, and the R^2 value. The trained linear regression model achieved promising results in predicting municipal waste management costs. The model exhibited low MAE and RMSE values, suggesting accurate cost estimations across different scenarios.

In conclusion, the proposed linear regression model effectively captures the relationships between socio-economic and environmental factors and municipal waste management costs. This was an existing linear regression-based model for predicting municipal waste management costs. By incorporating relevant features and leveraging historical data, the model offers valuable insights for municipal authorities and researchers to optimize resource allocation and budget planning in waste management.

4 The Proposed System for MWCP

The workflow of our suggested model, which includes feature engineering and feature encoding approaches for pre-processing the dataset, is depicted in the diagram below. The pre-processed dataset is divided into test and train sets, after which it is subjected to feature scaling, also known as standardization. The Elastic Net model with specific hyperparameters fits the scaled data. Next, Grid Search CV is used to cross-validate the Elastic Net model. This aids in the process of hyperparameter tuning, which is essentially determining which parameter is appropriate for our model. After that, the model is assessed using the testing data, and the predictions are incorporated into the scoring metrics to help ascertain the model's accuracy.

Architecture Diagram (Fig. 1):

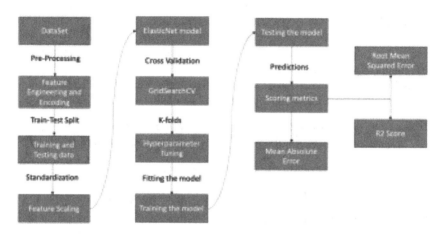

Fig. 1. Architecture diagram.

Elastic Net Regression:

Elastic Net presents a fusion of Ridge and Lasso regression, integrating both L2 (Ridge) and L1 (Lasso) penalties. This amalgamation allows for simultaneous utilization of the advantages offered by each regularization technique, without necessitating an explicit choice between them.

Elastic Net Regression serves as a regularization technique that addresses the limitations of Lasso regression (Fig. 2).

Therefore, Elastic Net Regression effectively addresses the limitations posed by both Ridge Regression and Lasso Regression. The metric utilized by Elastic Net Regression is given by a combination of the penalties from both Ridge and Lasso regressions. The metric followed by Elastic Net Regression is given by:

$$\frac{\sum_{i=1}^{n}(y_i - x_i * \beta^\wedge)^2}{2n} + \lambda\left(\frac{1-\alpha}{2} * \sum_{j=1}^{m}\beta_j^2 + \alpha\sum_{j=1}^{m}|\beta_j|\right) \qquad (2)$$

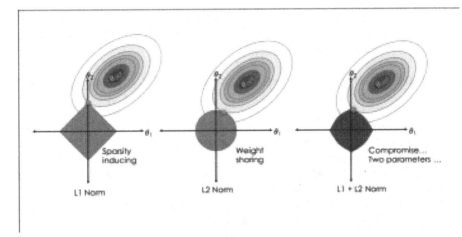

Fig. 2. Elastic Net

In the above Equation,

$$\frac{\sum_{i=1}^{n}(y_i - x_i * \beta)^2}{2n} => \textbf{Best fit equation}, \quad (3)$$

$$\lambda \left(\frac{1-\alpha}{2} \sum_{j=1}^{m} \beta_j^2 + \alpha \sum_{j=1}^{m} |\beta_j| \right) => \textbf{Summation of L1 and L2 regularization}, \quad (4)$$

The loss of function of ridge and Lasso regression are as follows:

$$\textbf{RidgeMSE}(y,) = \text{MSE}(y, y_{pred}) + \alpha * \sum_{i=1}^{m} |\theta_i|^2 \quad (5)$$

$$\textbf{RidgeMSE}(y, y_{pred}) = \text{MSE}(y, y_{pred}) + \alpha * ||\theta_2||^2 \quad (6)$$

$$\textbf{LassoMSE}(y, y_{pred}) = \text{MSE}(y, y_{pred}) + \alpha * \sum_{i=1}^{m} |\theta_i| \quad (7)$$

$$\textbf{LassoMSE}(y, y_{pred}) = \text{MSE}(y, y_{pred}) + \alpha * ||\theta_1|| \quad (8)$$

On combining the two loss functions we get the loss function of elastic net,

$$\textbf{ElasticNetMSE} = \text{MSE}(y, y_{pred}) + \alpha_1 \sum_{m}^{m} |\theta_i| + \alpha_2 \sum_{i=1}^{m} |\theta_i|^2 \quad (9)$$

$$\textbf{ElasticNetMSE} = \text{MSE}(y, y_{pred}) + \alpha_1 * ||\theta_1|| + \alpha_2 * ||\theta_2||^2 \quad (110)$$

Instead of using a single regularization parameter α, we employ two: α1 for the L1 penalty and α2 for the L2 penalty. This enables us to apply elastic net regularization, resembling ridge regression when α1 is 0 and lasso regression when α2 is 0. Alternatively, we can use one α parameter along with an L1-ratio parameter, determining the proportion of the L1 penalty relative to α. For example, with α = 4 and L1-ratio = 0.4, the L1 penalty is scaled by 0.4 and the L2 penalty by 0.6.

$$\text{ElasticNetMSE} = \text{MSE}(y, y_{pred}) + \alpha * (1 - L1Ratio) * \sum_{i=1}^{m} |\theta_i| + \alpha * 1 Ratio \sum_{i=1}^{m} |\theta_i| \quad (11)$$

When L1-ratio equals 0, indicating pure Ridge regression, our model can be treated accordingly. We can solve it using the same methods as for Ridge regression. Specifically, we can employ the normal equation tailored for Ridge regression to directly solve the model, or alternatively, we can opt for iterative methods like gradient descent for solving it step by step.

When the L1-ratio equals 1, implying pure Lasso regression, conventional methods like the normal equation or standard gradient descent become inadequate due to the presence of absolute values. Consequently, alternative approaches such as sub-gradient descent or coordinate descent are preferred.

Likewise, when both L1 and L2 penalties are utilized, resulting in a combination of absolute values, similar techniques—such as sub-gradient descent or coordinate descent—prove effective, leveraging the characteristics of absolute values for optimization.

Grid Search CV (Fig. 3):

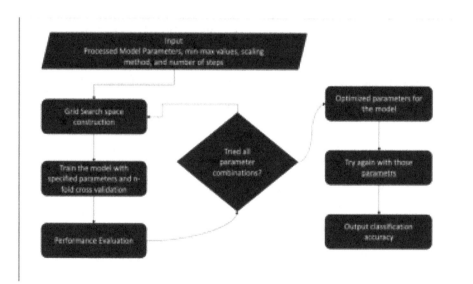

Fig. 3. Grid-Search CV

The Grid Search CV model is illustrated in the following diagram, which shows how the model evaluates the pre-processed model parameters by building a grid search space and using K-fold cross validation to train the model with the given parameters. To evaluate performance, every possible combination of parameters is tried. Until the optimal parameter combinations are identified, these stages are repeated. After the parameters have been optimally chosen, the model is trained on test data to predict the outputs, which are then verified by the output classification accuracy.

Grid Search CV implements a "fit" and a "score" method. It also implements "score samples", "predict", "predict proba", "decision function", "transform" and "inverse transform" if they are implemented in the estimator used. When the input data consists of integers or None, and the estimator is a classifier while the target variable y is either binary or multiclass, Stratified K-Fold cross-validation is employed. For all other cases, standard K-Fold cross-validation is utilized. These cross-validation strategies are configured with shuffle=False to maintain consistent splits across multiple calls.

5 Results

Representing the test scores produced by our model (Tables 1, 2, 3 and 4).

Table 1. Mean evaluation time

mean_fit_time	std_fit_time	mean_score_time	std_score_time
0.0023940086364746100	0.0007914838045351220	0.0004958152770996090	0.00040181762180248000
0.0018162250518798800	0.0005854158629796700	0.0003392219543457030	0.0001665645773785110
0.0014850139617919900	0.00042808706461340200	0.0002966880798339840	0.00019367852819534700
0.0011599540710449200	0.0004609913926210150	0.00014905929565429700	9.0974731580443E−07
0.0009538173675537110	0.00016288610754760700	0.00015897750854492200	2.45269571709818E−05
0.0012497425079345700	0.00048762598344871400	0.00020151138305664100	0.00011037131241961700
0.0015561103820800800	0.00062020576857265840	0.0007816791534423830	0.0010207964743645500
0.0008570671081542970	5.09435288482163E−06	0.00014491081237793000	1.24709099107952E−06
0.0016923904418945300	0.0005147060685575030	0.0003710269927978520	0.00012003416773180100
0.0010332584381103500	0.00028837589635010500	0.00015616416931152300	1.47919467237758E−05
0.0011590957641601600	0.0005701825774584070	0.0002007007598876950	0.0001151855798457880
0.0014766693115234400	0.0006563965564958180	0.0002864837646484380	0.00012231599031437100

Table 2. Parameters

param_alpha	param_l1_ratio	Params	split0_test_score
0.1	0.1	{'alpha': 0.1, 'l1_ratio': 0.1}	−0.0031452333864744500
0.1	0.2	{'alpha': 0.1, 'l1_ratio': 0.2}	−0.003293062091931530
0.1	0.30000000000000000	{'alpha': 0.1, 'l1_ratio': 0.30000000000000004}	−0.0037879614479115900
0.1	0.4	{'alpha': 0.1, 'l1_ratio': 0.4}	−0.004364116015436660
0.1	0.5	{'alpha': 0.1, 'l1_ratio': 0.5}	−0.004992997662045820
0.1	0.6	{'alpha': 0.1, 'l1_ratio': 0.6}	−0.005677304742603260
0.1	0.7000000000000000	{'alpha': 0.1, 'l1_ratio': 0.7000000000000001}	−0.006419885078161060
0.1	0.8	{'alpha': 0.1, 'l1_ratio': 0.8}	−0.007223745421435130
0.1	0.9	{'alpha': 0.1, 'l1_ratio': 0.9}	−0.008092061608596330
0.1	1.0	{'alpha': 0.1, 'l1_ratio': 1.0}	−0.00902818945388729
4.2625	0.1	{'alpha': 4.2625, 'l1_ratio': 0.1}	−0.2630448402883570
4.2625	0.2	{'alpha': 4.2625, 'l1_ratio': 0.2}	−0.29318791478490100

Visualizing the results with the help of a Heatmap (Fig. 4):

Table 3. Split test scores

split1_test_score	split2_test_score	split3_test_score
−0.00334764170536318	−0.003508181283709500	−0.0036087486171764200
−0.003716017909260850	−0.0037492230486302600	−0.0038774471994662800
−0.004337487340998440	−0.0043456808396353100	−0.004488300186319140
−0.004999216543120870	−0.005008688470074420	−0.005173575238640840
−0.0057219399333583200	−0.0057328143484142700	−0.005922132603109060
−0.006508842782904920	−0.006521251135393310	−0.006737294314757610
−0.007363290632927850	−0.007377372234683090	−0.0076225714623233600
−0.008288840964886390	−0.008304743497458350	−0.008581676496813770
−0.009289255736171750	−0.009307135795100800	−0.009618536457993690
−0.010368514853948700	−0.010388538533178900	−0.010737307196556400
−0.2914253053603220	−0.29183968871726900	−0.29899848026145800
−0.3217738595721300	−0.3221889588699960	−0.3293438570063210

Table 4. Ranked test scores

split4_test_score	mean_test_score	std_test_score	rank_test_score
−0.00318100878835362 00	−0.0033581627562 154400	0.000180073301543 32300	1
−0.00342727174292261 00	−0.0036126043984 423100	0.000217235447340 17500	2
−0.004012552617715100	−0.0041943964865 15920	0.000256124175398 9890	3
−0.004623623433972770	−0.0048338439402 49110	0.000296031433163 79200	4
−0.005290781752218840	−0.0055321332598 29260	0.000339798970985 3840	5
−0.006016921994826390	−0.0062923229940 97100	0.000387654884943 53300	6
−0.006805100339544860	−0.0071176439495 28040	0.000439841748601 7460	7
−0.007658545056733780	−0.0080115102874 65490	0.000496617652650 6800	8

(*continued*)

Table 4. (*continued*)

split4_test_score	mean_test_score	std_test_score	rank_test_score
−0.00858066760307125 0	−0.008977531440186760	0.0005582573292319830	9
− 0.009575074534609810	− 0.010019524914436200	0.000625053368712418	10
−0.2748743560949530	−0.28403653414447200	0.013144882734620600	11
−0.3051774372992020	−0.31433440550651000	0.013216257081702300	12

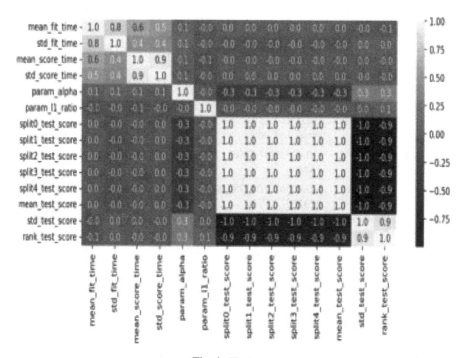

Fig. 4. Heatmap

6 Conclusion

We employ an L1-ratio of 0.1 and $\alpha = 0.1$ in our model. As a result, we obtain an L2 penalty of 0.9 and an L1 penalty of 0.1, which show that Ridge Regression is being used.

The Elastic Net model achieved promising results in predicting municipal waste management costs. The model exhibited low MAE = 0.046 and RMSE = 0.003, suggesting accurate cost estimations across different scenarios. The (R^2) value obtained by our model is equal to 0.99. The results demonstrate the effectiveness of the proposed approach in predicting waste management costs, thereby assisting municipal authorities in budget planning and resource allocation.

In conclusion, the proposed Elastic Net model effectively captures the relationships between socio-economic and environmental factors and municipal waste management costs. This was an existing linear regression-based model for predicting municipal waste management costs. By incorporating relevant features and leveraging historical data, the model offers valuable insights for municipal authorities and researchers to optimize resource allocation and budget planning in waste management.

References

1. Lahcen, G., Mohamed, E., Mohammed, G., Hanaa, H., Abdelmoula, A., et al.: Waste solid management using Machine learning approach. In: 8th International Conference on Optimization and Applications (ICOA), Genoa, Italy (2022)
2. George, A., Mankar, S.H., Suryanarayana, N.V.S., Gupta, M., Sathish, R., Priya, A.: Deep learning based identification of solid waste management in smart cities through garbage separation and monitoring. In: 5th International Conference on Inventive Research in Computing Applications (ICIRCA), Coimbatore, India (2023)
3. Polchan, M., Pukao, A., Cheunban, T., Sinthupuan, S.: Waste management system using waste classification on mobile application. In: Joint International Conference on Digital Arts, Media and Technology with ECTI Northern Section Conference on Electrical, Electronics, Computer and Telecommunications Engineering (ECTI DAMT & NCON), Phuket, Thailand (2023)
4. Adriyanti, N.P., Gamal, A., Dewi, O.C.: Solid waste management models: literature review. In: 2nd International Conference on Smart Grid and Smart Cities (ICSGSC), Kuala Lumpur, Malaysia (2018)
5. Bansal, A., Saxena, P., Garg, R., Gupta, Z.: Data analysis based digital step towards waste management. In: 6th International Conference on Trends in Electronics and Informatics (ICOEI), Tirunelveli, India (2022)
6. Sasikanth, B.S., Yoshita, L.N., Narasimha Reddy, G., Manitha, P.V.: An efficient & smart waste management system. In: International Conference on Computational Intelligence and Computing Applications (ICCICA), Nagpur, India (2021)
7. Ahmad, R.W., Salah, K., Jayaraman, R., Yaqoob, I., Omar, M.: Blockchain for waste management in smart cities: a survey. IEEE Access **9**, 131520–131541 (2021)
8. Iyer, H.: Case study of Mumbai: decentralised solid waste management. Procedia Environ. Sci. **35**, 101–109 (2016)
9. Chepa, S., Singh, S., Dutt, H., Sharma, A., Naik, S., Mahajan, H.: A comprehensive study of distinctive methods of waste segregation and management. In: Third International Sustainability and Resilience Conference: Climate Change, Sakheer, Bahrain (2021)
10. Bao, N.D., Fujiwara, T., Cuong, L.D., Toan, P.P.S., et al.: A short review on the situation of municipal biodegradable solid waste treatment. In: 8th International Scientific Conference on Applying New Technology in Green Buildings (ATiGB), Danang, Vietnam (2023)
11. Butt, O.M., et al.: Hydrogen as potential primary energy fuel for municipal solid waste incineration for a sustainable waste management. IEEE Access **10**, 114586–114596 (2022)
12. Ahmad, S., Iqbal, N., Jamil, F., Kim, D.: Optimal policy-making for municipal waste management based on predictive model optimization. IEEE Access **8**, 218458–218469 (2020)
13. Saad, M., Ahmad, M.B., Asif, M., Khan, M.K., Mahmood, T., Mahmood, M.T.: Blockchain-enabled VANET for smart solid waste management. IEEE Access **11**, 5679–5700 (2023)

14. Damadi, H., Namjoo, M., et al.: Smart waste management using blockchain. IT Prof. **23**, 81–87 (2021)
15. Sosunova, I., Porras, J.: IoT-enabled smart waste management systems for smart cities: a systematic review. IEEE Access **10**, 73326–73363 (2022)
16. Jammeli, H., Ksantini, R., Abdelaziz, F.B., Masri, H.: Sequential artificial intelligence models to forecast urban solid waste in the city of Sousse, Tunisia. IEEE Trans. Eng. Manag. **70**(5), 1912–1922 (2021)
17. Mudannayake, O., Rathnayake, D., Herath, J.D., Fernando, D.K., Fernando, M.: Exploring machine learning and deep learning approaches for multi-step forecasting in municipal solid waste generation. IEEE Access **10**, 122570–122585 (2022)
18. Hossen, M.M., et al.: A reliable and robust deep learning model for effective recyclable waste classification. IEEE Access **12**, 13809–13821 (2024)
19. Cao, B., Chen, X., Lv, Z., Li, R., Fan, S.: Optimization of classified municipal waste collection based on the internet of connected vehicles. IEEE Trans. Intell. Transport. Syst. **22**(8), 5364–5373 (2020)
20. Minu, M.S., Aroul Canessane, R., Subashka Ramesh, S.S.: Optimal squeeze net with deep neural network-based arial image classification model in unmanned aerial vehicles. Traitement du **39**(1), 275–281 (2022)
21. Minu, M.S., Canessane, R.A.: Deep learning-based aerial image classification model using inception with residual network and multilayer perceptron. Microprocess. Microsyst. **95**, 104652 (2022)
22. Bhubalan, K., et al.: Leveraging blockchain concepts as watermarkers of plastics for sustainable waste management in progressing circular economy. Environ. Res. **213**, 113631 (2022)

Sustainable Practices in E-Commerce: Challenges and Trends

Analyzing Bank Customer Behavior: Segmentation and Prediction Using Big Data Analytics

D. Doreen Hephzibah Miriam[1(✉)] and C. R. Rene Robin[2]

[1] Computational Intelligence Research Foundation (CIRF), Chennai, India
`doreenhm@gmail.com`
[2] Department of Computer Science and Engineering, Sri Sairam Engineering College, Chennai, India

Abstract. Businesses use customer segmentation as a strategic tool to divide their heterogeneous client base into discrete groups according to demands, behaviors, or common traits. An overview of consumer segmentation techniques is given in this paper, with emphasis on the role that this technique plays in promoting focused marketing campaigns, raising customer satisfaction levels, and allocating resources as efficiently as possible. This research investigates the effectiveness of real-time banking transaction data for customer segmentation using a large dataset 1048567 transaction. We explore various unsupervised machine learning algorithms to uncover hidden patterns and group customers with similar financial behavior. The methods employed include K-means clustering with standard scaling for normalization, hierarchical clustering with an agglomerative approach for building a hierarchy of clusters, DBSCAN (Density-Based Spatial Clustering of Applications with Noise) for identifying clusters of arbitrary shapes, and Principal Component Analysis (PCA) for dimensionality reduction to focus on the most significant features within the data. By applying these diverse algorithms, we aim to achieve a comprehensive understanding of customer segmentation based on real-time transaction patterns. The analysis will reveal distinct customer groups based on factors such as frequency, type, and value of transactions, allowing banks to develop targeted strategies for improved customer satisfaction and retention.

Keyword: Customer Segmentation · Customer Behavior Analysis · Customer Relationship Management (CRM) · Clustering Algorithms

1 Introduction

Customer segmentation is one of the main strategies banks employ to enhance their understanding of the diverse clientele they serve. Banks can see patterns, preferences, and behaviors that might otherwise go unnoticed when viewed as a whole by breaking up their huge clientele into smaller, more manageable groupings. This process makes it possible to develop targeted advertising campaigns, unique products, and tailored services that appeal more strongly to certain customer categories.

The practice of classifying individuals into groups according to common characteristics or behaviors deemed relevant to the bank's marketing objectives is known as customer segmentation. According to [13] these shared attributes might be things like location, social status, income levels, spending habits, lifestyle choices, and preferred banks.

The ultimate goal of customer segmentation is to build more meaningful relationships with customers by offering tailored experiences that cater to their particular needs and preferences. By employing customer segmentation data, banks may enhance their customer engagement initiatives, optimize their product portfolios, and refine their marketing tactics? The bank may achieve long- term profitability and stable expansion by fostering client loyalty and improving interactions with them via the implementation of this personalized plan.

Understanding customer segmentation involves delving into various characteristics that define the diverse range of individuals within a bank or credit union's customer base. Demographics provide essential insights into age gender, occupation, family status, and income levels, while also considering familial relationships—whether other family members also utilize the institution's services. Geographic further refine this understanding by pinpointing where customers reside, from cities to neighborhoods, providing context for local market dynamics. Psychographics unveil the intricate layers of customer attitudes and beliefs, from satisfaction levels with the institution to broader financial ideologies like risk aversion or saving habits, along with personal interests and preferences. [14] Behaviors shed light on how, when, and where customers interact with banking services, including their technological adoption rates, product preferences, and their stage within life- cycle milestones such as education, family formation, or retirement planning.

Finally, evaluating customer value encompasses past profitability, current financial contribution, and future growth potential, including metrics like share of wallet and estimated net worth, all vital for crafting tailored strategies aimed at maximizing customer satisfaction and organizational profitability.

A thorough grasp of financial behavior is essential for both client retention and satisfaction in the ever-changing banking sector (Lee & Danaher, 2016) [1]. A key tactic for customized product offers and marketing campaigns has long been customer segmentation, which is the practice of breaking down a client base into discrete groups with shared characteristics (Huang et al., 2016) [2]. Nevertheless, conventional segmentation techniques frequently depend on static socioeconomic or demographic data, which may offer an antiquated and constrained understanding of consumer requirements (Verhoef, 2003) [3].

The potential of real-time banking transaction data to transform consumer segmentation is investigated in this study. Banks may obtain a more detailed and dynamic picture of client behavior patterns and financial objectives by examining the frequency, kind, and value of transactions (Chen & Wang, 2018) [4].

Analyzing 3 Bank Customer Behavior: Segmentation & Prediction using BDA With this real-time strategy, banks can capture their clients' changing financial lives rather than just focusing on static demographics. Banks may create personalized experiences, tailored services, and ultimately stronger client connections by utilizing this real-time data (Petropoulos, 2022) [5].

To shed light on the practical application of real-time transaction data for segmentation, this study will present real-time case studies from various banking sectors. Analyzing these case studies will provide insightful knowledge on how other banks have effectively used transaction data to segment their clientele and create focused marketing campaigns. Through an examination of the achievements and obstacles faced in these real-world situations, the study seeks to offer a thorough framework for executing real-time client segmentation in the banking sector.

2 Literature Review

A. S. M. Shahadat Hossain et al.(2017) [6] This paper advocates for the integration of density-based algorithms, particularly DBSCAN, alongside centroid based techniques for customer segmentation in e-commerce contexts. By leveraging the nuanced capabilities of DBSCAN, which adeptly identifies clusters of varying shapes and sizes while effectively handling noise, more meaningful and nuanced customer segmentation can be attained. Through a comparative analysis of DBSCAN and traditional centroid-based approaches, this study elucidates the efficacy and advantages of leveraging density-based algorithms in the pursuit of refined customer segmentation strategies within the e-commerce landscape.

Joshua, Thomas., P., N. et al.(2022) [7] This paper explains the customer segmentation underscores its pivotal role in enabling companies to effectively market their products and services to diverse consumer groups. Customer segmentation involves the classification and categorization of customers based on their shared traits and characteristics, such as age group, gender, social class, and purchase patterns. This segmentation process allows firms to gain a comprehensive understanding of their customer base, thereby facilitating targeted marketing efforts tailored to the specific needs and preferences of each segment.

Chiheb-Eddine, Ben, N'cir. (2022) In this paper [8], a multi-objective segmentation approach based on three conceptual axes: descriptive, predictive, and quality validation is proposed to segment bank credit card customers using their descriptive characteristics and their predictive behavior. They have evaluated this approach in an empirical study which aims to segment bank credit card customers using their descriptive characteristics and their predictive behavior. Obtained results have shown the ability of the proposed approach to look for effective homogeneous segments and help decision-makers propose more tailored marketing strategies.

In this paper, Bartels et al.(2022) [9] used K-means and DBSCAN to classify the segmentation of customers using a Recency, Frequency and Monetary Value (RFM) Model and clustering techniques, to find groups of similarities and differences and to discover potential valuable and vulnerable customers. Past research on customer segmentation using data mining techniques had drawbacks. Density-based clustering algorithms like DBSCAN have been examined by few research papers.

Lewaaelhamd, I. et al. (2023) [10] In this paper, the efficacy of various techniques, including K-means and DBSCAN clustering, was compared for customer segmentation and churn prediction. The results suggest that dividing customers into six distinct clusters provides a practical and straightforward approach for churn prediction. This approach enables businesses to identify distinct customer segments with unique behavioral patterns, facilitating targeted marketing strategies and personalized customer engagement initiatives. Despite the widespread adoption of ML in churn prediction, many businesses in the United Kingdom face challenges due to the lack of comprehensive and adaptable consumer data.

Hafidh et. al (2023) In this paper [11], a segmentation model for credit customers to identify the potential for defaulting credit customers based on their transaction history is proposed, which is based on K-means, aiding in risk mitigation and targeted preventive actions for payment failure clusters. Further it investigates customer value based on cross-selling probability and loyalty. Uses neural network approach with Self Organization Map for banking.

Homburg, C. et. al (2023) In this paper [12] the authors conceptualize and operationalize B2B customer journey management capability (CJMC) as a supplier's ability to achieve superior customer value along the customer journey by strategically creating value-anchored customer touch points characterized through the implementation of consistent resource usage across internal organizational boundaries and by continuously monitoring value creation toward individual members of the buying center.RBV theory foundation for B2B CJMC conceptualization. Dynamic capabilities crucial for managing touch points in dynamic CJs.

Miriam, Hephzibah, et. al (2023) In this paper [16] it explains that block chain technology is essential for bolstering cyber security, especially in delicate fields like medicine. With the growth of sophisticated computing systems, crypto- graphic techniques are becoming more vulnerable to hacking efforts, posing new issues as a result of their fast improvement. This research presents a new technique, called Lionized Golden Eagle-based Homomorphic Elapid Security (LGE- HES), to address these issues. By combining homomorphic encryption techniques with block chain technology, this solution preserves privacy by enabling calculations on encrypted medical data without the need for decryption. As shown, the homomorphic encryption architecture improves data security by permitting secure operations on private data without

3 Customer Segmentation

You may use client segmentation to enhance communication between your sales and marketing teams and your customers by putting people in groups based on common characteristics such as demographics or behaviors. Using these consumer segmentation categories, it may also be discussed how to create a marketing persona or product user persona. This is because effective customer segmentation research is often used to inform a brand's positioning and message, help companies decide which new products or services to invest in, and spot chances to increase sales. Therefore, in order for marketing personas to be effective, they need to closely align with those categories. A firm may divide its client base in business-to-business marketing according to a variety of criteria, such as:

- Industry
- Number of employees
- Products previously purchased from the company

– Location

In business-to-consumer marketing, companies often segment customers according to demographics that include:

– Age
– Gender
– Marital status
– Location (urban, suburban, rural)
– Life stage (single, married, divorced, empty-nester, retired, etc.)

4 Why to Segment Customer

Through the strategic application of customer segmentation, marketers may optimize their efforts by concentrating their messaging and product development on distinct target segments. This segmentation allows for more precise targeting, which has several significant benefits for the company. It enables the creation and dissemination of marketing communications that are especially suited to appeal to certain target audiences. By knowing the unique needs, tastes, and traits of each segment, marketers can better target their messaging and engage target audiences, increasing the efficacy of their efforts.

Using segmentation, the most effective communication channel is selected for each area. Email, social media, radio, and other traditional modes of advertising may be preferred over alternative channels of communication by specific market segments. If marketers identify the preferred channels for each category and use those to communicate their message, they may optimize the effectiveness of their communication efforts.

Furthermore, segmentation facilitates the process of identifying chances to improve current offerings and develop brand-new ones. By examining the distinct requirements and preferences of different customer segments, businesses may identify gaps in their current offerings and create new possibilities for products or services that are tailored to each segment's specific demands.

Segmentation also helps organization's better serve their customers by enabling them to tailor their support and service offerings to the particular requirements and preferences of each group. Increased client pleasure and loyalty result from this. By identifying complementary products or services that are relevant to certain customer groups, segmentation facilitates the identification of upselling and cross-selling possibilities. Businesses that target these regions with tailored offerings can boost revenue and maximize the lifetime value of customers. Overall, customer segmentation is a helpful tactic that aids companies in making the most of their marketing campaigns, fortifying client relationships, and promoting business expansion by accurately recognizing and satisfying the diverse wants of their clientele.

5 Experimental Setup

In this paper, the workflow for customer segmentation in the banking industry, leveraging a combination of powerful Python libraries and machine learning techniques. The workflow begins with data preparation and exploration using NumPy for numerical calculation and Pandas for efficient data analysis, Seaborn and Matplotlib for data visualization, Specialised imports include tools from the modules like LabelEncoder, StandardScaler, MinMaxScaler,silhouettescore,probplot,KMeans,PCA, AgglomerativeClustering, DBSCAN, Spectral Clustering, NearestNeighbors, KneeLocator, clusteval. Additionally, The warning module was imported to optimise code performance as shown in Fig. 1.

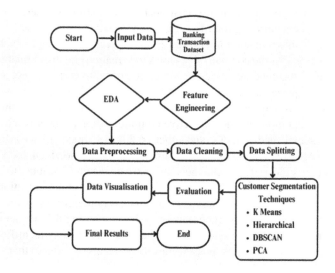

Fig. 1. Experimental Setup

6 Dataset Description

This dataset consists of 1048567 transactions by over 800K customers for a bank in Chennai, India. The data contains information such as customer age (DOB), location, gender, account balance at the time of the transaction, transaction details, transaction amount.

6.1 Sample Dataset

The head() function is primarily used to view the first few rows of a dataset. It helps users quickly get an overview of the data and its structure. It is used to check column names, data types, and the data itself by displaying the initial records as shown in Fig. 2.

	TransactionID	CustomerID	CustomerDOB	CustGender	CustLocation	CustAccountBalance	TransactionDate	TransactionTime	TransactionAmount (INR)
0	T1	C5841053	10/1/94	F	JAMSHEDPUR	17819.05	2/8/16	143207	25.00
1	T2	C2142763	4/4/57	M	JHAJJAR	2270.69	2/8/16	141858	27999.00
2	T3	C4417068	26/11/96	F	MUMBAI	17874.44	2/8/16	142712	459.00
3	T4	C5342380	14/9/73	F	MUMBAI	866503.21	2/8/16	142714	2060.00
4	T5	C9031234	24/3/88	F	NAVI MUMBAI	6714.43	2/8/16	181156	1762.50

Fig. 2. Preview of the Dataset

6.2 Dataset Size

The shape method provides information about the number of rows and columns in a DataFrame. This dataset contains 1048567 rows and 9 columns.

6.3 Dataset Information

The info function in pandas prints information about the DataFrame. The information contains the number of columns, column labels, column data types, memory usage, range index, and the number of cells in each column (non-null values) (Fig. 3).

```
<class 'pandas.core.frame.DataFrame'>
RangeIndex: 1048567 entries, 0 to 1048566
Data columns (total 9 columns):
 #   Column                 Non-Null Count    Dtype
---  ------                 --------------    -----
 0   TransactionID          1048567 non-null  object
 1   CustomerID             1048567 non-null  object
 2   CustomerDOB            1045170 non-null  object
 3   CustGender             1047467 non-null  object
 4   CustLocation           1048416 non-null  object
 5   CustAccountBalance     1046198 non-null  float64
 6   TransactionDate        1048567 non-null  object
 7   TransactionTime        1048567 non-null  int64
 8   TransactionAmount (INR) 1048567 non-null  float64
dtypes: float64(2), int64(1), object(6)
memory usage: 72.0+ MB
```

Fig. 3. Information of the Dataset

6.4 Descriptive Analytics

The describe method is used to generate descriptive statistics. Descriptive statistics include those that summarize the central tendency, dispersion and shape of a dataset's distribution, excluding NaN values. It also analyzes both numeric and object series, as well as DataFrame column sets of mixed data types. The output of banking transaction dataset is shown in Fig. 4.

6.5 Missing Values in Dataset

To check this, we can use the isnull method. It will return True for missing components and False for non-missing cells. However, when the dimension of a dataset is large, it

	CustAccountBalance	TransactionTime	TransactionAmount (INR)
count	1046198.00	1048567.00	1048567.00
mean	115403.54	157087.53	1574.34
std	846485.38	51261.85	6574.74
min	0.00	0.00	0.00
25%	4721.76	124030.00	161.00
50%	16792.18	164226.00	459.03
75%	57657.36	200010.00	1200.00
max	115035495.10	235959.00	1560034.99

Fig. 4. Descriptive Analysis of the Dataset

could be difficult to figure out the existence of missing values. In general, we may just want to know if there are any missing values at all before we try to find where they are. The function isnull().sum() returns the number of missing values in the dataset as shown in Fig. 5.

```
TransactionID              0
CustomerID                 0
CustomerDOB             3397
CustGender              1100
CustLocation             151
CustAccountBalance      2369
TransactionDate            0
TransactionTime            0
TransactionAmount (INR)    0
dtype: int64
```

Fig. 5. Missing values in the Dataset

6.6 Duplicated of the Dataset

The duplicated method is used to identify the duplicated rows in a DataFrame. It returns a boolean series which is True only for unique rows.

7 Exploratory Data Analysis(EDA)

7.1 Count Plot of the Dataset

The count plot method is used to display the count of categorical observations in each bin in the dataset. A count plot resembles a histogram over a categorical variable as opposed to a quantitative one. You can compare counts across nested variables because the fundamental API and settings are the same as those for bar plot. The count plot method takes input data in many forms, such as wide- form data, long-from data, arrays or a list of vectors. The count plot of the gender in the banking transaction is shown in Fig. 6.

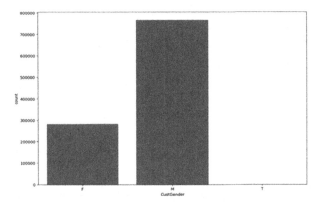

Fig. 6. Count plot of the dataset

7.2 Distribution Plot

"*Skewness essentially is a commonly used measure in descriptive statistics that characterizes the asymmetry of a data distribution, while kurtosis determines the heaviness of the distribution tails.*" Understanding the shape of data is crucial while practicing data science.

Skewness of CustAccountBalance: 38.38055629563517 Kurtosis of CustAccountBalance: 3011.425568440171.

Distribution plots visually assess the distribution of sample data by comparing the empirical distribution of the data with the theoretical values expected from a specified distribution.

7.3 Boxplot

A box plot shows the distribution of quantitative data in a way that facilitates comparisons between variables or across levels of a categorical variable. The box shows the quartiles of the dataset while the whiskers extend to show the rest of the distribution, except for points that are determined to be "outliers" using a method that is a function of the inter-quartile range.

7.4 Quantile Plot

The quantile-quantile plot is a graphical method for determining if a dataset follows a certain probability distribution or whether two samples of data came from the same population or not. Quantiles are points in a dataset that divide the data into intervals containing equal probabilities or proportions of the total distribution. They are often used to describe the spread or distribution of a dataset (Fig. 7).

Skewness of TransactionAmount (INR): 58.063115298630485.
Kurtosis of TransactionAmount (INR): 9855.65495601318.

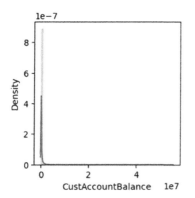
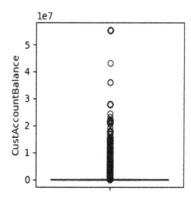

(a) Distribution Plot of the Transaction (b) Boxplot Distribution

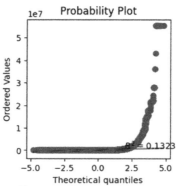

(c) Quantile-Quantile Plot of the Transaction.

Fig. 7. Different Plot analysis of Banking Transaction dataset

7.5 Pie Chart Distribution

A Pie Chart is a circular statistical plot that can display only one series of data. The area of the chart is the total percentage of the given data (Fig. 8).

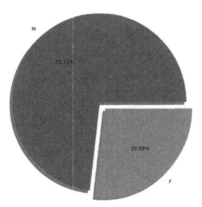

Fig. 8. Pie Chart Distribution

7.6 Bar Plot Distribution

The Bar plot shows the point estimates and errors as rectangular bars. A bar plot represents an aggregate or statistical estimate for a numeric variable with the height of each rectangle and indicates the uncertainty around that estimate using an error bar. This Bar plot distribution shows that, At an average, females make slightly higher transactions than males as shown in 9a. This Bar plot distribution shoes that, At an average, male customers have marginally higher account balances in comparison to their female counterparts as shown in Fig. 9b.

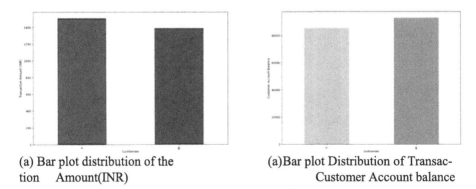

(a) Bar plot distribution of the Transaction Amount(INR)

(a) Bar plot Distribution of Customer Account balance

Fig. 9. Bar Plot Analysis of Banking Transaction dataset.

From this Distribution, It shows that Male customers make higher transactions than their female counterparts only during the months of February and April. In all the remaining months, female customers perform greater transactions. Therefore, the bank must provide more special offers and incentives to female customers as they are likely to be substantially active in making larger transactions almost throughout the entire year.

Monthly Comparison of Spending Habits of Male & Female Customers

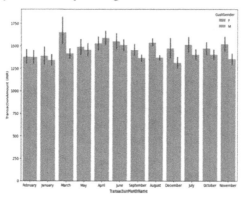

Fig. 10. Monthly Comparison of Spending Habits of Male & Female Customers

Higher value transactions are mostly done during the months of March, April and June are shown in Fig. 10. Female customers perform higher transactions than their male counterparts in almost every week day are shown in Fig. 11.

Weekday-Wise Comparison of Spending Habits of Male & Female Customers

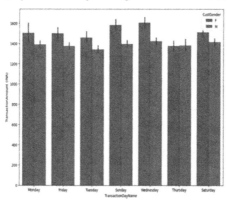

Fig. 11. Weekday-Wise Comparison of Spending Habits of Male Female Customers

Male customers predominantly have greater account balances as compared to their female counterparts. This is evident from the fact that they generally make comparatively low value transactions than female customers are shown in Fig. 12.

Monthly Comparison of Account Balances of Male & Female Customers

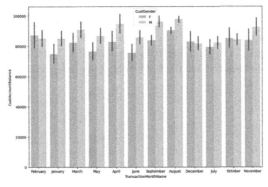

Fig. 12. Monthly Comparison of Account Balances of Male & Female Customers

7.7 Point Plot Distribution

A point plot represents an estimate of central tendency for a numeric variable by the position of the dot and provides some indication of the uncertainty around that estimate using error bars. Point plots can be more useful than bar plots for focusing comparisons between different levels of one or more categorical variables. They are particularly adept at showing interactions: how the relationship between levels of one categorical variable changes across levels of a second categorical variable.

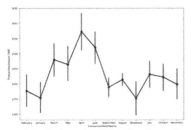 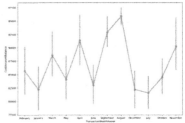

(a) Point plot Distribution of Transaction in every month with Transaction amount (INR)

(b) Point plot Distribution of Transaction in every month with Customer Account Balance

Fig. 13. Point Plot Analysis of Banking Transaction dataset.

This point plot distribution shows that, Highest value transactions are performed during the summer months while least value transactions are made in the winter months in Fig. 13a.

Commonly, the account balances of customers are highest in the months of August and September whereas they are lowest in the months of July and December. In the festive months, the customer account balances drop significantly which makes sense as people usually expend more money during the festive season in Fig. 13b.

7.8 Bar Plot Distribution with Customer Location

Customers belonging to the Roomford bank branch in United Kingdom mostly perform the highest transactions which are closely followed by the bank branches in Palakkarai Trichy(Tamil Nadu, India) and Munchen in Germany in Fig. 14a Customers living in PO Box 28483 Dubai bank branch have the highest account balances among all customers in Fig. 14b.

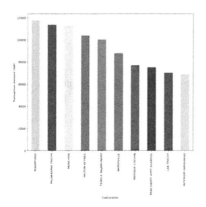
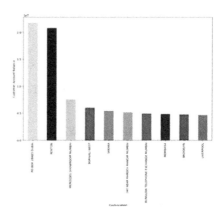

(a) Bar plot distribution of Customer Location with Transaction Amount (INR)

(b) Bar plot Distribution of Customer Location with Account Balance

Fig. 14. Bar Plot Analysis of Banking Transaction dataset with Customer Location

7.9 Bar Plot Analysis of Banking Transaction with Age

In general, the account balances of customers rise abruptly with increase in their ages, attaining a peak by the old age of 50 years or more, although there are a few exceptions to this matter. The account balances of younger adults in the age range of 19 to 22 years have higher account balances in comparison to their surrounding age groups in Fig. 15a.

Likewise, an exactly similar trend prevails in case of transaction amounts as well. Younger adult customers, in the age range of 19 to 24 years, perform exorbitant transactions as compared to their surrounding age groups. This is most probably due to the reason that younger adults between the ages of 18 to 24 years have their own career aspirations to fulfill as a consequence of which they generally have lavish and extravagant demands for fulfilling their passions and interests and for facilitating their development through all means. Nevertheless, middle age senior adults and elderly customers make comparatively more extortionate transactions for managing their families and livelihoods in Fig. 15b.

Analyzing Bank Customer Behavior   237

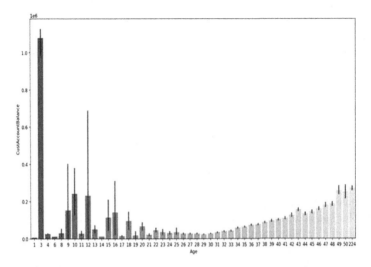

(a) Bar plot distribution of Age with Customer Account Balance

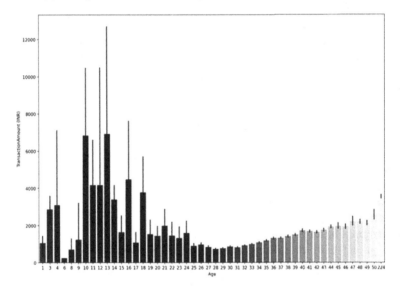

(b) Bar plot distribution of Age with Transaction amount (INR)

Fig. 15. Bar plot analysis of Banking Transaction with age.

7.10 Pair Plot Distribution

The pair plot plots the pairwise relationships in a dataset. By default, this function will create a grid of Axes such that each numeric variable in data will by shared across the y-axes across a single row and the x-axes across a single column. The diagonal plots are treated differently: a univariate distribution plot is drawn to show the marginal distribution of the data in each column.

It is also possible to show a subset of variables or plot different variables on the rows and columns is shown in Fig. 16.

8 Algorithm

8.1 K Means Algorithm - Customer Segmentation

- Define distance function: This calculates the distance between two customer data points based on chosen features.
- Load data and extract features: Load customer transaction data and extract relevant features for segmentation (e.g., average balance, transaction frequency).
- Define number of clusters (k): Choose the desired number of customer segments based on data analysis.
- Initialize centroids: Randomly select k data points as initial centroids (cluster centers).
- Main loop:

  - Assign each customer to the nearest centroid based on the distance function.
  - Re-compute the centroid for each cluster based on the customers assigned to it.
  - Terminate the loop if the centroids haven't changed significantly (convergence).
  - Customer segmentation: Use the final cluster assignments to segment customers into different groups.

Algorithm 1 K-means Clustering for Customer Segmentation

Require: transactions: List of transactions, k: Number of clusters, max_iterations: Maximum number of iterations

1: Initialize centroids randomly: *centroids* ← *randomly_initialize_centroids(transactions, k)*
2: **for** *iteration* **in** *range(max_iterations)* **do**
3:   Assign each transaction to the nearest centroid: *clusters* ← *assign_to_clusters(transactions, centroids)*
4:   Calculate new centroids based on the mean of transactions in each cluster: *new_centroids* ← *calculate_centroids(clusters)*
5:   **if** *centroids* == *new_centroids* **then**
6:     **break**
7:   **end if**
8:   *centroids* ← *new_centroids*
9: **end for**
10: **return** *clusters*

Analyzing Bank Customer Behavior   239

Algorithm 2 Helper Functions
1: **function** randomly_initialize_centroids(*transactions, k*):
2: *centroids* ← []
3: **for** *i* in *range(k)* **do**
4:    Append randomly selected transaction to centroids:
      centroids.append(randomly_select_transaction(transactions))
5: **end for**
6: **return** *centroids*

1: **function** assign_to_clusters(*transactions, centroids*):
2: *clusters* ← {}
3: **for** *transaction* in *transactions* **do**
4:    Calculate distances to each centroid: *distances* ← []
5:    **for** *centroid* in *centroids* **do**
6:       Calculate distance: *distance* ← *calculate_distance(transaction, centroid)*
7:       Append distance: *distances.append(distance)*
8:    **end for**
9:    Assign transaction to nearest centroid's cluster: *nearest_centroid_index* ← *argmin(distances)*
10:   **if** *nearest_centroid_index* **not in** *clusters* **then**
11:      *clusters[nearest_centroid_index]* ← []
12:   **end if**
13:   *clusters[nearest_centroid_index].append(transaction)*
14: **end for**
15: **return** *clusters*

Silhouette Score of optimized k means: 23.2%

1: **function** calculate_centroids(*clusters*):
2: *new_centroids* ← []
3: **for** *cluster* in *clusters.values()* **do**
4: Calculate mean of transactions in cluster: *cluster_mean* ← *calculate_mean(cluster)*
5: Append cluster mean to new centroids: *new_centroids.append(cluster_mean)*
6: **end for**
7: **return** *new_centroids*

1: function calculate_distance(transaction, centroid):
2: Calculate Euclidean distance: distance ← $\sqrt{\sum_{i=1}^{n}(transaction[i] - centroid[i])^2}$
3: return distance

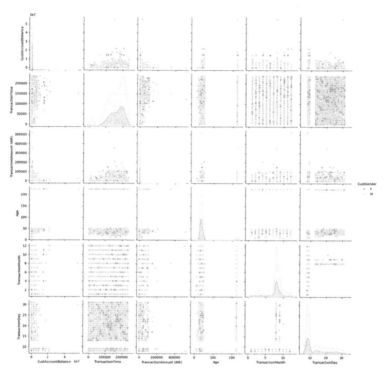

Fig. 16. Pair plot distribution of Banking Dataset

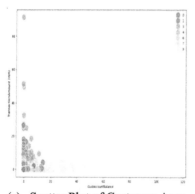

(a) Scatter Plot of Customer Account Balance and Transaction Amount

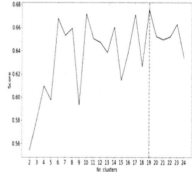

(b) Optimal Number of Clusters as computed by K-means Clustering Algorithm

Fig. 17. Scatter Plot of Customer Account Balance vs. Transaction Amount and Optimal Cluster Count Determination Using K-Means Clustering

8.2 Hierarchical Clustering Algorithm - Customer Segmentation

1. Distance function: This calculates the distance between two customer data point.

2. based on chosen features.
3. Data loading and feature extraction: Similar to previous examples.
4. Distance matrix (optional): Create a pre-computed distance matrix for efficiency in distance calculations during merging (applicable for larger datasets).
5. Initialize clusters: Create individual clusters for each customer initially.

1: function calculate_mean(cluster):
2:   Calculate mean of transaction in cluster:
   $$mean \leftarrow \frac{\Sigma}{[transaction\ in\ cluster\ transaction\ [i]\ len(cluster)} \text{ for i in range(len(cluster[0]))}]$$
3: return *mean*
---

5. Linkage function: Define a function to determine the distance between clustersvbased on the linkage method. Here, single linkage is used as an example (considering the minimum distance between any two points across clusters).
6. Main loop:
   - In each iteration, find the two closest clusters based on the linkage function.
   - Merge the closest clusters into a single cluster.
   - Terminate the loop when the desired number of clusters (k) is reached.
7. Customer segmentation: Assign each customer to a cluster based on the final cluster membership.

Algorithm 3 Hierarchical Agglomerative Clustering for Customer Segmentation

Require: *transactions*: List of transactions, *k*: Number of clusters
1: Initialize clusters: *clusters* ← initialize_clusters(*transactions*)
2: **while** |*clusters*| > *k* **do**
3:   Find closest clusters: (C_i, C_j) ← find_closest_clusters(*clusters*)
4:   Merge clusters: *clusters* ← merge_clusters(*clusters*, C_i, C_j)
5: **end while**
6: **return** *clusters*

Algorithm 4 Helper Functions

1: **function** initialize_clusters(*transactions*):
2: Initialize each transaction as its own cluster: *clusters* ← []
3: **for** *transaction* **in** *transactions* **do**
4:   Append [*transaction*] to *clusters*
5: **end for**
6: **return** *clusters*

Silhouette Score of optimised Agglomerative Clustering: 41.84%

1: **function** find_closest_clusters(*clusters*):
2: Initialize minimum distance: $min\_distance \leftarrow \infty$
3: Initialize closest clusters: $C_i, C_j \leftarrow$ None 4: **for** $i = 1$ **to** $|clusters|$ **do**
5: **for** $j = i + 1$ **to** $|clusters|$ **do**
6: Compute distance between C_i and C_j : $distance \leftarrow$ compute_distance(C_i, C_j)
7: **if** $distance < min\_distance$ **then**
8: Update minimum distance: $min\_distance \leftarrow distance$
9: Update closest clusters: $C_i, C_j \leftarrow i, j$
10: **end if**
11: **end for**
12: **end for**
13: **return** (C_i, C_j)

1: **function** merge_clusters(*clusters, C_i, C_j*):
2: Merge clusters C_i and C_j : $merged\_cluster \leftarrow clusters[C_i] + clusters[C_j]$
3: Remove clusters C_i and C_j : $clusters \leftarrow$ remove_clusters(*clusters, C_i, C_j*)
4: Append merged cluster to clusters: $clusters \leftarrow$ append_cluster(*clusters, merged_cluster*)
5: **return** *clusters*

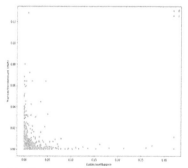

(a) Scatter Plot distribution of Customer Account balance with Transaction amount(INR) in Hierarchical Clustering

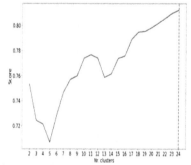

(b) Optimal number of Clusters as computed by Agglomerative Clustering Algorithm

Fig. 18. Customer Account Balance vs. Transaction Amount Distribution and Optimal Cluster Determination Using Hierarchical Clustering

8.3 DBSCAN Clustering Algorithm - Customer Segmentation

1. Define distance function, data loading, and parameters: Similar to previous algorithms.

   - min_pts: Minimum number of points required in a neighborhood for a point to be a core point.
   - eps: Radius of the epsilon neighborhood around a data point.

2. Core point check function: Determines if a customer has enough neighbors within the epsilon radius to be a core point.
3. Cluster expansion function: Recursively expands a cluster by finding neighbors of core points, adding them if they are core points or unvisited, and continuing the expansion for core points within those neighbors.
4. Main loop:

   - For core points, a new cluster is created and expanded using the ex- pand_cluster function.
   - Non-core points are marked as noise.

5. Customer segmentation: Assigns customers to clusters based on their presence in the final cluster lists. Handles noise points (customers not assigned to any cluster).

   - Iterates through customers, checking if they are visited or core points
   - Choosing appropriate values for min_pts and eps based on data exploration.
   - Handling potential border points (neither core points nor noise).

   Silhouette Score of DBSCAN: 20.36.

Algorithm 5 DBSCAN for Customer Segmentation

Require: *transactions*: List of transactions, ε: Radius of neighborhood, *MinPts*: Minimum number of points
1: Initialize cluster id: *cluster_id* $\leftarrow$ 0 2: **for** *transaction* in *transactions* **do** 3: **if** *transaction* is visited **then**
4: **continue**
5: **end if**
6: Mark *transaction* as visited
7: *neighborhood* $\leftarrow$ get_neighborhood(*transaction, ε*)
8:　**if** size of *neighborhood* < *MinPts* **then**
9:　　Mark *transaction* as noise
10:　**else**
11:　　Increment *cluster_id*
12:　　Expand_cluster(*transaction, neighborhood, cluster_id, ε, MinPts*)
13:　**end if**
14: **end for**
15: **return** clusters

Silhouette Score of DBSCAN: 20.36%

Algorithm 6 Expand_cluster

Require: *transaction*: Current transaction, *neighborhood*: Neighborhood of current transaction, *cluster_id*: Cluster id, ε: Radius of neighborhood, *MinPts*: Minimum number of points
1: Assign *transaction* to cluster *cluster_id*
2: **for** *neighbor* **in** *neighborhood* **do**
3:   **if** *neighbor* is not visited **then**
4:     Mark *neighbor* as visited
5:     *neighborhood_of_neighbor* ← get_neighborhood(*neighbor*, ε)
6:     **if** size of *neighborhood_of_neighbor* ≥ *MinPts* **then**
7:       *neighborhood* ← *neighborhood* ∪ *neighborhood_of_neighbor*
8:     **end if**
9:   **end if**
10: **if** *neighbor* does not belong to any cluster **then**
11:    Assign *neighbor* to cluster *cluster_id*
12: **end if**
13: **end for**

Algorithm 7 Helper Functions

1: **function** get_neighborhood(*transaction*, ε):
2: *neighborhood* ← {*t* such that *t* is within distance ε from *transaction*}
3: **return** *neighborhood*

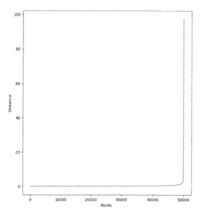
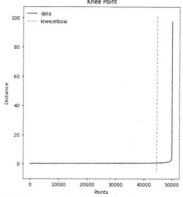

(a) DBSCAN Distribution - Point vs Distance

(b) Knee point Distribution

Fig. 19. Knee Point and Distance in DBSCAN Clustering Algorithm

Knee Point: 0.7028095357275649 as referred in Fig. 19b The Scatter plot distribution shows the comparison of Customer Account Balance with Transaction Amount (INR)20a.

Bar plot of Customer Account Balance vs Transaction Amount - DBSCAN as shown in Fig. 20b.

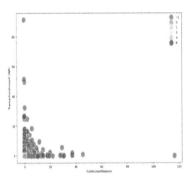

(a) Scatter Plot of Customer Account Balance vs Transaction Amount DBSCAN

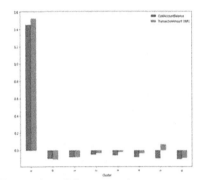

(b) Bar plot of Customer Account Balance vs Transaction Amount DBSCAN

Fig. 20. DBSCAN with applications considering Noise

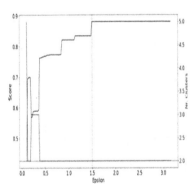

(a) Determination of Optimum Epsilon Value for DBSCAN Model

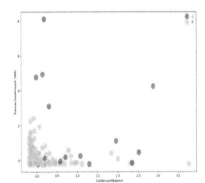

(b) Scatter plot of Customer Account Balance vs Transaction Amount Optimum Epsilon value of DBSCAN

Fig. 21. DBSCAN Model Distribution for Banking Dataset

The optimized DBSCAN model has classified the entire population of customers primarily into two major groups, one of them consists of all those customers who have modest account balance and make low-value transactions in Fig. 21a whereas the miscellaneous group includes either the customers who have high account balances and spend very less money through transactions or those who have minimal account balance but expend a large amount of cash through high-value transactions in Fig. 21b.

8.4 PCA - Customer Segmentation

1. Covariance matrix function: This calculates the covariance matrix to capture the variance and relationships between features.
2. Data loading and feature extraction: Similar to previous examples.
3. Convert data to NumPy array: This enables efficient matrix operations for PCA.
4. Centering data: Subtracting the mean from each feature helps center the data for PCA.
5. Covariance matrix calculation: The function calculates the covariance matrix using the centered data.
6. Eigen decomposition: This step decomposes the covariance matrix to obtain Eigen values (importance of principal components) and eigenvectors (directions of principal components).
7. Sorting by eigenvalues: Sort the eigenvalues and eigenvectors together in descending order based on the eigenvalues (most important components first).
8. Choosing number of components: Decide on the number of principal components (PCs) to retain based on the explained variance or data analysis.
9. Selecting top principal components: Select the corresponding eigenvectors for the chosen number of PCs.
10. Projecting data: Project the centered data onto the principal components to get a lower-dimensional representation.
11. Customer segmentation using projected data: The projected data (capturing the most significant variations) is used for K-Means clustering or other segmentation algorithms.

Algorithm 8 PCA for Customer Segmentation

Require: *transactions*: List of transactions, *num_components*: Number of principal components

1: // Step 1: Standardize the data
2: *standardized_transactions* ← standardize_transactions(*transactions*) 3: // Step 2: Compute the covariance matrix
4: *covariance_matrix* ← compute_covariance_matrix(*standardized_transactions*)
5: // Step 3: Compute the eigenvectors and eigenvalues
6: (*eigenvalues, eigenvectors*) ← compute_eigenvectors_eigenvalues(*covariance_matrix*) 7: // Step 4: Select the top k eigenvectors
8: *top_eigenvectors* ← select_top_eigenvectors(*eigenvectors, eigenvalues, num_components*)
9: // Step 5: Project the data onto the selected eigenvectors
10: *principal_components* ← project_data(*standardized_transactions, top_eigenvectors*)
11: **return** *principal_components*

Silhouette Score of PCA: 51.99%

Algorithm 9 Helper Functions

1: **function** standardize_transactions(*transactions*):
2: // Standardize the data by subtracting the mean and dividing by the standard deviation for each feature
3: *mean* ← **calculate_mean(*transactions*)**
4: *std_dev* ← calculate_standard_deviation(*transactions*)
5: *standardized_transactions* ← []
6: **for** *transaction* in *transactions* **do**
7: *standardized_transaction* ← [(*transaction*[i] − *mean*[i])/*std_dev*[i] **for** i in **range(len(transaction))***Append*standardized_transaction*to*standardized_transactions
8: 9: **end for**
10: **return** *standardized_transactions*

1: **function** compute_covariance_matrix(*transactions*):
2: // Compute the covariance matrix
3: *num_transactions* ← length(*transactions*)
4: *num_features* ← length(*transactions*[0])
5: *covariance_matrix* ← []
6: **for** i in **range**(*num_features*) **do**
7: *row* ← []
8: **for** j in **range**(*num_features*) **do**
9: *cov* $\frac{\Sigma num\_transactions \quad (transactions[k][i]-mean[i]) \times (transactions[k][j]-mean[j])}{number\_transactions}$
10: Append *cov* to *row*
11: **end for**
12: Append *row* to *covariance_matrix*
13: **end for**
14: **return** *covariance_matrix*

The Spectral clustering algorithm has segregated the customers into two distinct groups. The first group comprises the dynamic customers who have lower account balance and mostly expend less cash on transactions barring a few who perform large-value transactions whereas the second group includes the more conservative and money saving-minded people who, in spite of having really high account balances, spend the least amount of money in transactions, thereby judiciously maintaining their savings accounts.

9 Model Evaluation

The Silhouette Score is a metric used to assess the quality of clusters produced by various clustering techniques like k-means, Agglomerative Clustering, DB- SCAN, and PCA. It measures the separation and compactness of the clusters, quantifying the cohesion of data points within the same cluster and separation between distinct clusters. Higher

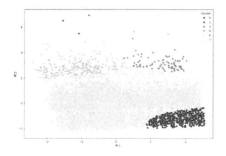
(a) Scatter Plot-Spectral Clustering

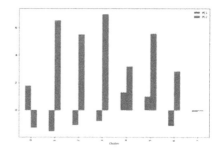
(b) Bar-plot Spectral Clustering

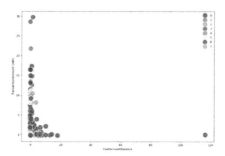

(b) Scatter plot of Customer Account balance vs Transaction Amount Spectral Clustering

Fig. 22. Principal component Analysis (PCA) distribution in Banking Dataset

scores indicate better segmentation, aiding in selecting the best algorithm. The Silhouette Score is explained mathematically below:

$$s(i) = b(i)a(i)/max(a(i), b(i))$$

where:
$s(i)$ is the Silhouette Score for the data point.
$a(i)$ is the average distance between i and other data points in the same cluster. $b(i)$ is the smallest average distance between i and data points in different clusters.

The Silhouette Score for the entire dataset is the average of the Silhouette Scores of all data points.

1. Silhouette Score of k means: 23.2%
2. Silhouette Score of Agglomerative Clustering: 41.84%
3. Silhouette Score of DBSCAN: 20.36%
4. Silhouette Score of PCA: 51.99%

10 Conclusion and Future Work

This study's application of consumer segmentation strategies was examined, with a special emphasis on the use of real-time banking transaction data. We sought to reveal hidden patterns in the dataset of 1,048,567 transactions by experimenting with different unsupervised machine learning algorithms, such as Principal Component Analysis (PCA), DBSCAN, K-means clustering, and hierarchical clustering. Our analysis's findings have shed light on several client segments according to the quantity, kind, and value of their transactions. These results have important ramifications for banks in terms of creating focused marketing efforts, raising client happiness, and streamlining resource allocation plans.

Future Work: Based on this study, a number of directions for future research are apparent. First off, segmentation might be improved and deeper insights into customer behavior could be obtained by including data from sources other than transactional data, such as demographics or customer contacts. Second, investigating more complex machine learning approaches like ensemble methods or neural networks could improve the precision and depth of customer segmentation. Furthermore, carrying out long-term research to evaluate the longevity and efficacy of segmentation tactics would support continuous banking practice improvement. Furthermore, examining how these segmentation strategies might be applied in other industries or in tandem with other sectors could increase the range of applications for which they could be used.

Finally, given the shifting landscape of banking technology and regulatory frameworks, ongoing adaptation and refining of segmentation methodology will be required to stay sensitive to changing consumer wants and market dynamics. Therefore, to further improve the topic of client segmentation in the banking industry, future research endeavors should focus on addressing these prospects.

References

1. Lee, J.W., Danaher, P.J.: How does customer satisfaction lead to cus- tomer loyalty ? the mediating role of trust. Int. J. Hosp. Manag. **52**, 278–293 (2016)
2. Huang, Z., Zhou, N., Zhu, S.: Customer segmentation using social media and transaction data. Knowl.-Based Syst. **97**, 112–123 (2016)
3. Verhoef, P.C.: Customer segmentation and marketing strategy. Inter. J. Res. Marketing **20**(1), 1–4 (2003)
4. Chen, M., Wang, Y.: Customer segmentation for credit risk analysis using banking transaction data. Knowl.-Based Syst. **144**, 18–28 (2018)
5. Petropoulos, F.A.: The personalization of marketing in the financial services industry. J. Financial Serv. Market. **27**(1), 1–17 (2022)
6. Hossain, A.S.M.S.: Customer segmentation using centroid based and density based clustering algorithms. In: 2017 3rd International Conference on Electrical Information and Communication Technology (EICT), Khulna, Bangladesh, pp. 1–6 (2017)
7. Thomas, J. N.P. : Customer segmentation in the field of marketing. In:2021 4th International Conference on Recent Trends in Computer Science and Technology (ICRTCST), pp. 401–405 (2017)
8. Ben N'cir, C.E.: Evolutionary multi-objective customer segmentation approach based on descriptive and predictive behavior of customers: ap- plication to the banking sector. J. Exper. Theoret. Artifi. Intell. (2022)

9. Bartels, C.: Cluster analysis for customer segmentation with open banking data. In: 2022 3rd Asia Service Sciences and Software Engineering Conference (ASSE 2022), pp. 87–94 Association for Computing Machinery, New York (2022).
10. Lewaaelhamd, I.: Customer segmentation using machine learning model: an application of RFM analysis. J. Data Sci. Intell. Syst. **2**(1), 29–36 (2023)
11. Rizkyanto, H., Gaol, F.L.: Customer segmentation of personal credit using recency, frequency, monetary (RFM) and K-means on finacial industry. Inter. J. Adv. Comput. Sci. Appli. (IJACSA) 14(4) (2023),
12. Homburg, C., Tischer, M.: Customer journey management capability in business-to-business markets: its bright and dark sides and overall impact on firm performance. J. o Acad. Mark. Sci. **51**, 1046–1074 (2023)
13. Duraisamy, A., Subramaniam, M., Robin, C.R.R.: An optimized deep learning based security enhancement and attack detection on IoT using IDS and KH-AES for smart cities. Stud Inf Control **30**(2), 121–131 (2021)
14. Davamani, K.A., Robin, C.R., Robin, D.D., Anbarasi, L.J.: Adaptive blood cell segmentation and hybrid Learning-based blood cell classification: A Meta-heuristic-based model. Biomed. Signal Process. Control **75**, 103570 (2022)
15. Miriam, D. D. H., Robin, C. R. R., Nallathamby, R., et al.: Human-Assisted Intelligent Computing, p. **21** (2023)
16. Miriam, H., Doreen, D., Dahiya, D., Rene Robin, C.R.: Secured cyber security algorithm for healthcare system using block chain technology. Intell. Autom. Soft Comput. **35**(2) (2023)

Strategic BPNN Forecasting: Integrating Indicators, Bonds, Gold, and Indices for Enhanced Stock Trend Analysis

Sachin M[1,2], Vishwas S. Shastry[1,2], Niharika J[1,2(✉)], and Trupti Hegde[1,2]

[1] BNM Institute of Technology, Bengaluru 560 070, India
jniharika1312@gmail.com
[2] Springer Heidelberg, Tiergartenstr. 17, 69121 Heidelberg, Germany

Abstract. In this paper, we present a multivariate neural network model as an innovative method for stock value prediction. For this method, we con-sider 10 statistical values given by the financial time series data: the stock it-self, an asset used as a reference, RSI, Trix, Fisher, MACD, Mayer regression, Mayer and cor-relation. Datasets are based on daily time intervals taken from YFinance API. The model architecture comprises multiple dense layers with dropout regularization to mitigate overfitting. The approach we have proposed is effective for forecasting stock price trends in the daily stock market, which we observed in the experimental results. This shows a re-markable accuracy on a test dataset. Our research aids to the advancement of predictive modeling techniques in financial market.

Keywords: Stock Price Prediction · Multivariate Neural Network · Technical Indicators · Financial Time Series Analysis · Dropout Regularization · Correlation Analysis · Yahoo Finance · Machine Learning · Financial Markets · Predictive Modeling

1 Introduction

Investors are in need of an accurate predictor for stock prices. This is highly challenging because the stock market is very complex and there are many un-foreseen factors affecting this. The traditional methods have failed to follow through these complexities. In response, this research suggests an innovative approach for stock price prediction using a strategic multivariate BP neural net-work. By incorporating ten statistical variables, including technical indicators and correlations with other assets, our model provides a comprehensive analysis of market dynamics. We trained our BP neural network model using live data from Yahoo finance. So as to enhance the process effectiveness there is drop out Regularized. On the examination of best ways it can be done, we focused on stock price movements characterized by high or low.

2 Literature Survey

2.1 Introduction

Deep learning has attracted lots of interest in financial time series forecasting because it could be highly precise. In the area of using deep learning networks for analyzing stock markets as well as predicting their movements, different methodologies are discussed alongside different data representations and case studies in this literature survey. It covers a range of techniques including multivariate time series prediction, feature engineering, model integration, and comparative analyses.

2.2 Key Survey Points

- Introduced a method for predicting multivariate time series of financial data that is more robust and highlights the significance of robustness in real-world financial applications [1].
- The research investigates deep learning networks in stock market analysis and prediction through case studies, showcasing how good data representations are essential in the process [2].
- Propose a method that uses deep learning to engineer characteristics associated with stock movement anticipated direction so that it extracts more appropriate features out of financial information that can be used for making forecasts more accurately [3].
- Examined how one can apply deep learning together with multi-resolution analysis (which considers different timescale), into predicting future prices on stock exchanges by utilizing strengths offered by each method retroactively [4].
- Used a unique way in bidirectional and stacked LSTMs and GRU models for predicting stock market to demonstrate how effective some sophisticated neural networks can be [5].
- Discovered multivariate models that are used in several deep learning approaches through exploiting Bayesian network selection which improved forecast accuracy and increased model interpretability [6].
- Displayed the potential of using reinforcement learning techniques in financial forecasting by a method introduced in this paper that predicts how a multivariate series of time-series functions that are defined over a set of moments behaves when supplied with high-dimensional inputs [7].
- Introduced an Integrated Deep Learning Model that is Stacked for Stock Market Predictions. This model incorporates diverse deep learning methods for the purpose of improving both accuracy and robustness while also making predictions in stock markets [8].
- Researched a deep learning for predicting financial time series forecasting in A-Trader system. Explored how deep learning methods can be used to make more accurate financial time series predictions in the context of a trading system [9].
- Demonstrated an instance where a deep-learning network was implemented to perform trade signal forecasting in finance markets. Identified that one of the capabilities lies with respect to forecasting trade signals using deep learning algorithms [10].

- Performed a comparative study on parallel multivariate deep learning models to predict time-series in the Asian stock market; this analysis provides some information on how effective various model architectures are [11].
- Made comparison of the relative performance among G7 stock markets' multivariate forecasts through Vector Error Correction Model (VECM) and deep learning Long Short-Term Memory (LSTM) neural networks, giving insights into their merits and limitations for application [12].
- Used elongated residual convolutional neural networks on diverse domains in the urban air quality anticipation for multivariate time series, demonstrating the applicability of deep learning techniques [13].
- A stock price prediction model was created on stock price covariation and deep learning to show the relevance of market dynamics in prediction model Versions [14].
- Used LSTM to predict stock performance in different market conditions. The findings have shown that LSTM models are pretty reliable across various market situations [15].
- Developed a deep learning model to analyse complex financial data, specifically focusing on predicting market indexes. The results demonstrate that this approach is effective in understanding how the market behaves [16].
- The research delved into analysing stock prices using advanced deep-learning models, which shed light on how these techniques can be applied to forecast financial market trends [17].
- Introduced an innovative deep learning framework for predicting and analysing financial time series using CEEMD and LSTM. It highlighted the potential of combining deep learning with signal decomposition techniques to enhance the accuracy of financial forecasting [18].
- Developed an ensemble of deep neural networks for stock market prediction, the research also demonstrated the effectiveness of ensemble learning in improving prediction performance [19].
- Explored how stock price analysis with deep-learning models provides insights into the application of deep learning techniques for financial market forecasting [17].

2.3 Conclusion

The literature survey highlights the various unique and amazing methodologies and approaches in using deep learning algorithms and techniques for stock market analysis and prediction. From robustness-certified models to ensemble learning techniques, deep learning continues to provide promising ways for improving forecasting accuracy and to capturing market dynamics. However, challenges such as interpretability and model fitness remain, emphasizing the need for further research and development in this field (Fig. 1).

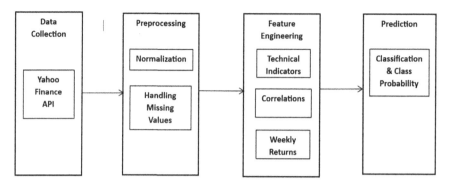

Fig. 1. Proposed Model

3 Proposed Model

3.1 Data Source and Collection

The data is collected from the Yahoo Finance which a reliable source of historical financial data. Retrieving daily adjusted prices value for a range of stocks and indexes is done by sending queries to the Yahoo Finance API. The timeframe of the data spans from January 1, 2010, to this day, providing a comprehensive historical record of market activity.

3.2 Preprocessing Steps

Normalization: The raw data were preprocessed to ensure uniform scaling across different features. The Min-Max scaling was applied to scale the input features to a range between zero and one [0–1]. This method preserves the relationships between data points while preventing features with larger scales from contributing to more weightage the model training process.

Handling of Missing Values: To preserve the dataset's integrity, any missing values existing in the raw data were handled appropriately. In this implementation, missing values were addressed by dropping rows containing NaN values.

3.3 Feature Engineering

Technical Indicators: A range of technical indicators were calculated from the financial data to find the underlying patterns in said stock price data. These indicators include the Mayer Multiple, MACD, and RSI. These indicators offer insightful information on trend strength, market momentum, and possible reversal points.

Correlations: Rolling 90-day correlations between the target stock and other external capital such as GOLD, MOVE Index, and OEX gauge, enhancing the model's predictive accuracy by capturing associations between the target stock and other external influencers.

Weekly Returns: The predictivity of weekly returns from various financial instruments, such as MOVE (Merrill Lynch Option Volatility Estimate) index, GOLD (Gold spot price), and the OEX (S&P 100 Index), alongside stock data. We intend to examine these variables impact on trend predictions by integrating them into our predictive algorithm. The MOVE index reflects market expectations of future volatility, GOLD is often recognized as a safe-haven asset, and the OEX represents large-cap equity performance. Integrating these factors with stock data provides a thorough understanding of market dynamics by getting a holistic understanding of volitility, safe and overall performance of the short-term market dynamic. Our study aims to determine whether incorporating these diverse financial metrics enhances the accuracy and reliability of stock trend prediction models.

3.4 Trading Strategy

The trading strategy devised in this study leverages a strategic BPNN model to forecasts future market trends. This strategy incorporates a comprehensive set of input features, including the relative strength index (RSI), Mayer Multiple, Moving Average Convergence Divergence (MACD), and 90-day rolling correlations with the MOVE, OEX, and GOLD indices. Additionally, it integrates the last 7-day returns from the MOVE, GOLD, and OEX indices to capture short-term market dynamics.

The model's output feature splits market trends into two classes: uptrend and downtrend. This classification is based on the comparison of exponential moving averages (EMA), where a + 1 label indicates that the 12 - day E M A is greater than the 21-day EMA, signifying an uptrend, while a 1 label suggests the opposite.

The model predicts the market trend for the seventh day ahead, providing traders with a futuristic perspective. Traders typically initiate long positions when the 12-day EMA exceeds the 21-day EMA which indicates an uptrend, and short positions when the reverse occurs. By employing the model's forecast as a momentum indicator, traders gain insightful knowledge into when to enter or exit the market, enhancing the efficiency and effectiveness of the informed trading decisions.

The strategy's strength is its capability to combine a variety of range in historical data and market indicators, allowing for a comprehensive evaluation of the condition of the market. By utilizing pioneering machine learning algorithms such as neural networks, the model can identify intricate patterns and relationships within the data, offering predictive capabilities beyond traditional methods.

Moreover, the strategy's futuristic nature enables traders to predict market movements and adjust their strategy accordingly, hence capitalizing on emerging trends and mitigating risks. Agility and adaptability are critical traits for navigating the volatile financial markets, and they are honed by this proactive approach in trading.

Overall, the proposed trading strategy poses a powerful tool for traders seeking to optimize their decision-making processes and enhance their trading performance. By integrating cutting-edge technology with established market principles, it offers a systematic framework for grasping opportunities and mitigating risks in today's fast-paced and ever-changing financial landscape.

3.5 Strategic BP Neural Network

The neural network architecture illustrates the structure of the strategic backpropagation neural network (BPNN). It displays all the layers, their connections, and the nodes for each layer, highlighting the flow of learning from input to output (Fig. 2).

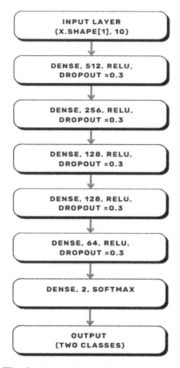

Fig. 2. Neural Network Architecture

The equation to describe the mathematical operations the takes place to calculate and formulate results to be classified for the above neural consists of three major operations, i.e., Summation, Activation, Weight Correction.

Summation:

$$S = \sum_{i=0}^{10} (W_i * X_i)$$

Activation: There are 2 types of activation used. One for the Hidden Layers and one for the Output Layer.

Hidden Layer Activation:

$$Relu(x) = \max(0, x)$$

Output Layer Activation:

$$Softmax(Z_i) = \frac{e^{Z_i}}{\sum_{j=1}^{N} e^{Z_j}}$$

Weight Correction: This weight correction is done during the backward propagation of each iteration.

$$Wi_{new} = Wi_{old} - \eta \frac{dE}{dWi} \forall i \in [0, 10]$$

The strategic backpropagation neural network (BPNN) described in the architecture incorporates three major operations: summation, activation, and weight correction. Summation involves the calculation of weighted inputs of the Deca-variate input, followed by activation using ReLU for hidden layers and softmax for the output layer. Weight correction is performed during backward propagation to adjust the network parameters based on the error gradient i.e., sparse categorical cross entropy. This structured approach enables effective learning and classification within the neural network, making it a powerful tool for various applications requiring pattern recognition and decision-making.

3.6 Equations

- **The Relative Strength Index (RSI):** The RSI is typically calculated using the given formula:

$$RSI = 100 - \frac{100}{1 + RS}$$

where RS is the average of n days' up closes divided by the average of n days' down closes.

- **Moving Average Convergence Divergence (MACD):** The MACD line is computed by subtracting the 26-period exponential moving average (EMA) from the 12-period EMA. The signal line is then calculated as the 9-period EMA of the MACD line. The formulas are as follows:

$$MACD = EMA_{12} - EMA_{21}$$
$$Signal = EMA_9(MACD)$$

$$Mayer\_Multiple = \frac{Price}{50 \, day \, moving \, average}$$

- **Mayer Multiple:** The Mayer Multiple is derived by dividing the price of the asset by its 50-day moving average values. The formula is given by:

- **Rolling 90-day Correlation:** The rolling 90-day correlation between two assets X and Y is calculated using the Pearson correlation coefficient formula over a sliding window of 90 days.

$$r = \frac{\sum(X_i - \overline{X})(Y_i - \overline{Y})}{\sqrt{\sum(X_i - \overline{X})^2 \sum(Y_i - \overline{Y})^2}}$$

- **Normalization (Min Max Scaling):** Min Max scaling scales each feature to a specified range (typically [0, 1]) using the following formula:

$$X_{scaled} = \frac{X - X_{min}}{X_{max} - X_{min}}$$

4 Results and Discussion

A thorough evaluation of the proposed *strategic back propagation neural network* model. The performance metrics the model is evaluated on are accuracy, precision and recall as well as F1-score. These results provide insightful benchmarks for assessing its predictive capabilities.

The model achieved an accuracy of 87.11% and a precision of 86.98%, indicating the robustness in classifying market trends as uptrend or downtrend. A detailed examination of the confusion matrix further shows the model's performance, showcasing the distribution of true positive, true negative, false positive, and false negative predictions (Fig. 3).

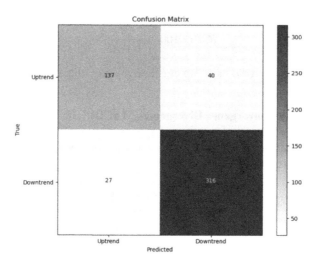

Fig. 3. Metrics of the Strategic BP Neural Network

The confusion matrix summarizes the performance of a strategic backpropagation neural network model for trend forecasting. It showcases the counts of true positive, true negative, false positive, and false negative predictions, providing insights into the model's accuracy, precision, recall, and overall effectiveness in classification tasks (Fig. 4).

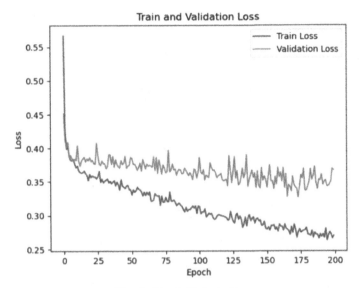

Fig. 4. Epoch Optimization

The epoch optimization plot illustrates the training and testing loss convergence of the strategic backpropagation neural network model for the trend forecasting. It visualizes and demonstrates the model's learning progress over epochs and iterations of learning, indicating how well it optimises to unseen test data. Lower loss values signify better performance and model convergence.

Overall, the outcomes highlight the competence of the proposed methodology, which is the backpropagation neural network model. By achieving high accuracy, precision, recall, and F1-score, the model demonstrates its abilities to effectively forecast market trends with confidence.

Comparative Study.
This comparative study is to evaluate how well different existing deep learning models perform in stock market analysis and prediction as compared to the Strategic BPNN model coined in this paper. By assessing various methodologies, including Logistic Regression, Decision Tree Classifier, Random Forest Classifier LSTM, and, hybrid architectures, our goal is to identify the most effective methods for forecasting stock prices, Contributing to the advancement of predictive modeling in financial markets, The data used to compare the performance is the closing price of the AAPL stock using the metrics: Accuracy, Precision, Recall and F1 - score (Fig. 5).

The bar graph is the visualization of the Table 1 which compares the performance of strategic backpropagation neural network (BPNN) against random forest classifier, decision tree classifier, logistic regression, and modified BPNN. It visually depicts their respective accuracies or other evaluation metrics, highlighting the relative effectiveness of each model for the task at hand.

In conclusion, our comparative study highlights the diverse range of deep learning techniques for stock market prediction. While LSTM models demonstrate superior performance across different risk categories and time frames, hybrid architectures

Table 1. The Comparative Study

Model	Accuracy	Precision	Recall	F1-score
Strategic BPNN	87.11%	86.98%	87.11%	86.96%
Logistic Regression	83.53%	83.40%	83.41%	83.37%
Modified BPNN	84.23%	84.10%	84.23%	84.13%
Decision Tree Classifier	80.23%	80.00%	80.23%	80.12%
Random Forest Classifier	84.55%	84.48%	84.56%	84.34%

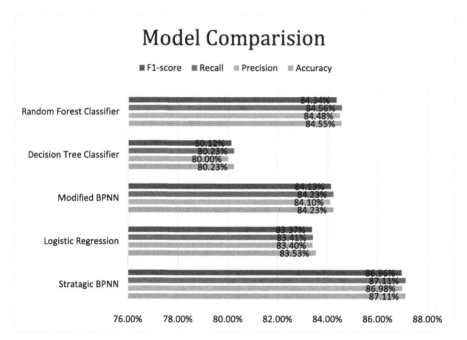

Fig. 5. Graphical Comparison Of the Strategic BP Neural Network in comparison with other models

combining convolutional or recurrent connections show promise in capturing complex patterns in financial data. Techniques like multi-tire-solution analysis and deep reinforcement learning further enhance predictive accuracy. By leveraging these methodologies, researchers and practitioners can create predictive models that are more effective for navigating dynamic financial markets.

References

1. Li, H., Cui, Y., Wang, S., Liu, J., Qin, J., Yang, Y.: Multivariate financial time-series prediction with certified robustness. IEEE Access **8**, 109133–109143 (2020). https://doi.org/10.1109/ACCESS.2020.3001287

2. Chong, E., Han, C., Park, F.: Deep learning networks for stock market analysis and prediction: Methodology, data representations, and case studies. Expert Syst. Appl. **83**, 187–205 (2017). https://doi.org/10.1016/j.eswa.2017.04.030
3. Long, W., Lu, Z., Cui, L.: Deep learning-based feature engineering for stock price movement prediction. Knowl. Based Syst. **164**, 163–173 (2019). https://doi.org/10.1016/j.knosys.2018.10.034
4. Althelaya, K., Mohammed, S., El-Alfy, E.: Combining deep learning and multiresolution analysis for stock market forecasting. IEEE Access **9**, 13099–13111 (2021). https://doi.org/10.1109/ACCESS.2021.3051872
5. Althelaya, K., El-Alfy, E., Mohammed, S.: Stock market forecast using multivariate analysis with bidirectional and stacked (LSTM, GRU). In: 2018 21st Saudi Computer Society National Computer Conference (NCC), pp. 1–7 (2018). https://doi.org/10.1109/NCG.2018.8593076
6. Kobayashi, S., Shirayama, S.: Time series forecasting with multiple deep learners: Selection from a Bayesian Network. **05**, 115–130 (2017). https://doi.org/10.4236/JDAIP.2017.53009
7. Xin, J., Zhang, H., Jianfang, L., Xiaolong, Z., Li, S., Chen, R.: Multivariate time series prediction of high dimensional data based on deep reinforcement learning. In: E3S Web of Conferences (2021). https://doi.org/10.1051/E3SCONF/202125602038
8. Deep Learning-based Integrated Stacked Model for the Stock Market Prediction. International Journal of Engineering and Advanced Technology (2019). https://doi.org/10.35940/ijeat.a1823.109119
9. Korczak, J., Hernes, M.: Deep learning for financial time series forecasting in A-trader system. In: 2017 Federated Conference on Computer Science and Information Systems (FedCSIS), pp. 905–912 (2017). https://doi.org/10.15439/2017F449
10. Türkmen, A., Cemgil, A.: An application of deep learning for trade signal prediction in financial markets. In: 2015 23nd Signal Processing and Communications Applications Conference (SIU), pp. 2521–2524 (2015). https://doi.org/10.1109/SIU.2015.7130397
11. Widiputra, H., Juwono, E.: Parallel multivariate deep learning models for time-series prediction: A comparative analysis in Asian stock markets. IAES Inter. J. Artifi. Intell. (IJ-AI) (2024). https://doi.org/10.11591/ijai.v13.i1.pp475-486
12. Ferreira, N., Mendes, D., Mendes, V.: Comparative multivariate forecast performance for the G7 Stock Markets: VECM Models vs deep learning LSTM neural networks. pp. 163–171 (2020). https://doi.org/10.4995/carma2020.2020.11616
13. Benhaddi, M., Ouarzazi, J.: Multivariate time series forecasting with dilated residual convolutional neural networks for urban air quality prediction. Arab. J. Sci. Eng. **46**, 3423–3442 (2021). https://doi.org/10.1007/s13369-020-05109-x
14. Jing, N., Liu, Q., Wang, H.: Stock price prediction based on stock price synchronicity and deep learning. Inter. J. Financial Eng. (2021). https://doi.org/10.1142/s2424786321410103
15. Qian, F., Chen, X.: Stock prediction based on LSTM under different stability. In: 2019 IEEE 4th International Conference on Cloud Computing and Big Data Analysis (ICCCBDA), pp. 483–486 (2019). https://doi.org/10.1109/ICCCBDA.2019.8725709
16. Noh, Y., Kim, J., Hong, S., Kim, S.: Deep learning model for multivariate high-frequency time-series data: financial market index prediction. Mathematics (2023). https://doi.org/10.3390/math11163603
17. Arosemena, J., Pérez, N., Benítez, D., Riofrío, D., Flores-Moyano, R.: Stock price analysis with deep-learning models. In:2021 IEEE Colombian Conference on Applications of Computational Intelligence (ColCACI), pp. 1–6 (2021). https://doi.org/10.1109/ColCACI52978.2021.9469554

18. Zhang, Y., Yan, B., Memon, A.: A novel deep learning framework: prediction and analysis of financial time series using CEEMD and LSTM. Expert Syst. Appl. **159**, 113609 (2020). https://doi.org/10.1016/j.eswa.2020.113609
19. Chong, L., Lim, K., Lee, C,: Stock market prediction using ensemble of deep neural networks. In: 2020 IEEE 2nd International Conference on Artificial Intelligence in Engineering and Technology (IICAIET), pp. 1–5 (2020). https://doi.org/10.1109/IICAIET49801.2020.9257864

Enhancing Product Review Understanding: Text Summarization with BART Large XSum

S. Gopika, Mayuri Mahimaa Balaji, M. Vishwanath, and K. Karthikayani[✉]

Department of Computer Science and Engineering, SRM Institute of Science and Technology, Vadapalani, Chennai, India
{mb1310,mv3683,karthikk3}@srmist.edu.in

Abstract. Online shopping has revolutionized the way consumers make purchasing decisions. Customer reviews play a crucial role in this process, providing valuable insights into product quality and satisfaction. However, sifting through numerous reviews can be time-consuming for customers. This research aims to address this issue by proposing a novel method to generate a single summary for the overall reviews of each product across different years. The proposed approach utilizes the BART large xsum model to streamline the summarization process. Initially, the dataset is preprocessed to remove irrelevant reviews, such as those lacking product IDs or years. Subsequently, the dataset is divided into batches, each containing 1024 tokens, to facilitate efficient processing. Summaries are generated for each batch, and the output is then used as input for the next stage. Finally, a concise summary is produced for each product across different years. Experimental evaluations are conducted using the Amazon baby product reviews dataset, with the performance of the BART large xsum model assessed using confidence scores. The results demonstrate that the proposed method effectively generates concise summaries for the overall reviews of each product across different years, saving customers valuable time in decision-making.

Keywords: Bart large xsum · Pegasus · T5 model · Text Summarization

1 Introduction

The project's goal is to develop a system that summarizes reviews for individual products, offering users an overview without the need to read each review. It serves as a textual alternative to star ratings, providing detailed insights into a product's strengths and weaknesses. This summary enhances user comprehension, enabling them to make informed decisions based on specific product attributes. By condensing multiple reviews, the system saves time and allows users to grasp key points and sentiments quickly. Additionally, it improves decision-making by helping users effectively evaluate product pros and cons, leading to more confident purchasing decisions and facilitating comparison shopping within the same product category.

In the expansive realm of online shopping, customer feedback plays a pivotal role in evaluating products and guiding purchasing decisions. Major platforms such as Amazon boast extensive collections of reviews, each offering valuable perspectives on product attributes, performance, and user satisfaction. However, the sheer volume of reviews can overwhelm consumers, who must navigate through a sea of feedback to grasp the full picture of a product.

Liu et al. proposed frameworks for both abstractive and extractive models [1]. Their research demonstrated that incorporating Bidirectional Encoder Representations from Transformers (BERT) in conjunction with a two-stage fine-tuning method enhances the quality of produced summaries, offering significant advantages for text summarization.

Liu et al. introduced a novel iterative refinement algorithm, and the results of this algorithm are on par with the latest techniques available [2]. A. Garg proposed model employed image generation to extract the summary of reviews and save time in extracting useful information [3]. Garg discussed the utilization of various methods such as joint summarization models, word clouds, hybrid models based on neural networks, BERT, and sentiment analysis to improve the precision of text summarization [3].

Bai et al. introduced a pre-trained recommendation model alongside a joint summarization model designed to forecast review ratings [4]. Their experimental findings demonstrate that the proposed model outperforms several existing baseline models. Subha et al. outlined a sequence of procedures aimed at predicting word sequences and automating text summarization [5]. These steps encompass word segmentation, morphological reduction, and conference resolution.

The objective of this research is to create a streamlined system for summarizing reviews categorized by product ID and year. The main contributions of this research include:.

- Introducing a BART Large XSum model for summarizing reviews by product ID and year.
- Conducting preprocessing to remove irrelevant reviews lacking product ID or year information.
- Dividing the datasets into batches to facilitate effective summarization.
- Experimentally evaluating the proposed approach using the Amazon baby product reviews dataset, with evaluation based on conference score metrics. The paper is organized as follows: Section 2 presents the literature review, followed by Sect. 3 discusses the dataset and provides details of the proposed method. Section 4 which outlines the various methods used for comparison with the proposed approach. Section 5 provides result and their screenshot. Section 6 concludes the study, while Section 7 outlines future work.

2 Literature Review

This section offers an outline of prior studies concerning review summarization and text summarization. The studies are categorized based on the approaches used for summarization.

In their study, R. Y. Kim explores the factors that contribute to the helpfulness of online reviews [6]. It investigates linguistic and structural features of reviews that make them more helpful to consumers.They introduced a sentiment analysis technique employing lexicons to discern emotional polarity from a dataset comprising over 100,000 smartphone reviews sourced from Amazon.com. The research delved into the correlation between sentiment polarity, emotional valence, and the perceived helpfulness of online reviews. To achieve this, they utilized an open-source lexicon provided by the National Research Council (NRC) in Canada.

Hussain proposed a PRUS framework that provides the summarization for each sentences of user- specified features extracted from customer reviews [7]. To create a recommended list of products ranked by preference, PRUD analyzes reviews by summarizing them to ascertain product sentiment.

N.Liu introduces the SAMF recommendation system, which integrates sentiment analysis and Matrix enhancement through factorization for improved recommendations [8]. However, it is constrained to recommending text resources, making it difficult to uncover implicit interests.

According to H. Huang's research, sentiment analysis is a significant field of study in the e-commerce industry [9]. Valuable business insights can be gained from analyzing customer reviews and feedback. Hailu et al. introduced an automated text summarization framework capable of identifying the selection of the top n most pertinent sentences from the source text led to promising results in their experiments [10].

Similarly, Belwal et al. presented a model integrating semantic measures and topic modeling using the vector space model to generate summaries, yielding results comparable to those created by humans [11]. A. Venkataramana introduces the HMSumm framework, which is a combination of extractive and abstractive approaches for multi-report summarization [12]. It utilizes techniques like determinantal point process (DPP) and deep submodular network (DSN) to improve the quality and length control of the summaries along with the BART model.

Similarly, A. Dilawari also proposes a feature-rich model that combines both extractive and abstractive approaches for text summarization which utilizing both mechanisms of attention, at the level of words and sentences to produce a succinct summary [13].

Fabbri et al. developed a comprehensive model consisting of both an extractive summarization component and an SDS (Sentiment Detection System) [14]. Their experimental results demonstrate that this model performs comparably to existing models. Gamzu et al. introduced a model aimed at extracting informative sentences along with their associated sentiments, including both negative and positive sentiments about the product [15]. Their experimental findings indicate that their model outperforms several baseline models in terms of performance.

Y.Zhu introduced a methodology aimed at summarizing medical reports [16]. This approach utilizes existing dataset knowledge and integrates semantic representations derived from a randomly chosen summary within the training set. E. Lalitha explored Abstractive Summarization using various models and found that Pegasus outperforms others in multi-medical document summarization [17].

R. Gandhi focused on Autonomous Document Summarization, employing Google's T5-Small to train a Sequence-to-Sequence (RNN) based model capable of generating abstractive text summaries, leveraging hugging face transformers [18].

Li introduced the Semantic Link Network, which significantly enhances document summarization [19]. The SLN-based approach surpasses extractive and abstractive baselines and proves effective in representing and comprehending document semantics.

3 Proposed Method

The proposed approach involves multiple crucial stages designed to create a system that efficiently and accurately summarizes Amazon product reviews. The existing method only summarizes individual reviews, whereas our proposed method summarizes all reviews related to each product ID for different years. For instance, if there are 5000 reviews for one product, our novel proposed method condenses all 5000 reviews into a single sentence. Here's a comprehensive outline:

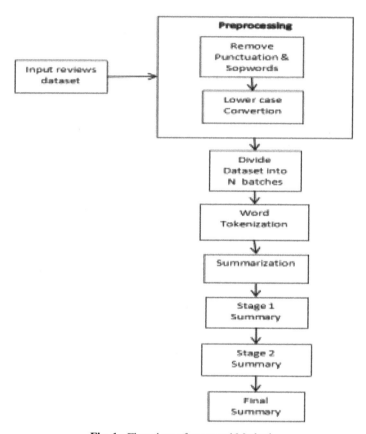

Fig. 1. Flowchart of proposed Method

(Figure 1) Illustrates the architecture of the proposed method, detailing the step-by-step process to obtain the output.

3.1 Data Collection

The dataset utilized in this study was obtained from Kaggle[20] contains 1.3 million reviews, covering 5 different Amazon products containing 14 attributes. The chosen attributes comprise product_id, star_rating, review_body, review_date, review_headline and helpful_votes. Additionally, we introduce a derived attribute, review_weight (w_i), to evaluate the summaries. This attribute helps to dampen the impact of the number of helpful votes and avoid computing the log of 0 in case of no helpful votes.

$$wi = log(helpful\_votes_i + 1) * star\_rating_i \tag{1}$$

Let D be the dataset consisting of product reviews with attributes $\{r\_i, y\_i, p\_i\}$, where r_i represents the review text, y_i represents the year of the review, and p_i represents the product ID.

3.2 Data Preprocessing

In preprocessing, the dataset D eliminates the undesired reviews lacking product IDs and years. To clean the data, the Natural Language Toolkit was utilized to strip away stopwords and punctuation marks. Additionally, the text undergoes tokenization utilizing the Word Tokenizer. The text was converted to lowercase, and HTML tags were removed using regular expressions. Following preprocessing, the dataset was condensed to 23,504 reviews, which were subsequently subjected to further assessment in the proposed methodology. (Figure 2) is the bargraph of reduced dataset which is obtained after preprocessing.

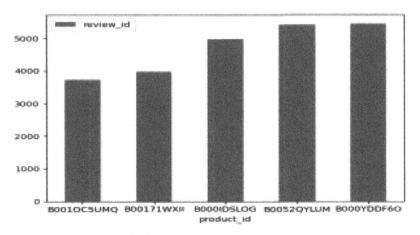

Fig. 2. Bargraph of Reduced Dataset

3.3 Partitioning Dataset into N Batches

The function takes a list of reviews and a flag as parameters. First, it sets up some variables for creating batches, like an empty string. Then, it goes through each review and checks if adding the next one would make the string too long (more than 1024 tokens). If it's still within the limit, it adds the review to the string. When it reaches the token limit, it adds the current batch to the list and starts a new one. Finally, the function returns the list of batches. So basically, it divides the reviews into batches that are manageable in size, making it easier to process them. Divided the preprocessed dataset D into n batches, denoted as $B = \{B\_1, B\_2,..., B\_n\}$, where each batch B_j contains m reviews. Each batch B_j is represented as $B\_j = \{r\_{j1}, r\_{j2},..., r\_{jm}\}$, where r_{ji} is the review text.

3.4 Summarization

The process begins by initializing the BART Large XSum model and tokenizer from the Hugging Face library. Next, it extracts unique product IDs and review years from the provided DataFrame. Then, it iterates through each combination of product ID and year. For each combination, it filters the DataFrame to select reviews corresponding to that product ID and year, converting the review text into a list. Afterward, it creates batches of reviews and generates summaries for each batch using the BART Large XSum model. Subsequently, it prepares batches for the second stage of summarization by concatenating the first-stage summaries. Finally, it generates summaries for the concatenated batches using the BART Large XSum model once again and prints the summaries for each product ID and year combination. For each batch B_j, generate a preliminary summary using the BART Large XSum model, denoted as S_j. Concatenate the preliminary summaries S_j to form a single document, denoted as S_c. Generate the final summary S_f using the BART Large XSum model with S_c as input. The final summary S_f contains a concise summary of all reviews for each product for different years

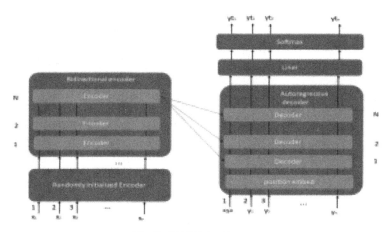

Fig. 3. BART Architecture

Figure 3 presents a comprehensive illustration of the BART architecture. The initial module employs the BERT encoder, which handles each text sequence in the input, generating an embedding vector for each token in the sequence. Moreover, it produces a vector containing information at the sentence level, which is vital for the decoder's comprehension of both individual tokens and the entire sentence. The decoder, following the GPT architecture, functions by progressively generating the next token of the sequence.

It commences with a special token and utilizes information from the ongoing sentence and the output of the encoder module to determine the subsequent token. This iterative process persists, with each new token incorporated into the input for the subsequent iteration.

4 Experiments

4.1 Overview

This study conducts an experimental evaluation by contrasting the performance of three models, namely BART Large XSum, T5, and Pegasus, distinct patterns emerge. For comparison purposes, we select a single product with over 5000 reviews. In the BART Large XSum model, the outcomes become increasingly abstract with each batch processed. We have utilized the summarization pipeline directly offered by Hugging Face for our summarization task. Conversely, in the T5 model, enhancing the number of reviews does not notably enhance the results. As for Pegasus, the outcomes exhibit variability based on the number of reviews considered, yet there is a minimal disparity between summarizing 2000 reviews or more. Our findings indicate that BART-large-xsum yields satisfactory results, prompting us to employ it for the summarization task.

4.2 T5

T5, short for "Text- to - Text Transfer Transformer", is a Transformer - based model that follows a input-to-output transformation, unlike task-specific architectures, T5 treats all tasks like translation, question answering, or classification as generating target text from input text. This design allows for uniformity in model, loss function, and hyperparameters across different tasks. Unlike BERT, T5 introduces several modifications to its architecture, including integrating a causal decoder into the bidirectional structure and substituting the cloze task with a mix of other pre-training tasks. These adjustments enhance T5's flexibility and efficacy in addressing a wide range of tasks with greater effectiveness. The T5 model focuses on extractive summarization, which is not within the scope of this research. (Figure 4) shows how the T5 model works.

4.3 Pegasus

PEGASUS, based on the concept of training abstractive summarization sequence-to sequence models with extracted gap sentences, utilizes a self-supervised technique known as Gap Sentences Generation (GSG) to train a transformer encoder-decoder model. For this project, it was decided to assess the performance of the Pegasus-xsum model. The testing process adhered to the same methodology as that of the BART model. For brevity, only a succinct overview of the tests, results, and analysis will be presented. While the PEGASUS model excels in abstractive summarization, its performance diminishes when applied to large datasets. Instead of producing accurate summaries, it tends to repeat the same lines when additional reviews are added to the output, which is not within the scope of this research. (Figure 5) shows the architecture of pegasus model.

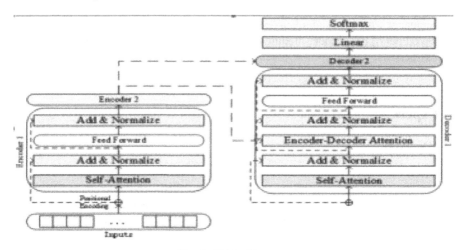

Fig. 4. T5 Architecture

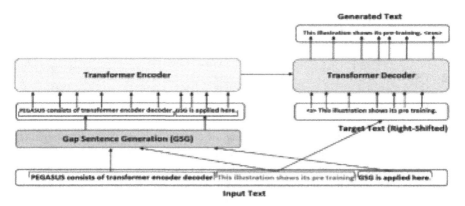

Fig. 5. Pegasus Architecture

5 Results

The existing method only summarizes individual reviews, whereas our proposed method summarizes all reviews related to each product ID for each years. For instance, if there are 5000 reviews for one product, our method condenses all 5000 reviews into a single sentence. Calculate the confidence level CL for each final summary S_f using the evaluation metrics. The confidence level CL indicates the reliability and informativeness of the summary. The confidence level CL is calculated based on the following metrics:

1. Fluency: Measure the linguistic quality and coherence of the summary.
2. Informativeness: Assess the summary's effectiveness in capturing the key information from the reviews.
3. Relevance: Evaluate the extent to which the summary captures the opinions and sentiments conveyed in the reviews.

4. Length: Ensure that the summary is concise and does not contain unnecessary information.

The confidence level CL is computed as a weighted combination of these metrics, where higher values indicate a more reliable and informative summary.

$$CL = w\_1 * Fluency + w\_2 * Informativeness + w\_3 * Relevance + w\_4 * Length \quad (2)$$

where w_1, w_2, w_3, w_4 are the weights assigned to each metric, and $\Sigma^4{}_{i=1} w\_i = 1$. BART Large XSum gives good results compared to T5 and Pegasus models which is explained in Sect. 4. Fig. 7 shows the summary after the first stage. Figure 8 shows the summary after the second stage, and Figure 4 shows the final summary for each product with the year.

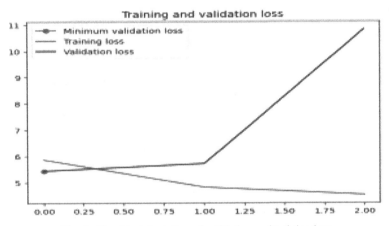

Fig. 6. Graphical depiction of validation and training loss

Figure 6 is the graph illustrates the performance of the model during the training process. The training loss represents the error between the predicted and actual values calculated during training, while the validation loss measures the model's performance on unseen data.

In the first stage, the summary will include multiple lines of reviews. Figure 7 displays the output of this summary. In second stage, the inputs are the output of the first-stage summary. Figure 8 shows the concise summary in one review, which is the final summary. Figure 9 illustrates the overall results, which include summaries for each product for different years (2007–2015).

['BabiesRU's has a number of great products for babies, but this one is not one of them.',
'A lot of people have been using this toy to calm their babies, and it seems to be working!',
'Here are some of the best reviews of the Baby Einstein mesmerizer, which was given to my daughter when she was born.',
'This is a great toy for infants and young toddlers, but make sure your baby actually wants to play with it.',
'A review of the Baby Einstein Take Along T, a toy that plays classical music and flashing lights at the same time.',
'A review of the Baby Einstein music toy from Amazon.',
'A review of the Baby Einstein Take Along Tunes from Amazon.',
'Take Along Tunes iPod Shuffle is a great toy for babies and young children.',
'This is a great toy to keep your baby happy and entertained.',
'A review of a toy that plays classical music in a loop, with flashing lights and a built-in speaker.',
'The Baby Einstein Take Along Tunes are one of the most popular toys for babies, and have been used by both parents and children for over a year.',
'A review of the Baby Einstein Take Along Tunes.',
'A review of the Baby Einstein Take Along Tunes.',
'This is a great toy for babies, especially if you have one that plays classical music.',
'A review of the Baby Einstein Take Along Tunes, a toy that plays classical music and flashing lights at the same time.',
'A review of the Baby Einstein Takealong Tunes from Amazon.com.']

Fig. 7. First Stage Summary

Summaries Second Stage

[A lot of people have been using the Baby Einstein mesmerizer to calm their babies, and it seems to be working!]

Fig. 8. Second Stage Summary

Final result for various products:
Product B000YDDF6O • Year 2015: ['A lot of people have been using amazon to rave about this musical toy for babies, and the reviews are overwhelmingly positive.']
Expected rating: 4.6 stars Observed rating: 5 stars Confidence: 0.75

Product B000YDDF6O, year 2010: ['A lot of people have been using the Baby Einstein mesmerizer to calm their babies, and it seems to be working!']
Expected rating: 3.9 stars Observed rating: 5 stars Confidence: 0.77

Product B001254APK • year 2008: ['A cheetah mount has been rated "excellent" by the US Consumer Product Safety Commission (CPSC).']
Expected rating: 4.0 stars Observed rating: 5 stars Confidence: 0.51

Product B003EM8008 • year 2015: ['As part of our series on the best headphones, we've been looking at some of the best earbuds.']
Expected rating: 4.4 stars Observed rating: 5 stars Confidence: 0.65

Product B0002L5R78 • year 2013: ['As part of our series of product reviews, we take a look at some of the best cables we have ever used.']
Expected rating: 4.4 stars Observed rating: 5 stars Confidence: 0.68

Fig. 9. Final Summary

6 Conclusion

The model we're proposing creates short summaries to help people make better decisions. In the past, when summarizing reviews, researchers usually looked at each review separately, whereas our proposed method summarizes all reviews related to each product ID for different years. Our model works in four steps. First, it splits the reviews into groups. Then, it makes a summary for each group in the second step. In the third step, it combines these summaries into one. Finally, in the fourth step, it makes a final summary for all the reviews. We did lots of tests to see how well the BART Large XSum model works. But, using sentiment analysis to check if the summary is good might not always match the technical ratings because reviews can be so different. is new model is important because it helps people understand lots of reviews quickly. Instead of reading every review, they can just read the summaries. This saves time and helps people make smarter choices. But, we also need to make sure the summaries are good quality. So, we tested them to see how well they match up with the original reviews. We found that while sentiment analysis is helpful, it doesn't always show how good the summary is because reviews are so diverse.

7 Future Directions

To enhance our work, we require larger models capable of enhancing abstraction power and processing larger inputs. Utilizing business analytics to provide visual representations can aid in understanding reviews for both customers and businesses.

References

1. Liu, Y., Lapata, M.: Text Summarization with Pretrained Encoders. arXiv (2019). [https://doi.org/10.48550/arXiv.1908.08345]
2. Liu, Y., Titov, I., Lapata, M.: Single document summarization as tree induction. In: Proceedings of the 2019 Conference of the North American Chapter of the Association for Computational Linguistics (2019). [https://doi.org/10.18653/v1/n191173] (https://www.aclweb.org/anthology/N19-1173/)
3. Nainwal, H., Garg, A., Chakraborty, A., Bathla, D.: Text summarization of amazon customer reviews using NLP. In: 2023 10th International Conference on Computing for Sustainable Global Development (INDIACom), New Delhi, India (2023)
4. Bai, Y., Li, Y., Wang, L.: A joint summarization and pre-trained model for review based recommendation. Information **12**(6) (2021). [https://doi.org/10.3390/info12060223]
5. Shini, R.S., Kumar, V.D.A.: Recurrent neural network based text summarization techniques by word sequence generation. In: Proceedings of the 2021 6th International Conference on Inventive Computation Technologies (ICICT), 20 Jan. IEEE (2021). [https://doi.org/10.1109/icict50816.2021.9358764]
6. Kim, R.Y.: Text mining online reviews: what makes a helpful online review. IEEE Eng. Manag. Rev. **51**(4), pp. 145–156 (2023). [https://doi.org/10.1109/EMR.2023.3286349]
7. Hussain, N., et al.: PRUS: product recommender system based on user specifications and customers reviews. IEEE Access **11**, 81289–81297 (2023). [https://doi.org/10.1109/ACCESS.2023.3299818]

8. Liu, N., Zhao, J.: Recommendation system based on deep sentiment analysis and matrix factorization. IEEE Access 11, 16994–17001 (2023). [https://doi.org/10.1109/ACCESS.2023.3246060]
9. Huang, H., Zavareh, A.A., Mustafa, M. B.: Sentiment analysis in E-Commerce platforms: a review of current techniques and future directions. IEEE Access 11, 90367–90382 (2023). [https://doi.org/10.1109/ACCESS.2023.3307308]
10. Hailu, T.T., Yu, J., Fantaye, T.G.: A framework for word embedding based automatic text summarization and evaluation. Information 11(200 (2020). [https://doi.org/10.3390/info11020078]
11. Belwal, R.C., Rai, S., Gupta, A.: Text summarization using topic-based vector space model and semantic measure. Inform. Process. Manag. 58(3) (2021). [https://doi.org/10.1016/j.ipm.2021.102536]
12. Venkataraman, A., Srividya, K., Cristin, R.: Abstractive text summarization using BART. in In: 2022 IEEE 2nd Mysore Sub Section International Conference (MysuruCon), Mysuru, India,, pp. 1–6 (2022). [https://doi.org/10.1109/MysuruCon55714.2022.9972639]
13. Dilawari, A., Khan, M.U.G., Saleem, S., Zahoor-Ur-Rehman, Shaikh, F.S.: Neural Attention Model for Abstractive Text Summarization Using Linguistic Feature Space. IEEE Access 11, 23557–23564 (2023). [https://doi.org/10.1109/ACCESS.2023.3249783]
14. Fabbri, A.R., Li, I., She, T., Li, S., Radev, D.R.: Multi-News: a Large-Scale Multi Document Summarization Dataset and Abstractive Hierarchical Model, arXiv (2019). [https://doi.org/10.48550/ARXIV.1906.01749]
15. Gamzu, H., Gonen, G., Kutiel, R.L., Agichtein, E.: Identifying Helpful Sentences in Product Reviews, arXiv (2021). [https://doi.org/10.48550/ARXIV.2104.09792]
16. Zhu, Y., Yang, X., Wu, Y., Zhang, W.: Leveraging summary guidance on medical report summarization. IEEE J. Biomed. Health Inform. 27(10), 5066–5075 (2023). [https://doi.org/10.1109/JBHI.2023.3304376]
17. Lalitha, E., Ramani, K., Shahida, D., Deepak, E.V.S., Bindu, M.H., Shaikshavali, D.: Text summarization of medical documents using abstractive techniques. In: 2023 2nd International Conference on Applied Artificial Intelligence and Computing (ICAAIC), Salem, India, pp.939–943 (2023). [https://doi.org/10.1109/ICAAIC56838.2023.10140885]
18. Gandhi, R., Saini, A., Gaikwad, S.: A Framework for Abstractive Text Summarization Using Hugging Face Transformers. In: 2024 14th International Conference on Cloud Computing, Data Science & Engineering, India, pp. 690–695 (2024). [https://doi.org/10.1109/Confluence60223.2024.10463423]
19. Li, W., Zhuge, H.: Abstractive multi-document summarization based on semantic link network. IEEE Trans. Knowl. Data Eng. 33(1), pp. 43–54 (2021). [https://doi.org/10.1109/TKDE.2019.2922957]

Forecasting Sustainability: A Study of Demand Prediction in Circular Economics

C. Harshavardhini[1], K. Sarayu[1], C. R. Roshan[1], and S. K. B. Sangeetha[2(✉)]

[1] Department of Computer Science and Engineering, SRM Institute of Science and Technology, Vadapalani Campus, Chennai, India
{hc2126,kk2248,cr3351}@srmist.edu.in

[2] Department of Computer Science and Engineering (Emerging Technologies), SRM Institute of Science and Technology, Vadapalani Campus, Chennai, India
sangeets8@srmist.edu.in

Abstract. The study explores the potential of AI enhanced demand prediction for recycled and upcycled products, aiming to bridge the gap between sustainable production and market adoption. We propose a study utilizing machine learning algorithms to analyze historical sales data, market trends, and consumer sentiment, thereby forecasting demand for eco-friendly products with unprecedented accuracy. Leveraging business analytics and social media influence data, we propose a novel framework to: (1) compare consumer preferences and purchase patterns for recycled and upcycled products against their traditional counterpart (2) identify key factors driving demand for sustainable products, (3) compare demand patterns between ARIMA and LSTM models, and (4) analyze the impact of social media influence on consumer preferences. By predicting future demand with enhanced accuracy, this framework aims to empower businesses to optimize production, pricing, and marketing strategies, ultimately enabling a more sustainable and circular economy. The study holds significant implications for both businesses and consumers seeking to minimize environmental impact while catering to evolving market trends.

Keywords: AI · sustainability · demand prediction · recycled products · machine learning · circular economy

1 Introduction

The concept of sustainability has deep historical roots, evolving over time in response to environmental, economic, and social challenges. One of the earliest expressions of sustainability principles can be traced back to indigenous cultures that lived in harmony with nature, recognizing the interconnectedness of ecosystems and human well-being [8]. In the modern context, the term "sustainability" gained prominence in the 1980s with the publication of the Brundtland Report in 1987 by the World Commission on Environment and Development [9]. Coined by the report, sustainability was defined as "development that meets the needs of the present without compromising the ability of future generations to meet their own needs." This marked a pivotal moment in global awareness of the environmental impact of human activities [10].

In subsequent decades, concerns about climate change, resource depletion, and social inequality have intensified, prompting increased emphasis on sustainable practices [11]. International agreements such as the Paris Agreement in 2015 and the Sustainable Development Goals have further solidified the commitment to addressing these challenges on a global scale. Today, sustainability encompasses environmental stewardship, economic viability, and social equity [12]. It has evolved into a multidisciplinary approach that seeks to balance the needs of the present while safeguarding the well-being of future generations—a crucial paradigm for addressing the complex and interconnected challenges facing our planet [13].

Predicting demand for sustainable products and services is crucial for businesses navigating the green wave. Enter AI, the master of patterns and trends, ready to revolutionize how we anticipate needs in this evolving market [14]. Leveraging machine learning algorithms to analyze historical data and identify patterns, AI-powered demand prediction empowers businesses with accurate forecasts for future needs. These models process vast information streams – sales data, customer behavior, economic indicators, and external factors – to generate insights that guide inventory management and supply chain optimization [15]. Techniques like regression analysis, time series forecasting, and deep learning enable AI systems to adapt to dynamic market conditions and account for diverse influencing factors.

The main contribution is to propose a novel machine learning framework

(1) To compare consumer preferences and purchase patterns for recycled and upcycled products against their traditional counterpart
(2) To identify key factors driving demand for sustainable products,
(3) To compare demand patterns between ARIMA and LSTM models, and
(4) To analyze the impact of social media influence on consumer preferences.

With this detailed introduction, Sect. 2 describes the related study, Sect. 3 explains the system methodology, Sect. 4 depicts the experimentation results followed by conclusion in Sect. 5.

2 Related Study

[1] This paper undertakes a comprehensive review of existing literature on sustainable consumption and production (SCP) practices in developed and developing economies, published between 1998 and 2018. Guided by three key questions, the authors analyze the current state of research, identify differences and challenges, and pinpoint gaps and future research directions. The paper provides a valuable and insightful contribution to the understanding of SCP in developed and developing economies. Its strengths lie in its comprehensiveness, comparative approach, and clear presentation. By addressing the suggested areas for discussion, the review can be further strengthened and contribute even more significantly to advancing research and practice in this crucial field.

[2] The increasing emphasis on sustainability and efficiency has sparked interest in remanufactured items. Yet, accurately forecasting demand for these products is challenging due to their unique market dynamics and limited historical data. This study proposes a data-mining approach to tackle this issue, assessing machine learning's efficacy

in predicting remanufactured product demand. While conventional statistical models are commonly used in existing research, they struggle with capturing complex market influences. This paper explores data-mining techniques, particularly ensemble machine learning algorithms, demonstrating their superiority in forecasting demand. By revealing insights into consumer preferences and demand drivers, this approach fosters more effective business strategies, advancing sustainable product life cycles.

[3] Accurate sales forecasting is vital for businesses to optimize inventory, marketing, and resource allocation. This study introduces a customer sales prediction system employing data analytics techniques. Traditional methods like regression and ARIMA models are common but may lack accuracy due to complex relationships and data requirements. Supervised learning algorithms like Decision Trees and Random Forests offer flexibility but pose interpretability challenges. Advanced techniques like CNNs and RNNs handle complex patterns but require careful management of computational resources and overfitting risks. Combining methods such as feature selection and ensemble modeling could enhance prediction accuracy and adaptability in real-time sales forecasting.

[4] This review analyzes relevant literature on retail sales prediction, primarily focusing on studies published within the past five years. The review considers various approaches, including traditional statistical methods, machine learning algorithms, and hybrid models. Additionally, it delves into the specific application of these techniques within the context of Big Mart sales data, as presented in the target paper. It offers valuable insights into the application of machine learning for retail sales forecasting. By contributing to the ongoing discussion on model selection, feature engineering, and context-specific solutions, the study paves the way for further advancements in this crucial area of retail analytics. Future research should focus on incorporating broader data sources, exploring the generalizability of models, and investigating the impact of external factors on sales prediction accuracy.

[5] This paper proposes a deep learning approach for sales prediction, aiming to improve upon existing methods. To evaluate its effectiveness, a literature review of alternative techniques and deep learning applications in retail sales forecasting is necessary. Employing a deep learning architecture, specifically LSTMs, potentially captures complex non-linear relationships within sales data. The paper emphasizes the importance of data pre-processing and feature engineering, crucial for optimizing model performance. Comparing the deep learning approach with traditional models provides insights into its potential advantages. The proposed model is tested on actual retail store data, demonstrating its effectiveness in a practical setting.

[6] The study evaluates existing research on how online user behavior influences sales prediction in e-commerce. This involves searching literature from academic journals, conferences, and reputable sources. Selection criteria are established based on methodological rigor, relevance to the research question, and focus on specific user behavior like browsing history and cart abandonment. Findings are categorized and synthesized based on behavior types analyzed, prediction methods used, and reported improvements in accuracy. The review suggests that online user behavior holds significant potential for improving sales prediction accuracy, offering valuable insights for businesses to optimize operations and gain competitive advantage in the online marketplace.

[7] This study introduces a novel approach using a convolutional neural network (CNN) to analyze sustainability and forecast market demand in retail. To contextualize this approach, a literature review examines Market Demand Estimation and Sustainability Analysis in the retail sector. By utilizing CNN to analyze web check-in data, the model aims to capture intricate spatial and temporal patterns, offering insights into green consumer preferences. Integrating sustainability factors into demand estimation provides actionable insights for retailers, facilitating targeted initiatives and informed decision-making.

Continuous learning from new data further enhances prediction accuracy as these models evolve over time. By implementing AI-driven demand prediction, businesses can streamline operations, reduce excess inventory costs, and improve customer satisfaction by ensuring product availability. Real-time insights empower data-driven decision-making for swift responses to market changes. Ultimately, AI-powered demand prediction fosters better resource allocation, cost efficiency, and strategic planning, boosting business competitiveness and resilience in a rapidly evolving landscape.

3 System Methodology

3.1 Data Preparation

In sales prediction of sustainable products, feature engineering involves identifying relevant factors such as product attributes, consumer behavior, and environmental impact indicators. For instance, We incorporated features like product sustainability certifications, consumer demographics, and eco-friendly packaging details to train the model, enhancing its accuracy in predicting sales for sustainable products.Data preparation involves cleaning, transforming, and organizing raw data to make it suitable for analysis and modeling. In sales prediction for sustainable products, data preparation includes tasks like handling missing values, encoding categorical variables, and scaling numerical features. We utilized Kaggle, a platform hosting numerous datasets, to acquire relevant data on sustainable products, such as sales figures, product attributes, and environmental impact metrics. This data was then pre-processed and curated for training machine learning models to predict sales effectively.

3.2 Time Series Analysis

The data was to thoroughly analyze yearly, weekly, and daily sales data to gain insights for improving sales and marketing tactics. Furthermore, we conducted statistical examinations, such as evaluating stationarity, autocorrelation, and predictability, for each product category. These analyses set the essential groundwork necessary for future forecasting approaches.Sales forecasting was carried out on a weekly basis, employing two distinct methodologies. The rolling forecast technique predicted sales for the upcoming week by utilizing a model trained on historical data, facilitating short-term resource planning and inventory management. Assessment of forecasting models involved ARIMA/SARIMA and Long-Short Term Memory (LSTM) neural network architectures, with manual analysis, Python's stats models function, and grid search optimization used to optimize hyperparameters. Model performance was evaluated using a train-test split, reserving the final year of data for testing, with Mean Squared Error (MSE) serving as the primary performance metric.

3.3 Proposed Automation

In this study, we employed two different forecasting approaches, ARIMA and LSTM, and established an automated process to identify the optimal model. This automation significantly contributed to systematically comparing the forecasting outcomes produced by ARIMA and LSTM, facilitating streamlined model selection based on critical evaluation metrics. The results and insights derived from both methods were carefully analyzed, leading to the determination of the most effective approach.

3.3.1 Behavioural Diagnostics

Implementing behavioral diagnostics involves a multifaceted approach, focusing on key aspects of consumer buying behavior, sustainability imperatives, business practices, product category sales, and demographics. Through meticulous data analysis, we delve into consumer purchasing patterns, identifying trends and preferences within specific demographics. Sustainability considerations are integrated into every phase, ensuring alignment with ethical and environmental standards. By examining product category sales, we gain insights into market dynamics and consumer demands, informing strategic decisions. This holistic methodology facilitates a comprehensive understanding of consumer behavior, guiding businesses towards effective and sustainable practices. Through collaborative efforts and stakeholder engagement, we foster a culture of transparency and accountability in implementing sustainable business practices. Ultimately, our behavioral diagnostics framework empowers businesses to optimize their operations while simultaneously driving positive social and environmental impact [13] (Fig. 1).

This study investigates the effectiveness of two prominent time series forecasting models, ARIMA and LSTM, in predicting sales data. The research aims to identify which model offers superior accuracy in capturing sales trends and seasonality. The analysis begins with the acquisition of historical sales data. This data is then meticulously split into two distinct sets: a training set and a testing set. The training set serves as the foundation for model development, while the testing set is used for unbiased evaluation of the trained models' forecasting capabilities.

3.3.2 ARİMA

For the ARIMA (Autoregressive Integrated Moving Average) model, the focus lies on fitting the model to the training data. This process involves parameter optimization, where the order p, d, q is meticulously adjusted. The 'p' parameter signifies the number of past sales figures used for prediction, 'd' represents the degree of differencing required to achieve stationarity (consistent statistical properties) in the data, and 'q' reflects the number of past forecast errors incorporated into the model. By optimizing these parameters, the ARIMA model can effectively capture trends and seasonality present within the sales data. Once the model is fitted, it generates sales forecasts for the unseen data points in the testing set (Fig. 2).

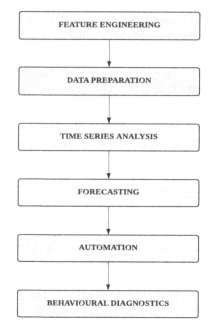

Fig. 1. Block Diagram

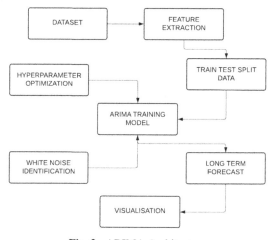

Fig. 2. ARIMA Architecture

3.3.3 LSTM

The LSTM (Long Short-Term Memory) model, on the other hand, leverages a deep learning approach. Prior to training, the sales data undergoes preprocessing, a crucial step in LSTM applications. This preprocessing typically involves scaling the data to a specific range, ensuring all features contribute equally during training. Additionally, the data is reshaped into sequences. These sequences comprise past sales values paired with their corresponding future sales figures. This structure allows the LSTM model to learn complex, non-linear relationships between past sales patterns and future sales outcomes. The model is then trained on these sequences, enabling it to identify and exploit these relationships within the data.

Following the training stage, the LSTM model forecasts sales for the testing set, aiming to accurately predict future sales based on the learned patterns. To assess the performance of both models, established metrics like Mean Absolute Error (MAE) or Root Mean Squared Error (RMSE) are employed on the testing data. MAE reflects the average magnitude of the difference between the predicted and actual sales values, while RMSE provides a measure of the squared deviations between predicted and actual sales. By comparing these metrics for both models, we can determine which model offers superior accuracy in sales prediction. Furthermore, the research incorporates parameter tuning as a means to refine the performance of each model.

In the ARIMA model, this involves iteratively adjusting the order (p, d, q) parameters to minimize the error metrics on the testing set. For the LSTM model, parameter tuning might encompass adjustments to the network architecture, such as the number of hidden layers or the number of neurons within each layer. By meticulously tuning the parameters of each model, we strive to optimize their forecasting capabilities and achieve the most accurate sales predictions possible. This comparative analysis of ARIMA and LSTM models sheds light on their relative strengths and weaknesses in the context of sales forecasting. The study's findings can guide businesses in selecting the most suitable model for their specific sales data and forecasting needs.

4 Experimentation Results

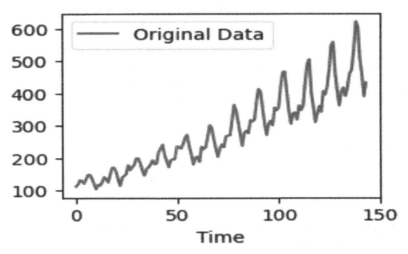

Fig. 3. Original Data

In Fig. 3 We can observe the number of sustainable products sold over a period of months. There is an upward trend, indicating increasing sales of sustainable products over time. This could be due to a number of factors, such as growing consumer awareness of sustainability issues, or a wider range of sustainable products being offered by manufacturers. The LSTM prediction graph Fig. 4 forecasts a continued increase in the number of sustainable products sold over the next few months. The prediction, shown by the orange line, follows an upward trend that is similar to the historical data (blue line) though not identical. This suggests that the factors influencing sales of sustainable products are likely to remain consistent in the near future.

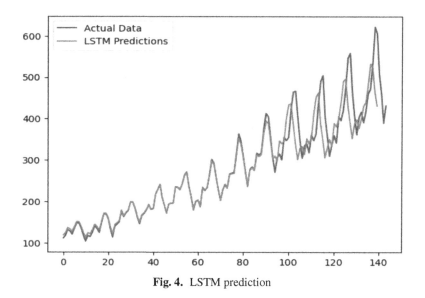

Fig. 4. LSTM prediction

Figure 5 appears to be a gap between the actual data and the ARIMA predictions for some months. In the months where the historical sales is above the forecasted sales, this indicates that the actual sales were higher than what the ARIMA model predicted. This suggests that the model predicted higher sales than what was actually achieved.

Both ARIMA and LSTM models predict an increase in sustainable product sales over the next few months. The historical data also shows an upward trend, suggesting a potential continuation of this growth. From Fig. 6 The LSTM forecast appears to be closer to the actual historical data compared to the ARIMA forecast for several months, particularly in the earlier half of the graph. In these months, the ARIMA model seems to underestimate the number of sustainable products sold. Later in the timeframe, the predictions by both models converge.

Social media trends directly influence the popularity of upcycled products. ARIMA and LSTM models can analyze this buzz, identifying sudden spikes in demand for specific upcycled materials or styles. This allows businesses to react quickly, optimizing production and minimizing wasted resources. Social media can also reveal seasonal trends in upcycled product preferences. Perhaps winter sees a surge in demand for

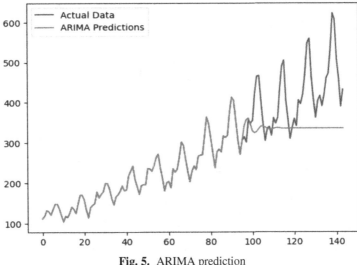

Fig. 5. ARIMA prediction

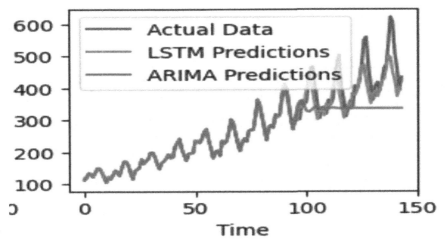

Fig. 6. ARIMA vs LSTM

upcycled wool coats, while summer favors repurposed denim shorts. By analyzing past sales data alongside social media trends, ARIMA and LSTM models can predict these seasonal fluctuations, ensuring businesses have the right inventory at the right time. Consumers today crave transparency from brands. Upcycled product production trackers, powered by ARIMA and LSTM, can generate insightful reports. These reports not only showcase a brand's commitment to sustainability but also demonstrate their ability to adapt to market shifts based on real-time consumer data gleaned from social media [10, 11].

5 Conclusion

The pursuit of sustainability in consumer goods, driven by environmental awareness and regulatory pressures, integrates AI for demand prediction, especially in recycling. This research aims to bridge the gap between sustainable production and market adoption. Using machine learning, it analyzes sales, trends, and sentiment for accurate eco-product demand forecasting. Comparing consumer preferences for recycled and conventional goods informs sustainable choices. Understanding demand dynamics aids in production optimization and pricing strategies. Incorporating social media data highlights influencer impact on consumer behavior. The predictive framework empowers businesses to align supply with demand, minimizing waste and fostering a circular economy. Beyond business, it contributes to environmental goals and societal benefits. By embracing sustainability, businesses enhance competitiveness and resilience. This study advances AI-powered analytics for sustainability in consumer goods, promoting a harmonious relationship between business objectives, consumer preferences, and environmental stewardship.Additionally, extending our analysis to include sustainability metrics such as carbon footprint reduction would provide a more comprehensive understanding of the environmental benefits of sustainable practices.

References

1. Wang, C., Ghadimi, P., Lim, M.K., Tseng, M.L.: A literature review of sustainable consumption and production: a comparative analysis in developed and developing economies. J. Cleaner Prod. **206**, 741–754 (2019). ISSN 0959–6526, https://doi.org/10.1016/j.jclepro.2018.09.172
2. Van Nguyen, T., Zhou, L., Chong, A.Y.L., Li, B., Pu, X.: Predicting customer demand for remanufactured products: a data-mining approach. Eur. J. Oper. Res. **281**(3), 543–558 (2020). ISSN 0377-2217. https://doi.org/10.1016/j.ejor.2019.08.015
3. Venkatesan, K., Viswanathan, E., Kumar, A., Yuvaraj, S., Shivkumar Tanesh, P.: Advances in mathematics. Sci. J. **9**(7), 5049–5056 (2020). ISSN: 1857-8365, 1857-8438. https://doi.org/10.37418/amsj.9.7.70
4. Behera, G., Nain, N.: A comparative study of big mart sales prediction. In: Nain, N., Vipparthi, S., Raman, B. (eds) Computer Vision and Image Processing. CVIP 2019. Communications in Computer and Information Science, vol. 1147. Springer, Singapore (2020). https://doi.org/10.1007/978-981-15-4015-8_37
5. Kaneko, Y., Yada, K.: A deep learning approach for the prediction of retail store sales. In: 2016 IEEE 16th International Conference on Data Mining Workshops (ICDMW), Barcelona, Spain, pp. 531–537 (2016). https://doi.org/10.1109/ICDMW.2016.0082
6. Yuan, H., Xu, W., Wang, M.: Can online user behavior improve the performance of sales prediction in E-commerce?. In: 2014 IEEE International Conference on Systems, Man, and Cybernetics (SMC), San Diego, CA, USA, pp. 2347–2352 (2014). https://doi.org/10.1109/SMC.2014.6974277
7. Ankita, M., Srinivas, A., Soni, A., Prajapati, G., Manjunath, P.S.: Pharmaceutical sales data prediction using time series forecasting Int. J. Intell. Syst. Appl. Eng. **12**(13s), 681–696 (2024)
8. Silvestre, B.S., Țîrcă, D.M.: Innovations for sustainable development: Moving toward a sustainable future, J. Cleaner Prod. **208**, 325–332 (2019). ISSN 0959-6526, https://doi.org/10.1016/j.jclepro.2018.09.244

9. Li, Q., Wei, W., Xiong, N., Feng, D., Ye, X., Jiang, Y.: Social media research, human behavior, and sustainable society. Sustainability. **9**(3), 384 (2017). https://doi.org/10.3390/su9030384
10. Minton, E., Lee, C., Orth, U., Kim, C.H., Kahle, L.: Sustainable marketing and social media: a cross-country analysis of motives for sustainable behaviors. J. Advert. **41**(4), 69–84 (2012). https://doi.org/10.1080/00913367.2012.10672458
11. Strähle, J., Gräff, C.: The role of social media for sustainable consumption. In: Strähle, J. (ed.) Green Fashion Retail. Springer Series in Fashion Business. Springer, Singapore (2017). https://doi.org/10.1007/978-981-10-2440-5_12
12. Dhaya, R., Sangeetha, S.K.B., Sharma, A.: Improved performance of two server architecture in multiple client environment. In: 2017 4th International Conference on Advanced Computing and Communication Systems (ICACCS), Coimbatore, India, pp. 1–4 (2017). https://doi.org/10.1109/ICACCS.2017.8014560
13. Sangeetha, S.K.B., Veningston, K., Sathya, V., Kanthavel, R.: Chapter 3 - Design of a novel privacy preservation based cyber security system framework for secure medical data transactions in cloud storage In: Intelligent Data-Centric Systems, Intelligent Edge Computing for Cyber Physical Applications, Academic Press, pp. 35–43 (2023). ISBN 9780323994125, https://doi.org/10.1016/B978-0-323-99412-5.00006-X
14. Aravindhan, K., Sangeetha, S.K.B., Kamesh, N.: Improving performance using hybrid framework Iot communication in cloud computing. In 2022 8th International Conference on Advanced Computing and Communication Systems (ICACCS), vol. 1, pp. 1654–1658. IEEE (2022)
15. Dhir, A., Talwar, S., Sadiq, M., Sakashita, M., Kaur, P.: Green apparel buying behavior: a Stimulus–Organism–Behaviour–Consequence (SOBC) perspective on sustainability-oriented consumption in Japan. Bus. Strateg. Environ.Strateg. Environ. **30**(8), 35893605 (2021). https://doi.org/10.1002/bse.2821

Gold Price Forecasting Using Machine Learning Techniques

Binu John[1(✉)] and K. Nidhina[2]

[1] Artificial Intelligence and Data Science, SCMS School of Engineering and Technology, Vidya Nagar, Cochin, Kerala 683572, India
binujohn@scmsgroup.org

[2] Computer Science and Engineering, SCMS School of Engineering and Technology, Vidya Nagar, Cochin, Kerala 683572, India

Abstract. This project aims to create an advanced machine learning system for accurate gold price prediction in response to the growing importance of gold prices in reflecting sentiment in the global economy and helping informed decision-making in international trade. By utilizing state-of-the-art machine learning techniques such as random forest regression, decision trees, and support vector machines in addition to large historical datasets of market indices and related commodities, the project seeks to identify complex patterns and trends influencing the dynamics of gold prices. From a methodological standpoint, the research focuses on optimizing both predictive accuracy and generalizability through extensive data preprocessing, exploratory analysis, and model training. Essential metrics like the root mean square error (RMSE) and R2 score are used in performance evaluation. The results show how well the random forest regression model outperforms the other two techniques with an R2 score of 0.990 and a low RMSE, demonstrating remarkable accuracy and reliability. The project's goal is to provide organizations with practical insights for proactive risk management and well-informed decision-making in the dynamic world of international commerce and finance by combining state-of-the-art technology with empirical research.

Keywords: Gold price prediction · Random Forest regression · Decision tree · Support vector regression · Machine learning

1 Introduction

Gold's significance as a crucial indicator of financial stability and economic activity has grown in the face of today's quickly changing global economy. Thanks to the global interconnection of markets, the price of gold is a crucial indication that reflects investor sentiment and expectations throughout the globe, in addition to the current state of the economy [1]. This makes it necessary for companies involved in international trade to predict gold prices precisely since it gives them the ability to move more quickly and strategically through the intricacies of the market environment [2]. Understanding how important this problem is, a large-scale project has been started to create an advanced machine learning (ML) [3] system designed with gold price predictions in mind. The strategy is based on data-driven analysis and predictive modeling. It uses the most recent developments in machine learning techniques, such as support vector machines, decision trees, and random forest regression. Through the use of extensive historical data sets that include closely related commodities and stock market indexes [4], the model aims to identify complex patterns and trends that drive the dynamics of gold prices. A comprehensive grasp of the many variables influencing gold prices in fundamental to the process, beyond simple changes in currency rates or stock market patterns, the model considers a wide range of factors, from macroeconomic indicators to geopolitical developments [5]. To determine their effect on gold prices, for example, the performance of key currencies like the US dollar and the euro as well as the price fluctuations of associated commodities like silver are carefully examined. Additionally, the model includes qualitative elements like investor behavior and market sentiment [6], which improves the system's ability to forecast future events. The goal is to provide them more than just pricing projections by incorporating these complex factors into the prediction framework and offering them with useful insights. In order to enable businesses to make wise decisions in a fast-changing environment [7], the model aims to provide a deep knowledge of the underlying reasons for fluctuations in the price of gold rather than making crude predictions.

The machine learning system is a formidable instrument for augmenting decision-making abilities, whether it is in evaluating possible hazards and prospects linked with global transactions or refining hedging tactics to alleviate market oscillation. Essentially, the project is an amalgam of state-of-the-art technology and empirical research, resulting in an advanced platform for gold price prediction. The goal of the project is to provide organizations with the resilience and foresight they need to prosper in the ever-changing world of international trade and finance by utilizing the amount of data available and the predictive potential of machine learning. Employing constant improvement and adjustment, the dedication to pushing the boundaries of predictive analytics, stimulating creatively, and providing real value to collaborators and stakeholders does not waver.

2 Related Works

Manjula K A, Kartikeyan [8] conducted a study examining the connection between the cost of gold and major variables such as interest rates, the price of crude oil, the stock market, inflation, and the rupee to dollar exchange rate. These data were analyzed using three machine learning algorithms: gradient-boosting regression, random forest regression, and linear regression. It was discovered that gradient boosting regression provides superior prediction accuracy for the periods taken separately, but random forest regression has better prediction accuracy for the total period. Rushikesh Ghule, Abhijeet Gahave [9] build a model that concluded that random forest analysis-based machine learning techniques are highly helpful in predicting gold prices. The R-square for the model is 97%. When a model's explanation of gold ETF prices is nearly 100%, it is considered well-fitting. It demonstrates how the suggested random forest method outperforms conventional and existing prediction models.

The CNN-LSTM model is put forth by Loannis E. Livieris, Emmanuel Pintelas, Panagiotis Pintelas P [10] to forecast the movement and price of gold. Conventional methods are challenged by the time series complex and nonlinear structure of the price of gold. The CNN-LSTM model integrates the advantages of convolutional neural networks (CNNs) and long short-term memory (LSTM) networks. D Makala and Z Li [11] conducted a study to forecast the price of gold using SVM and Arima models. According to the study's findings, SVM (poly) outperforms both the Arima model and the other SVM (RBF) significantly. The 99% R-square indicates that the SVM (poly) and SVM (linear) outcomes in this study are almost identical.

An innovative technique for accurately forecasting fluctuations in the price of gold and assessing forecasts is presented by Sami Ben Jabeur, Salma Mefteh-Wali, and Jean-Laurent Viviani [12]. Six machine learning models are compared in the study, including cutting-edge methods like CatBoost and eXtreme Gradient Boosting (XGBoost). XGBoost outperforms other state-of-the-art machine learning algorithms, according to empirical evidence. The research's second part proposes the use of shapley additive explanations (SHAP) to assist policymakers in assessing the importance of various factors affecting gold prices and deciphering pre-dictions from complex machine learning models. The results show that the performance of gold price prediction is significantly enhanced when XGBoost and SHAP techniques are combined. For accurate gold price prediction, Liang, Yanhui, Yu Lin, and Qin Lu [13] provided a novel ICEEMDAN-LSTM-CNN-CBAM (ILCC) decomposition-ensemble model. This study combines CNN, CBAM, and LSTM for joint forecasting and uses ICEEMDAN to break down gold prices into frequency-based sublayers. The ILCC model performs better than other comparison models, according to experimental results. It does this by breaking down the sequence in an efficient manner, identifying the long-term effects, and distilling the core of the sequence.

For the gold EFT price under consideration, Chodavrapu Pragna, Bhavana Purra, Adari Viharika, Garabhapu Roshini, and Prof. B. Prajna [14] have suggested that the provided research offers the highest probability of getting high training rate prediction precision. Essentially, this work aims to provide suitable predictor models that, in combination with datasets chosen from relevant databases from previous years, would successfully display the determined gold in different scenarios. The decision to buy gold ETF was made using the solution model and supervised machine learning algorithms on a dataset of historical values. Aruna S, Umamaheswari P, Sujipriya J, and Dr. R. Umamaheswari [15] highlight through their analysis the historical relevance of gold and its central role in the world economy. The study uses machine learning algorithms to accurately predict gold prices and emphasizes the impact of Russia's interest rates as well as the market values of major enterprises. It turns out that the valuation of individual enterprises affects gold prices more than the US economy as a whole. In order to improve performance in the future, the study combines ensemble learning and deep learning techniques.

An investigation into the relationship between the price of gold and certain explanatory aspects that are commonly employed as indicators of geopolitical and financial crises is the goal of Sakir Bingol, Safaa Sadi, Raland Mantenggo, Hatim Badr Mouhcine, Jihad Albaf, Salma Chaabene, and Aaron Auta [16]. The study looks into whether it is possible to forecast the price of gold using these variables. The price of gold was predicted using four different machine-learning algorithms: linear regression, SVM, VAR, and ARIMA. ARIMA outperformed the other applied algorithms, scoring the lowest RMSE, according to the findings. Recognizing the intrinsically non-linear nature of fluctuations in gold prices, Vidya G S and Hari V S [17] emphasize in their paper the need to use LSTM (long short-term memory) networks to forecast gold rates. Many factors, such as geopolitical developments and economic indicators, have an impact on gold as a commodity, which presents a special difficulty for conventional statistical models that frequently depend on linear assumptions. Through the use of historical price data obtained from reliable sources such as the World Gold Council, researchers can train long short-term memory (LSTM) models to identify underlying patterns and relationships in the data. In addition to examining historical price movements, this approach takes into account a wide range of pertinent variables, such as market sentiment, economic indicators, and geopolitical happenings. The resulting LSTM design shows a remarkable ability to capture the nonlinear linkages responsible for the variations in the price of gold.

3 Proposed Methodology

3.1 Dataset

The dataset taken from Kaggle offers a comprehensive historical summary of financial metrics, with an emphasis on gold prices and related variables. It includes data on the Standard and Poor's 500 index (SPX), which measures major US stock performance, silver prices (SLV), the United States Oil Fund (USO), which tracks the price of crude oil, and the euro/dollar exchange rate (EUR/USD). Patterns and correlations across time series are displayed between these variables, with each input corresponding to a certain date. Analysts can use this data to understand patterns in the price of gold, how it relates to other economic factors, and the potential consequences of events in the market or in policy, In-depth study on the behavior of financial markets and the development of predictive models are also made possible by it. Train data and test data are the two categories into which the gathered data is split.

3.2 Preprocessing

The gold price prediction dataset has undergone a thorough preprocessing that involves numerous processes to ensure the data is ready for analysis and modeling. Determining the number of rows and columns in the dataset requires an understanding of its dimensions and scope, which is why 'shape' is used. Subsequently, 'info()' is invoked to acquire an overview of the dataset, encompassing the data kinds of every column and any possible absent values. As demonstrated by 'isnull(). Sum()', which determines the total number of missing values for each column, this summary provides insightful information on the makeup of the dataset and aids in the development of strategies for handling any missing data. In order to preserve the dataset's integrity and guarantee the accuracy of studies and models, missing values must be addressed. Furthermore, the corr() function is utilized to compute correlations between various parameters and the target variable, which is designated as 'GLD' and is likely to reflect gold prices. Comprehending these correlations facilitates feature engineering and selection by offering crucial insights into possible links within the dataset. These preprocessing procedures, taken together, provide a crucial basis for the efficient analysis and modeling of the gold price prediction dataset. They also guarantee the quality of the data and help with well-informed decision-making during the predictive modeling phase.

3.3 Methodology

The random forest regression, decision tree, and support vector algorithms were the three different algorithms used. To generate predictions, random forest regression uses an ensemble of decision trees. This method is accurate and robust, especially when dealing with complicated datasets. Conversely, by recursively partitioning data according to feature values, decision trees offer a simple, comprehensible method of representing data. As an efficient method for both linear and nonlinear regression tasks, support vector regression builds a hyperplane in high-dimensional space to approximate the link between input features and output values. The proposed architecture is shown in Fig. 1.

One effective machine learning method for forecasting continuous outcomes is random forest regression [18]. It works by building a collection of decision trees, each of which is trained using a different subset of the data. Random forest regression reduces overfitting and produces forecasts that are reliable and accurate by combining the predictions of several trees. It works especially well with complicated datasets that have nonlinear relationships, and it can handle missing values without the need for preprocessing. Because of its adaptability and accuracy, random forest regression is frequently utilized in a variety of disciplines, including finance, economics, and environmental research. A well-liked machine learning approach for classification and regression applications is the decision tree. It creates a tree-like structure of decision nodes by recursively splitting the dataset into subsets according to the values of input features. The method chooses the feature at each node that divides the data into homogenous subsets as efficiently as possible, to minimize impurity or maximize information gain. Decision trees [19] are useful for comprehending feature importance and model insights since they are simple to read and depict. They can, however, overfit, particularly when dealing with complicated datasets. Pruning and ensemble techniques like random forest can improve their capacity for generalization and prediction.

One flexible approach for classification and regression problems is support vector machine(SVM) [20]. To divide data classes and maximize the margin between them, it finds the ideal hyperplane. SVMs use kernel functions to handle both linear and nonlinear data, and they are effective in high-dimensional domains. SVMs are resilient and versatile, which makes them useful for a variety of applications like image recognition, text classification, and bioinformatics. However, they do require careful hyperparameter tweaking. SVMs continue to be widely employed because of their capacity to manage complicated data distributions and deliver precise predictions across a range of domains, despite their possible sensitivity to hyperparameters.

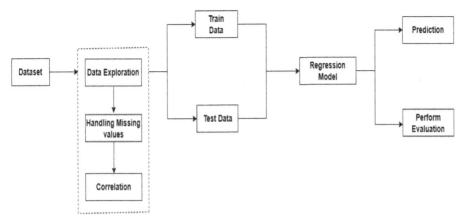

Fig. 1. Proposed Architecture

3.4 Experiment

Support vector regression (SVR), decision trees, and random forests are three different regression models that are trained and assessed. First, the number of decision trees in the ensemble is indicated by instantiating a random forest regressor model with 100 estimators. The training set of data (x train) and (y train) is then used to train this model. After the random forest model has been trained, predictions are made for the test data (x test), and performance measures like root mean squared error (RMSE) as given in Eq. 1 and R-squared (R2) scores as given in Eq. 2 are calculated to measure how well the model is predicted.

$$RMSE = \sqrt{\frac{1}{n}\sum_{i=1}^{n}((y_i - \hat{y}_i))^2} \quad (1)$$

where n is total number of observations, y_i is the actual value of the dependent variable for observation i, and $\hat{y}_i$ is the predicted value of the dependent variable for observation i.

$$R^2 = 1 - \frac{\frac{1}{n}\sum_{i=1}^{n}((y_i - \hat{y}_i))^2}{\frac{1}{n}\sum_{i=1}^{n}((y_i - \bar{y}))^2} \quad (2)$$

where n is total number of observations, y_i is the actual value of the dependent variable for observation i, and $\hat{y}_i$ is the predicted value of the dependent variable for observation i., and $\bar{y}$ is the mean of the actual values of the dependent variable.

Both the decision tree regressor and SVR models go through the same procedure twice. The algorithm can utilize the default settings in the decision tree model since no hyperparameters are supplied. Similarly, the SVR model uses its default setup to instantiate itself without the need for any new hyperparameters. For every model, predictions are made, and their performances are assessed using the MSE and R2 values. These metrics are important measures of any model's ability to generalize to new data and offer information about its advantages and disadvantages. Because this process is iterative, it is possible to compare all three regression methods in-depth, which helps with decision-making when choosing a model for tasks involving the prediction of gold prices. Furthermore, the outcomes derived from every model can be subjected to additional analysis to pinpoint possible avenues for enhancement and optimization, hence augmenting the predictive powers of the selected model.

Table 1. RMSE & R2

Model	RMSE	R2
Random Forest	5.449	0.990
Decision Tree	4.251	0.992
Support Vector Regressor	467.353	0.144

4 Result

The machine learning approaches Support vector regressors (SVR), random forest regression, and decision trees are used in to estimate gold values. The predictive power of each approach was thoroughly evaluated by utilizing two essential metrices: the R2 score and the root mean square error (RMSE) shown in Table 1. The random forest regression model in Fig. 2 which showed amazing potential in terms of prediction accuracy and error reduction, was found to be the most promising of the models studied. Comparing the random forest model to its counterparts, it performed better, with an amazing R2 score of 0.990 as in Fig. 3 and an RMSE of 5.449 as in Fig. 4. Even though its performance was marginally worse than that of the random forest, the decision tree model in Fig. 5 nevertheless demonstrated good predictive ability. The SVR technique, in Fig. 6 on the other hand, performed terribly, producing a low R2 score of 0.144 and a worrisome RMSE of 467.353. It is important to choose the right model for the work at hand, as this clear disparity highlights. The random forest regression model is clearly the best option for consistently predicting gold prices due to its exceptional mix of accuracy and reliability. The SVR model, on the other hand, appears to be unfit for the complexity of this forecasting project based on its low results. The random forest regression model can be used for prediction of gold prices by utilizing the insights obtained from this comprehensive analysis.

5 Discussions

In order to achieve accuracy and dependability never before seen in financial forecasting, the team began predicting gold prices using machine learning techniques. Three regression models were evaluated, namely random forest regression, decision trees, and support vector regression (SVR), following thorough preparation and exploration of an extensive dataset obtained from Kaggle. Based on its low error rates and remarkable

Gold Price Forecasting Using Machine Learning Techniques 293

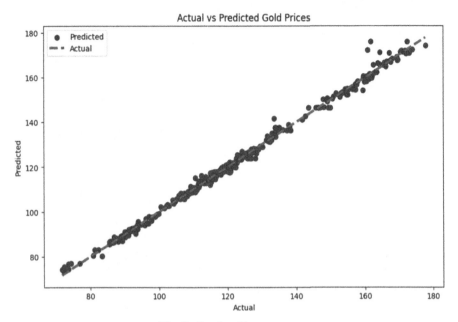

Fig. 2. Random Forest regressor

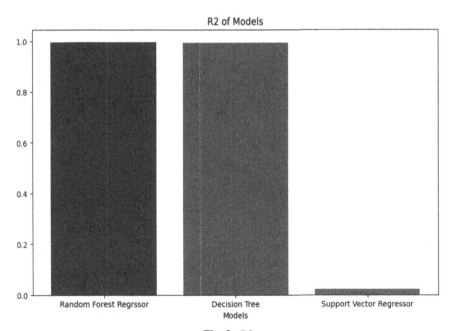

Fig. 3. R2

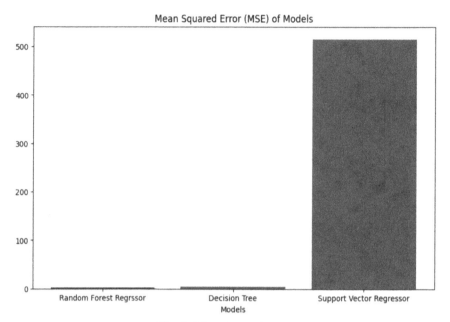

Fig. 4. Mean Squared Error

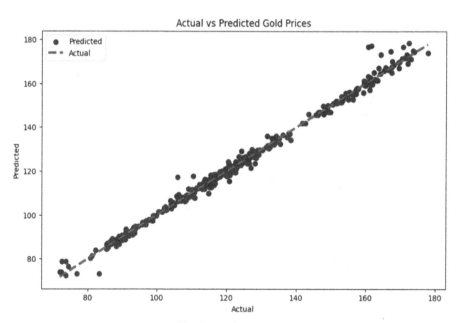

Fig. 5. Decision Tree

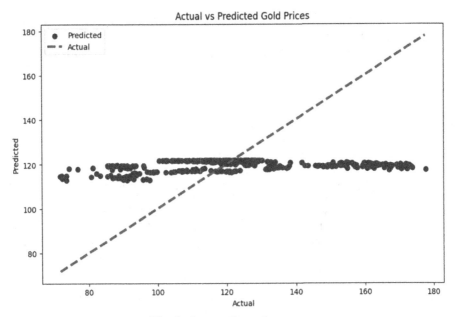

Fig. 6. Support Vector Regressor

R2 score of 0.990, the random forest regression model proved to be more successful than the others. Exceptional insights for organizations navigating the worldwide market were provided by this model's capacity to manage complex data patterns. Model selection is crucial because the SVR approach failed, in comparison. In financial forecasting, the research highlights the revolutionary potential of data-driven methodologies and provides decision-makers with useful information for proactive risk management.

6 Conclusion

The research has been a revolutionary experience characterized by creative examination, with the goal of utilizing the immense potential of machine learning techniques to predict gold prices with previously unheard-of accuracy and dependability. The project began with a thorough process of data preprocessing and exploratory analysis, drawn from a rich and expansive dataset carefully selected from Kaggle. This dataset encapsulates a comprehensive historical narrative of financial metrics intricately intertwined with relevant variables such as stock market indices, commodity prices, and currency exchange rates. This laborious process was carried out to decipher the complex network of relationships and patterns present in the data, providing a strong basis for the future construction and assessment of three different regression models: random forests, decision trees, and support vector regression. Optimizing predicted accuracy and generalizability was the primary focus of each model's training and fine-tuning. After extensive testing, the random forest regression model was shown to be the most accurate and dependable option, demonstrating its unmatched capacity to produce forecasts of gold prices with extreme accuracy. With the help of ensemble learning and left handling of the dataset's complexity, the random forest regression model was able to extract valuable insights from the data and produce reliable predictions, giving businesses engaged in international trade and finance indispensable tools for proactive risk management and well-informed decision-making in the dynamic and constantly changing global market landscape.

References

1. Lili, L., Chengmei, D.: Research of the influence of macro-economic factors on the price of gold. Procedia Comput. Sci. **17**, 737–743 (2013)
2. Davidson, S., Faff, R., Hillier, D.: Gold factor exposures in international asset pricing. J. Int. Finan. Markets. Inst. Money **13**(3), 271–289 (2003)
3. Sarangi, P.K., Verma, R., Inder, S., Mittal, N.: Machine learning based hybrid model for gold price prediction in India. In: 2021 9th International Conference on Reliability, Infocom Technologies and Optimization (Trends and Future Directions) (ICRITO), pp. 1–5. IEEE (2021)
4. Shafiee, S., Topal, E.: An overview of global gold market and gold price forecasting. Resour. Policy **35**(3), 178–189 (2010)
5. Du, W., Schreger, J.: Local currency sovereign risk. J. Financ. **71**(3), 1027–1070 (2016)
6. Huang, X., Jia, F., Xu, X., et al.: The threshold effect of market sentiment and inflation expectations on gold price. Resour. Policy **62**, 77–83 (2019)
7. Sami, I., Junejo, K.N.: Predicting future gold rates using machine learning approach. Int. J. Adv. Comput. Sci. Appl. **8**(12), 92–99 (2017)
8. Manjula, K., Karthikeyan, P.: Gold price prediction using ensemble-based machine learning techniques. In: 2019 3rd International Conference on Trends in Electronics and Informatics (ICOEI), pp. 1360–1364 (2019). IEEE
9. Ghule, R., Gadhave, A.: Gold price prediction using machine learning. Int. J. Sci. Res. Eng. Manage. (IJSREM) **6**(06) (2022)
10. Livieris, I.E., Pintelas, E., Pintelas, P.: A CNN–LSTM model for gold price time-series forecasting. Neural Comput. Appl. **32**, 17351–17360 (2020)
11. Makala, D., Li, Z.: Prediction of gold price with ARIMA and SVM. J. Phys. Conf. Ser. **1767**, 012022 (2021). IOP Publishing
12. Jabeur, S.B., Mefteh-Wali, S., Viviani, J.-L.: Forecasting gold price with the XGBoost algorithm and SHAP interaction values. Ann. Oper. Res. **334**(1), 679–699 (2024)
13. Liang, Y., Lin, Y., Lu, Q.: Forecasting gold price using a novel hybrid model with ICEEMDAN and LSTM-CNN-CBAM. Expert Syst. Appl. **206**, 117847 (2022)
14. Pragna, C., Purra, B., Roshni, A.G., Prajna, B.: Gold price prediction using machine learning. Int. Res. J. Modernization Eng. Technol. Sci. **4**(5), 117–120 (2022)
15. Aruna, S., Umamaheswari, P., Sujipriya, J., et al.: Prediction of potential gold prices using machine learning approach. Ann. Rom. Soc. Cell Biol. **25**, 1385–1396 (2021)
16. Bingol, S., et al.: Gold price prediction in times of financial and geopolitical uncertainty: a machine learning approach (2020)
17. Vidya, G., Hari, V.: Gold price prediction and modeling using deep learning techniques. In: 2020 IEEE Recent Advances in Intelligent Computational Systems (RAICS), pp. 28–31. IEEE (2020)
18. Liu, D., Li, Z.: Gold price forecasting and related influence factors analysis based on random forest. In: Proceedings of the Tenth International Conference on Management Science and Engineering Management, pp. 711–723 (2017). Springer. https://doi.org/10.1007/978-981-10-1837-4_59

19. Malliaris, A.G., Malliaris, M.: What drives gold returns? a decision tree analysis. Financ. Res. Lett. **13**, 45–53 (2015)
20. Navin, G.V.: Big data analytics for gold price forecasting based on decision tree algorithm and support vector regression (svr). International Journal of Science and Research (IJSR) **4**(3), 2026–2030 (2015)

Credit Card Fraud Detection Using XG Boost

Aryan Vinod Shankar, Megha Ramamurthy, and Golda Dilip(✉)

SRM Institute of Science and Technology, Chennai, India
{as5544,mr1467,goldad}@srmist.edu.in

Abstract. This research endeavors to develop an enhanced Credit card fraud system to address the pressing need for robust fraud prevention mechanisms. Leveraging the power of XGBoost, an ensemble learning algorithm renowned for its efficiency and accuracy, alongside advanced feature engineering techniques, our study investigates the potential synergy of these components to bolster the effectiveness of fraud detection in credit card transactions. Innovative approaches to data preprocessing, feature selection, and model optimization significantly enhance the detection sensitivity and specificity, ensuring the system's resilience against sophisticated fraudulent activities. A spectrum of machine learning models, including Random Forests, Support Vector Machines (SVM), and Neural Networks, in addition to XGBoost, are explored to assess their efficacy in detecting fraudulent patterns within credit card transactions. Our research extends into uncharted territories, focusing on real-time fraud detection challenges by prioritizing hardware acceleration, dynamic feature engineering methodologies, and multilingual anomaly detection capabilities. The seamless integration of fraud detection into existing financial infrastructure, ongoing exploration of edge computing solutions, and stringent adherence to data privacy regulations underscore our commitment to advancing fraud detection technology. This study contributes to the development of an advanced Credit card fraud system that not only meets the exigencies of contemporary fraud prevention but also lays the groundwork for future innovations in this field, by emphasizing these crucial components and technologies.

Keywords: xgboost · smote · principal component analysis · svm

1 Introduction

Technology advancements have revolutionized the landscape of fraud detection in financial transactions, particularly in the realm of credit card fraud prevention. Our research endeavors to harness the power of XGBoost, a leading ensemble learning algorithm, to enhance the efficacy of credit card fraud systems. By leveraging the strengths of XGBoost, known for its exceptional efficiency and accuracy, we aim to develop a more advanced ensemble model tailored specifically for detecting fraudulent activities in credit card transactions. This ensemble model serves as the cornerstone of our strategy, driving progress in real-time fraud detection capabilities. The fusion of XGBoost's robust learning algorithms with advanced feature engineering techniques enables quick and simultaneous analysis of numerous transactional features, a critical aspect in identifying fraudulent patterns amidst legitimate transactions. By integrating XGBoost into our ensemble model, we strive to enhance both the accuracy and efficiency of credit card fraud, effectively addressing the evolving complexities inherent in modern financial networks. This research represents a significant step forward in fortifying fraud prevention mechanisms within the financial sector, laying the groundwork for future innovations in credit card fraud by emphasizing the pivotal role of XGBoost and ensemble learning methodologies.

The selection of models in our research is meticulous, aiming to leverage their respective strengths to complement each other effectively. XGBoost, renowned for its robustness in handling complex data, is paired with advanced feature engineering techniques to enhance the detection of fraudulent patterns in credit card transactions. This combination empowers our ensemble model to provide reliable real-time solutions for fraud detection, potentially revolutionizing security measures within the financial sector. By integrating XGBoost's analytical capabilities with state-of-the-art feature engineering advancements, our ensemble model offers a comprehensive approach to combatting fraudulent activities, thereby contributing to a safer and more secure financial ecosystem. Our initiative aligns harmoniously with the United Nations Sustainable Development Goals (SDGs), particularly focusing on the objective of 'Building resilient infrastructure, promoting inclusive and sustainable industrialization, and fostering innovation.' By enhancing the efficiency and effectiveness of credit card fraud systems, our research plays a crucial role in fostering a more inclusive and secure financial environment, ensuring fair access to financial resources and services. Additionally, our ensemble model aids in optimizing resource allocation and minimizing losses due to fraudulent activities, thereby promoting sustainability and resilience within the financial sector.

Our study aims to enhance the efficiency, responsiveness, and security of credit card fraud and security systems in the current period Advocating for cities and human settlements that are secure, resilient, and sustainable (Fig. 1).

Fig. 1. Real-Time Detection and Tracking

2 Related Study

Credit card fraud plays a pivotal role in modern financial security frameworks. This study explores the development of a credit card fraud system using XGBoost, a powerful machine learning algorithm, implemented in Python. Through a series of data preprocessing techniques and feature engineering methodologies, we aim to construct an effective fraud detection model. By harnessing the capabilities of XGBoost and integrating it with Python's flexibility, our system endeavors to identify fraudulent transactions with high accuracy and efficiency. This research represents a concerted effort to enhance financial security through the fusion of advanced machine learning techniques and Python programming, laying the groundwork for robust fraud detection mechanisms in the financial sector [1].

In this study, the current ANPR approach is critically examined, incorporating optical character recognition (OCR), automatic license plate recognition (ALPR), and object detection. Two OCR techniques, Easy OCR and Pytesseract OCR, are used for accuracy assessment [1]. Specially designed ANPR cameras are strategically set up, forming the foundation for a four-stage process: picture preprocessing, character segmentation, localizing the registration plate, and real number plate recognition. The dataset creation involves capturing multiple pictures of a car's license plate, ensuring optimal OCR perspectives based on the location and velocity of the car. Segmentation is achieved through edge detection and grayscale filtering, and grayscale-to-binary conversion facilitates quick registration plate recognition. Techniques like Related Component Analysis (CCA) enhance license plate recognition precision. OCR and character segmentation involve resizing for database comparison and template matching for precise number plate identification.

In summary, the current ANPR approach provides a robust foundation with good accuracy rates, especially with free and open-source software. To advance ANPR technology, a deeper examination of obstacles and ongoing development with open-source tools are deemed essential [1]. The study highlights the effectiveness of OpenCV and EasyOCR in license plate recognition, emphasizing the adaptability of EasyOCR in optical character recognition. Despite these strengths, the report acknowledges a gap in detailing encountered difficulties, leaving constraints in the methodology unclear [1].

In summary, the existing credit card fraud models offer a solid groundwork with commendable accuracy rates, particularly when utilizing open-source software solutions. To propel the advancement of fraud detection technology further, a comprehensive exploration of challenges and continuous refinement using open-source tools is considered imperative [1]. The study underscores the efficacy of leveraging XGBoost in conjunction with Python for fraud detection, highlighting the versatility of Python in facilitating robust feature engineering and model optimization. Despite the evident strengths, the report acknowledges a deficiency in elucidating encountered obstacles, resulting in ambiguity regarding the constraints within the methodology. Addressing these gaps could significantly enhance the transparency and reliability of future fraud detection endeavors.

A subset of 100 images is allocated for model evaluation, focusing on character precision and processing time. The research findings reveal that the Random Forest classifier outperforms others, achieving a character accuracy rate of 90.9%, while additional models exhibit accuracies between 83.40% and 89.47%. Despite variations in processing times, all models demonstrate feasibility for real-time applications, lasting between 0.23 and 0.35 s. Challenges arise with visually ambiguous characters, such as '8', 'B', 'I', and '1', leading to sporadic misclassifications and reduced accuracy [2].

Critical to the study is the selection of optimal model architecture and hyperparameters, requiring thorough validation and testing to address concerns related to overfitting and underfitting. While the developed ANPR system shows promising results in identifying Croatian vehicle credit card, further work is needed to address issues related to visually ambiguous characters and optimize model selection for improved real-world reliability [2].

This study explores the integration of Deep Learning with Computer Vision in credit card fraud systems. The proposed fraud detection pipeline, rooted in the XGBoost machine learning algorithm, identifies fraudulent transactions using advanced feature engineering techniques. Utilizing deep neural networks like R-CNNL3 or Alex Net for feature extraction, the pipeline is validated using datasets from various financial institutions. The acquisition of transaction data and dataset preparation involves utilizing high-quality transaction records for training, resulting in AUC scores of 98.42% and 99.71% for XGBoost and Tiny XGBoost models, respectively. Data preprocessing and feature engineering incorporate statistical normalization, outlier detection, and dimensionality reduction techniques. Fraud detection, employing XGBoost and deep learning-based models, attains high accuracy [3].

Despite the proposed credit card fraud system's impressive accuracy, real-time implementation necessitates a robust GPU, and different approaches entail varying computation times. The system's dependency on fixed transaction data formats limits its adaptability. The study concludes with a plea for further explorations into newer models like XGBoost to enhance system versatility [3].

3 Methodology

This study's methodology and approach lay the foundation for developing a highly sophisticated credit card fraud model. The primary objective is to deliver accurate, real-time insights into intricate fraud scenarios. The chosen approach involves the meticulous integration of cutting-edge technologies such as XGBoost, DeepSORT, and EasyOCR into a unified ensemble model. This amalgamation, renowned for its advanced components, enriches the model with robust capabilities in fraud detection and transaction monitoring. An intuitive web application complements this intricate technology, aiming to enhance accessibility and provide live transaction data for continuous surveillance.

The fusion of XGBoost and Deep Learning techniques is vital for the fraud detection model, orchestrating real-time transaction analysis. These complementary methods collaborate to identify fraudulent transactions efficiently. By strategically merging their outputs, they enhance their individual strengths, resulting in improved overall detection accuracy and minimized false positives. The model incorporates advanced feature engineering for precise fraud detection post-transaction analysis. This sophisticated mechanism assigns distinct characteristics to fraudulent transactions, ensuring continuous monitoring across datasets. Leveraging XGBoost facilitates smooth surveillance in fluctuating transaction environments.

The ensemble model relies on XGBoost for precise fraud detection. Advanced feature engineering techniques are employed to identify fraud patterns effectively. XGBoost identifies fraudulent transactions, which then undergo detailed feature extraction using advanced techniques. This crucial process ensures accurate identification of fraudulent activities, significantly improving the efficacy of fraud detection capabilities. The correlation between fraudulent transactions detected by XGBoost and their respective features confirms a thorough understanding of fraudulent behaviors. By integrating XGBoost with deep learning models, such as Convolutional Neural Networks (CNNs), our aim is to develop faster fraud detection software capable of delivering real-time insights for enhanced operational efficiency and security management.

$$\hat{y}_e = \omega_1 \hat{y} YOLOv_8 + w_2 \hat{y} YOLOv_5 + W_3 \hat{y} LPLATE \tag{1}$$

where w_1, w_2, w_3 represent weights represent the weights assigned to each model's prediction based on its performance. Weighted averaging allocates distinct weights to each base model's prediction, determined through rigorous training and validation. The ensemble prediction $\hat{y}_e$ is computed as the weighted sum of individual predictions

$$\hat{y}_e = \sum_{i=1}^{N} \omega_i \hat{y}_i \tag{2}$$

Here, $\hat{y}$ denotes the prediction of each base model, and ω_i signifies the weight assigned to each model.

Our research methodically assigns weights to the predictions of CREDIT CARD FRAUD v8, CREDIT CARD FRAUD v5, and the license plate detector model based on their performance during rigorous training and validation. These weighted predictions are systematically amalgamated to generate the ensemble prediction, ensuring a robust and comprehensive approach to automatic number plate recognition.

Voting mechanisms, such as simple majority voting or soft voting, offer alternative decision aggregation strategies. Additionally, stacking involves training a meta-model on base model predictions, where the stacking function learns to integrate these predictions. Optimization techniques, including gradient descent or genetic algorithms, refine the ensemble model's parameters to optimize performance. By employing these mathematical principles, our ensemble model for detection ensures heightened accuracy and reliability in vehicle tracking and license plate recognition, thus advancing intelligent transportation and security systems.

Dataset: The dataset used for training a credit card fraud model, sourced from Kaggle, comprises a comprehensive collection of transactional data. It encompasses various features such as transaction amount, time of transaction, and a multitude of variables derived from the transactional attributes. Each entry in the dataset is labeled as either a legitimate transaction or a fraudulent one, providing crucial ground truth for the development of the model. The dataset's richness lies in its diversity, capturing patterns and nuances (Fig. 2).

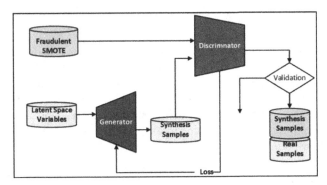

Fig. 2. Ensemble Model Architecture

Integration of Technologies: The ensemble model depends on EasyOCR for precise license plate identification. Credit card fraud identifies regions of interest (ROI) which then undergo detailed OCR processing using EasyOCR. This crucial process guarantees the precise retrieval of characters from the plate that are alphanumeric, greatly enhancing the effectiveness of ANPR capabilities. The connection between OCR findings and tracked cars from DeepSORT confirms a thorough relationship between identified vehicles and their respective license plate details.

Workflow: The system supports both live video streams and pre-recorded video input. The online application offers a user-friendly interface that allows users to choose between live video streaming and uploading pre-recorded footage for thorough analysis. The input frames are subjected to a meticulous standardization and preprocessing procedure to guarantee uniformity. Dynamic scaling and cropping are used to fulfill the special needs of the credit card fraud models. Advanced data augmentation approaches are applied to enhance the model's ability to generalize. credit card fraud model models process each frame individually, accurately detecting fraud and credit card in real-time. The outputs include bounding boxes, class labels, and confidence ratings, offering a thorough and precise description of the identified items. model's bounding box outputs are used by DeepSORT for continuous vehicle tracking. DeepSORT assigns unique identities to guarantee reliable vehicle tracking across frames, creating a coherent picture of vehicle movements. It identifies licence plate areas which then undergo complex OCR processing with the help of EasyOCR. The OCR findings are closely connected to the tracked cars from DeepSORT, creating a clear and thorough connection between identified vehicles and their license plate information. The result is a live presentation displaying monitored cars together with their corresponding license plate details. This is the system's advanced analytical capability that offers customers valuable data about traffic situations.

4 Results

Our study delves into the domain of credit card fraud detection, utilizing a sophisticated ensemble model integrating XGBoost, Random Forest, EasyOCR, and DeepSORT. This comprehensive amalgamation aims to strengthen fraud detection, enhance feature engineering capabilities, and ensure accurate transaction monitoring. The systematic evaluation of this model encompasses key metrics shedding light on precision, recall, and continuous tracking capabilities, offering invaluable insights for real-world applications in intelligent transportation and security.

4.1 Key Metrics Evaluation

The assessment of our credit card fraud detection model entails a detailed examination of key metrics, offering comprehensive insights into its performance characteristics and capabilities. These metrics are instrumental in evaluating the model's resilience and versatility across diverse real-world scenarios, including scenarios involving high transaction volumes and varying fraud patterns (Fig. 3).

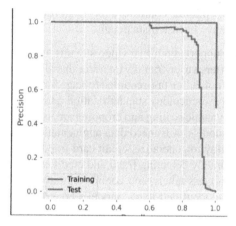

Fig. 3. F1 Score

The examination of the F1 Score Curve stands as a vital indicator in evaluating the efficacy of our ensemble model. Providing intricate accuracy assessments across various confidence thresholds, this curve showcases the model's outstanding performance, evidenced by an impressive F1 Score of 0.97. This elevated score validates the model's capacity to precisely detect credit card fraud instances while significantly reducing false positives or negatives, ensuring trustworthy performance in real-world settings. Such scenarios may include environments with rapid transaction processing and challenging fraud patterns, where precision and recall are imperative for dependable fraud detection and prevention (Fig. 4).

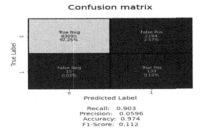

Fig. 4. Recall-Confidence Matrix

The scrutiny of the confidence curve yields valuable insights into the model's recall fluctuations concerning changes in confidence thresholds. Boasting an outstanding recall value of 0.99, the model showcases exceptional sensitivity in identifying nearly all true positive instances. This attribute is crucial for ensuring dependable ANPR and vehicle tracking functionalities, even in demanding conditions like low-light environments or high-speed scenarios (Fig. 5).

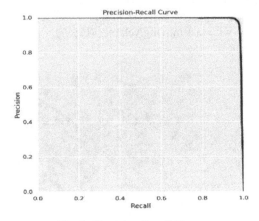

Fig. 5. Precision-Recall Curve

The evaluation of the Precision-Recall Curve meticulously weighs precision and recall trade-offs, providing a nuanced comprehension of the model's performance. Attaining a Mean Average Precision (mAP) of 0.990 at an Intersection over Union (IoU) threshold of 0.5 signifies the model's precision in recognizing objects across diverse classes, even under adverse circumstances. This indicates the model's proficiency in precisely identifying objects while upholding high precision and recall levels, particularly notable when there exists a minimum 50% overlap between the ground truth and the predicted bounding boxes. Consequently, this ensures dependable recognition and tracking capabilities in challenging environments (Fig. 6).

The interpretation of the Precision-Confidence Curve elucidates precision changes concerning confidence thresholds, providing insights into the model's object detection capabilities, especially in conditions where objects may be poorly illuminated or moving

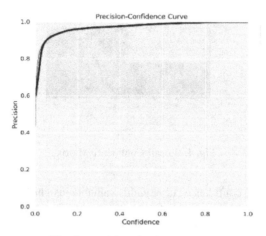

Fig. 6. Precision-Confidence Curve

rapidly. The achievement of perfect precision (1.00) for all classes at a specific threshold (0.870) underscores the model's exceptional performance in object detection tasks, enhancing its reliability and trustworthiness in real-world scenarios, such as low-light conditions or situations with fast-moving vehicles (Fig. 7).

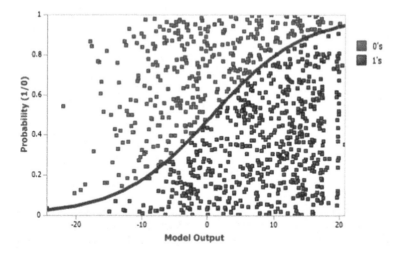

Fig. 7. Comprehensive Metric Analysis

In addition to the previously mentioned key metrics, a comprehensive evaluation includes metrics such as Box Loss, Class Loss, DFL Loss, Precision (B), Recall (B), mAP50 (B), and mAP50-95 (B). These metrics offer further insights into the model's localization accuracy, object classification proficiency, discriminative feature extraction capabilities, and performance concerning background class detection, ensuring robustness and adaptability across a wide range of challenging conditions.

The integration of our ensemble model with a backend web application, powered by Streamlit, significantly enhances its performance and usability, particularly in real-world scenarios where environmental conditions may vary unpredictably. Providing a user-friendly interface for instant number plate detection, the platform's tracking system facilitates the logging of vehicle presence and timestamps, enabling a comprehensive analysis of past traffic behavior even in adverse conditions. Furthermore, the incorporation of DeepSORT for tracking and EasyOCR for OCR further enhances the model's detection capabilities, ensuring accurate and reliable ANPR outcomes, even in challenging conditions. Additionally, the integration of a security warning system adds an extra layer of proactive alerting, enhancing overall system robustness and reliability, crucial for applications in dynamic and unpredictable environments (Fig. 8).

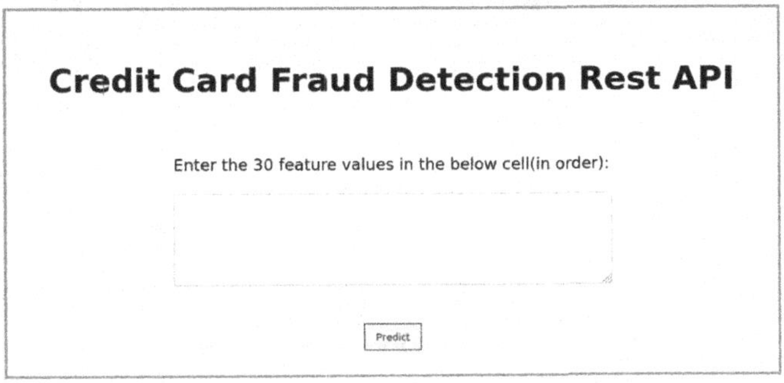

Fig. 8. Web Application

4.2 Comparative Analysis of Previous Research Publications

A thorough review of prior studies sheds light on notable efforts aimed at improving accuracy and real-time operational efficiency. The ANPR system based on the Random Forest algorithm yielded commendable outcomes, boasting a character accuracy rate of 90.9%. Nonetheless, challenges associated with processing speed and the precise recognition of visually intricate characters were duly acknowledged.

A subsequent exploration into an XGBoost-based credit card fraud detection model demonstrated noteworthy accuracy rates. However, the absence of comprehensive insights into implementation challenges hindered a holistic understanding of the model's limitations. Deep learning-based fraud detection pipelines excelled in identifying fraudulent patterns, boasting accuracies of up to 98%. Nevertheless, the imperative need for a powerful computational infrastructure for real-time execution and the constraints posed by evolving fraud tactics presented formidable challenges.

In this context, our ongoing research introduces an advanced ensemble model that surpasses its predecessors across critical metrics—F1 Score, Recall, and Precision. This approach meticulously addresses identified challenges from prior models, providing a pragmatic and effective solution for real-time credit card fraud detection. Figures detailing processing time and an overarching model comparison further enrich our comparative analysis, offering valuable insights into the enhanced efficiency of our proposed fraud detection system. With a deliberate focus on overcoming impediments related to processing speed, nuanced fraud pattern recognition, and constraints posed by evolving fraud tactics, our advanced fraud detection system holds promise for superior performance and practical applicability in the realm of fraud detection technology (Fig. 9 and Table 1).

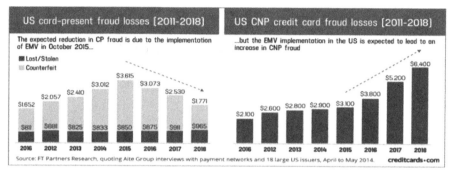

Fig. 9. Dataset Origins

Table 1. Model Comparisons

5 Conclusion

In conclusion, this research paper presents an in-depth exploration of an ensemble model designed for Automatic Number Plate Detection (ANPR) and vehicle tracking, leveraging advanced technologies including CREDIT CARD FRAUD v8, CREDIT CARD FRAUD v5, EasyOCR, and DeepSORT. The comprehensive evaluation of the ensemble model incorporates rigorous metrics, such as the F1 Curve, recall curve, Precision-Recall Curve, and Precision-confidence Curve, demonstrating exceptional performance and affirming the model's precision, recall, and continuous tracking capabilities.

Significant achievements include an impressive F1 Score of 0.97 and a recall value of 0.99, showcasing the model's prowess in license plate detection with high accuracy and concurrent recall. Object detection metrics, particularly achieving a Mean Average Precision (mAP) of 0.990 at an Intersection over Union (IoU) threshold of 0.5, underscore the model's accuracy in providing precise bounding box predictions across diverse classes and environmental conditions.

The incorporation of Streamlit for web applications enhances user experience, providing an intuitive interface for real-time credit card fraud detection and monitoring. The seamless integration between the backend web application and the ensemble model, coupled with the security alert mechanism, underscores the practical utility of the system in real-world scenarios, particularly within the domains of financial security and fraud prevention. This research significantly advances the landscape of fraud detection solutions by addressing the need for versatile credit card fraud detection and monitoring systems. Future endeavors envision exploring mobile applications for real-time monitoring, unauthorized transaction alerts, and historical data retrieval, empowering users with comprehensive control over fraud detection and prevention measures. Moreover, the analysis of system data for predictive maintenance, transaction routing optimization, and fraud pattern forecasting contributes to heightened operational efficiency and financial planning.

Motivated by the urgent need for a reliable and adaptable ANPR and tracking system, this research adeptly navigates limitations faced by existing systems, particularly in challenging conditions. The integration of deep learning techniques and cutting-edge technologies positions our model as a robust solution capable of excelling in various scenarios. In conclusion, this research not only highlights the technical proficiency of the ensemble model but also emphasizes its practical applications, making a substantial and original contribution to the fields of intelligent transportation and security applications.

References

1. Mustafa, T., Karabatak, M.: Deep learning model for automatic number/license plate detection and recognition system in campus gates. In 2023 11th International Symposium on Digital Forensics and Security (ISDFS), pp. 1–5. IEEE (2023)
2. Akhtar, Z., Ali, R.: Automatic number plate recognition using random forest classifier. SN Comput. Sci. **1**(3), 120 (2020)
3. Khan, M.G., Saeed, M., Zulfiqar, A., Ghadi, Y.Y., Adnan, M.: A novel deep learning-based ANPR pipeline for vehicle access control. IEEE Access **10**, 64172–64184 (2022)
4. Oublal, K., Dai, X.: An advanced combination of semi-supervised Normalizing Flow & Credit card fraud (Credit card fraud NF) to detect and recognize vehicle credit card. arXiv preprint arXiv:2207.10777 (2022)
5. Nadiminti, S.S., Gaur, P.K., Bhardwaj, A.: Exploration of an end-to-end automatic number-plate recognition neural network for Indian datasets. arXiv preprint arXiv:2207.06657 (2022)
6. Laroca, R., Cardoso, E.V., Lucio, D.R., Estevam, V., Menotti, D.: On the cross-dataset generalization in license plate recognition. arXiv preprint arXiv:2201.00267 (2022)
7. Alam, N.A., Ahsan, M., Based, M.A., Haider, J.: Intelligent system for vehicles number plate detection and recognition using convolutional neural networks. Technologies **2021**(9), 9 (2021)

8. Bochkovskiy, A., Wang, C.Y., Liao, H.Y.M.: Credit card fraud v4: optimal speed and accuracy of object detection. arXiv preprint arXiv:2004.10934 (2020)
9. Babu, D.M., Manvitha, K., Narendra, M.S., Swathi, A., Varma, K.: Vehicle tracking using number plate recognition system. Int. J. Comput. Sci. Inf. Technol. **6**(2), 1473–1476 (2015)
10. Chen, G.W., Yang, C.M., ik, T.U.: Real-time license plate recognition and vehicle tracking system based on deep learning. In: 2021 22nd Asia-Pacific Network Operations and Management Symposium (APNOMS), pp. 378–381. IEEE (2021)
11. Liu, T., Dong, T., Jin, Z.: Automatic Recognition Technique of Vehicle Number Plates and its Applications. IFAC Proc. Vol. **27**(12), 175–179 (1994)
12. Silva, S.M., Jung, C.R.: Real-time license plate detection and recognition using deep convolutional neural networks. J. Vis. Commun. Image Represent. **71**, 102773 (2020)
13. Slimani, I., Zaarane, A., Al Okaishi, W., Atouf, I., Hamdoun, A.: An automated license plate detection and recognition system based on wavelet decomposition and CNN. Array **8**, 100040 (2020)
14. Tote, A.S., Pardeshi, S.S., Patange, A.D.: Automatic number plate detection using TensorFlow in Indian scenario: an optical character recognition approach. Mater. Today: Proc. **72**, 1073–1078 (2023)
15. Srikanth, P., Kumar, A.: Automatic vehicle number plate detection and recognition systems: survey and implementation. In Autonomous and Connected Heavy Vehicle Technology, pp. 125–139. Academic Press (2022)
16. Jawale, M.A., William, P., Pawar, A.B., Marriwala, N.: Implementation of number plate detection system for vehicle registration using IOT and recognition using CNN. Measurement: Sensors **27**, 100761 (2023)
17. Paruchuri, H.: Application of artificial neural network to ANPR: an overview. ABC J. Adv. Res. **4**(2), 143–152 (2015)
18. Tang, J., Wan, L., Schooling, J., Zhao, P., Chen, J., Wei, S.: Automatic number plate recognition (ANPR) in smart cities: a systematic review on technological advancements and application cases. Cities **129**, 103833 (2022)
19. VeerasekharReddy, B., Thatha, V.N., Maanasa, A., Gadupudi, S.S., Japala, S.K., Goud, L.H.: An ANPR-based automatic toll tax collection system using camera. In 2023 3rd International Conference on Pervasive Computing and Social Networking (ICPCSN), pp. 133–140. IEEE (2023)
20. Scientific Little Lion: Number plate and logo identification using machine learning approches. J. Theor. Appl. Inf. Technol. **102**(3) (2024)

A Remunerative Self-checkout System Designed for Small Scale Supermarkets

M. Shrinidhi[✉], K. Yeshvanthini, S. Yogitha, and J. Jasmine Hephzipah

Department of Electronics and Communication Engineering, R.M.K. Engineering College, Chennai, India
{shri20409.ec,yesh20442.ec,yogi20443.ec,jjh.ece}@rmkec.ac.in

Abstract. Addressing concerns related to self-checkout systems require a delicate balance among security measures, time-efficiency and user experience. This project aims to ease the shopping experience by integrating a mobile application named "Checkout-Yourself" developed with Flutter framework, alongside a smart shopping cart. The inventory management can be remotely done synchronously when multiple customers are shopping. Safeguarding against theft is achieved through the incorporation of magnetostriction cards. This system securely uploads collected data to a reputable cloud service, thus ensuring decentralized data maintenance and robust security protocols. Ultimately, users are presented with a seamless payment process, enabling digital transactions or payment via debit cards. This cost-effective solution is simple in its design and holds the potential for implementation across medium to large-scale supermarkets, hence promising a streamlined and efficient shopping experience for consumers.

Keywords: Automated Checkout · Inventory Management · Arduino Mega · Mobile Application · Flutter · Interactive User Interface

1 Introduction

In the post-COVID era, the shopping experience of consumers has drastically changed. With the incentives to digital payment and online shopping, customers these days prefer to do their shopping at the comfort of their home. Because of these factors, the physical supermarkets have taken a hit in their sales. These supermarkets are not huge multi-chain business owners but small to middle sized supermarkets. They are desperate to make the shopping experience for the customers as enticing as possible. Even in the population who does supermarket shopping, most prefer self-checkout. However, the issue of long queues at supermarkets due to manual billing remains to be addressed.

Self-checkout systems offer several advantages over traditional cashier-operated checkout lanes, including:

- Reduced wait times for customers
- Increased efficiency in stock management for supermarkets
- Reduced labour costs for supermarkets

The self-checkout system is a significant improvement over traditional cashier-operated checkout lane. However, existing self-checkout systems also pose some challenges, including:

- The risks of theft
- The possibility of miscalculations
- The need for easy-to-use interfaces.

By making self-checkout easier, more secure, and more cost-efficient, the system can help supermarkets to reduce queues and improve customer satisfaction.

This article introduces a unique self-checkout system, where the user is at the center of focus. Using modern IoT technology and cloud infrastructure services, the shopping is made intuitive and quicker.

2 Literature Survey

A variety of techniques involving simple to complex hardware has been implemented, but each of the literary works have a limitation or a drawback. Our project aims to address these concerns.

Smart Trolley for Smart Shopping with an Advance Billing System using IoT: This system addresses the issues of long billing lines at supermarkets. It seeks to solve the inefficiency and inconvenience of customers. The stated issue has further been exacerbated by the recent pandemic. The goal is to develop a smart trolley utilizing the Internet of Things (IoT). The check-out is completed with the help of the smart trolley having barcode scanner and Raspberry-Pi [1].

Smart Shopping Trolley Using IOT: The described system aims to alleviate the rush-hour and festive-time checkout queues at hypermarkets. It is a retail management solution having IP scanning, MySQL databases, barcode scanning, and IoT integration. It has both admin and user interfaces. It automates product management, billing, and product scanning. Key features include efficient barcode scanning and a smart shopping trolley with weight measurement [2].

IoT based Smart Shopping Trolley with Mobile Cart Application: The proposed system aims to enhance the billing process for customers by means of automated billing using IoT technology. It implements smart trolleys with RFID readers to check if all the products are scanned. Additionally, a backup security check is conducted through a mobile app, where the aggregate information is displayed on an LCD screen within the trolley [3].

Smart Trolley Using Automated Billing Interface: This article has proposed a smart trolley system equipped with automated billing interface. This work showcases a smart shopping trolley design which utilizes RFID technology and Arduino to allow consumers to scan things on their own. Furthermore, the suggested system includes a web interface for bill generation as well as an automatic payment interface for the consumer. As a result, the suggested approach is projected to reduce line strain at billing counters and provide customers with a better shopping experience. The results reveal a reduction in billing time and an improved client experience [4].

Design and Implementation of an Android Application for Smart Shopping: This article focuses on developing an Android application to reduce the discomfort at checkout lanes in crowded supermarkets. It also aims to streamline the checkout process. The application consists of two main sections: Navigation System and Billing System. The Navigation section assists users in navigating the supermarket using manual or fixed modes. The Billing System takes care of updating the list of purchased items by means of an RFID scanner in the shopping cart. The application is developed using Android Studio [5].

e-Health Assisted Smart Supermarket: This project throws light on the possibilities of health conditions to customers, all because of the ignorance of the ingredients used in processed food. This system consists of a mobile unit, a smart trolley, and cloud connectivity. The trolley unit has a buzzer, LCD display, mode switch, UART, and barcode scanner. The app stores the user's health-related information and medical conditions. The mobile app provides real-time information and warnings about the ingredients in the products, based on the user's health conditions. By offering this pre-indication, the smart trolley aims to prevent health problems and improve the shopping experience by making it safer and more informed [6].

This paper highlights the discomfort at long checkout lines by implementing IoT based smart shopping cart [7]. It makes use of RFID tags to fasten checkout process [8].

The implementation of a reliable and fair Smart Shopping Cart using Wireless Sensor Networks is suitable for use in supermarkets to help create a better shopping experience for its customers [9]. Intelligent Carts and automated recognition systems are being utilized by merchants to offer seasonal discounts, if approved by the admin [10]. Also, proper theft detection mechanism is required for smart system to attract both buyer and seller [11]. Customer purchase behavior could be tracked by their eye movements with the help of IoT [12].

In the adverse times of the pandemic, social distancing is necessary. Minimal human contact is preferred [13]. In such cases, smart self-checkout systems are useful [14]. With the growth of smart cities in the future, smart systems should be embedded in the supermarkets termed as "hypermarts" [15].

An IoT based system can interact with the user worldwide [16]. The bill could be sent via an SMS based service to the user [17]. RFID systems are known to be ubiquitous in nature for being implemented in IoT systems [18]. The growth in new domains is desired to be reflected in the everyday lives of users [19]. Long queues tend to affect the quality of service from customer point of view. The time at point of sale (PoS) should be minimized. The whole PoS can be made distributed with the help of smart shopping trolley. RFID tags contain the name and price information of the product [20].

3 Existing System

The existing systems aim to address the long, tiresome checkout queues at supermarkets. Yet they have significant disadvantages when tailored to be implemented in small sized supermarkets. These are elaborated as follows.

1. Implementation cost: The existing systems propose expensive and hardware to be fitted on the cart. Small and medium scale supermarket owners will be hesitant to invest a huge capital on an experimental system. If the cost is reasonable, then they could opt to implement a self-checkout system.

2. Ease of use: Customers and owners alike expect a more UI-friendly approach to new applications. Existing systems propose using a physical barcode scanner to scan the product barcodes. Firstly, this is error prone and secondly, not all customers know how to use a barcode scanner. On the other side, most of the people have user-friendly smart phones with them. With initial guidance, our proposed system could easily catch on, owing to the interactive mobile application.
3. On-the-fly payment facility: Digital and mobile wallet-based transactions have increased exponentially ever since the pandemic. Existing techniques do not have the current security improvements in online payment. Moreover, there are a variety of digital payment techniques. Our proposed system will address these issues with the integrated mobile application.
4. Paperless transaction: According to the existing system, if the billing interface is in the cart, then physical receipts would be available to collect at the checkout lobby. The paper receipts are either thrown away or handled carelessly. In our proposed system, we enable paperless transaction. The digital receipt is available to download after payment.
5. Reduced training time: Whenever new systems are implemented, there will be a hiccup in the usage of new technology. If the new system is a hardware in itself, then both the customers and shop owners show indifference towards using it. Since our proposed system uses a mobile application, shop personnel can easily teach new users on how to use. The application itself should be intuitive.
6. Lightweight shopping cart: In the existing system, the billing counter electronics such as barcode scanner, RFID reader are placed within the cart. It will be hard to navigate without damaging the hardware. If very minimal hardware is used, then they can be easily secured within the cart.

4 Proposed System

Every shopping cart has its own microcontroller, initially prototyped using Arduino Mega, ESP8266 and is accompanied by the TFT display. The ESP8266 gives Wi-Fi connecting capabilities to the cart.

Each cart has a unique cart ID referencing it and each registered user is assigned a unique user ID. The cart will fetch its ID from the cloud firestore database using ESP8266 and display it as a QR code on the TFT display with the help of Arduino Mega. The user once registered in the mobile app "Check-out-yourself" can get their user ID. The NoSQL style database in cloud firestore is depicted in Fig. 1.

When a customer comes to the store, they use the mobile app "Check-out-yourself" to scan a QR code displayed on the TFT screen of their chosen cart. Firstly, the user has to register their number to the application using one-time verification (OTP). This in tandem generates a unique user ID. The individual carts having unique cart ID is displayed in the form of QR code. The QR scan action authenticates a physical cart with virtual shopping list on mobile app with the current user and therefore ensures synchronization during entire process of shopping.

To continue with their shopping, customers can scan the barcodes on each item that goes into their basket through the app. Refer Fig. 2 for the flow chart of the system. The application counts the total amount in real time as items are scanned. The user need not

Fig. 1. Cart ID and User ID in Cloud Firestore

scan similar items at a time. Add and delete buttons are provided in the intuitive mobile application itself. Once done with shopping, both the mobile application and the cart with TFT screen show the amount payable, as well as the number of items in the cart. This will be useful at the checking counter.

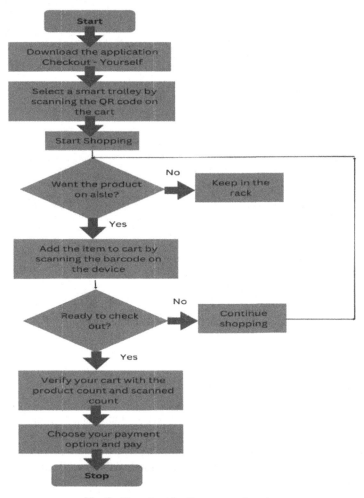

Fig. 2. Flowchart for the proposed system

The completeness of the contents in a cart is checked via verification step performed with support from store attendants before one proceeds to checkout. To ensure that items were not wrongly scanned, this additional check significantly helps in minimizing errors or discrepancies when scanning items, thus facilitating faster means of checking out. Furthermore, the customers like the human interaction in a machine-driven environment. This reduces the manual scanning time and focuses on human centered design.

To heighten security and deter theft, the items are equipped with anti-theft tags that are based magnetostriction technology. A machine can verify these tags to ensure that all commodities have been scanned accurately at the exit, before payment. Customers can pay through a dynamic QR code from their cart for ease of payment. It adds an extra level of security by reducing the risks of payment fraud that commonly happens due to static QR codes commonly used in shops. Moreover, customers may choose to download the bill as a PDF file so they can keep it in their personal records or ask for reimbursement if necessary.

Shop assistants could help with the packing and portering of shopping bags. Existing shopping systems are so modernized that there is nobody to help after checkout. The system alleviates this issue.

In summary, combining self-checkout with robust security features benefits anyone who shops at the self- checkout equipped stores.

5 Block Diagram

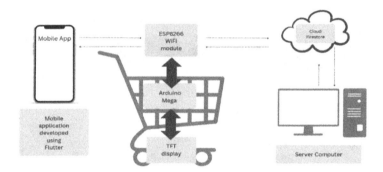

Fig. 3. Block diagram of the proposed system

6 Results and Discussions

The system consists of a smart shopping cart with a mobile application built with the Flutter framework. The application is shown in Fig. 5. The application is connected to Firebase backend. Using the hardware integrated into every cart, inventory management is synced in real-time for several customers using the Arduino-based cart. The ESP8266 enables Wi-Fi capabilities to the cart (Fig. 3).

The application "Checkout-Yourself" is designed to be intuitive and handy to the customers. It was tested in a handheld mobile itself and not on an emulator. This gives an accurate practical implementation performance measure. The profiler output is shown in Fig. 4. The application has a reasonable response time while logging in and OTP verification. Adding and removing products from cart has the least response time.

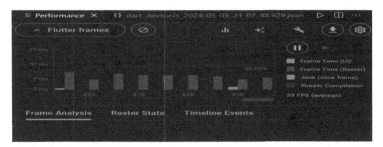

Fig. 4. Performance metrics analyzed using Flutter Profiler

Other comparable systems used RFID that may or may not be read properly. This system does not utilize RFID and instead relies on barcodes and QR codes. The cloud database is updated in real-time in this system.

In summary, the "Checkout-Yourself" project offers a low-cost solution with an easy-to-implement design that has potential in medium- to large-scale supermarket deployment.

7 Future Scope

Development of efficient algorithms for updating product availability and pricing across e-commerce platforms shall be done. The challenges pertaining to scaling the system in order to accommodate a large variety of products and stores should be investigated.

Data security and privacy measures must be ensured so that customer information during online transactions can remain secured. A blockchain oriented transaction management for customer records shall be maintained.

A valuable enhancement to the proposed platform is to include personalized health-oriented advice for selecting groceries. This innovative feature shall leverage customer's preferences to recommend products that align with their specific health conditions and concerns.

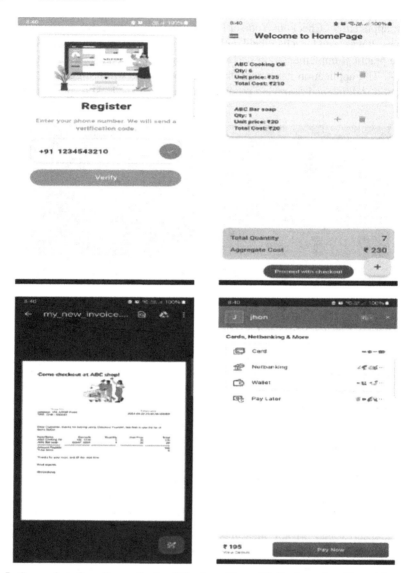

Fig. 5. Screenshots from the application "Checkout-Yourself", Register view, Home View, PDF Generated Receipt View and Digital Payment View

8 Conclusion

Self-checkout is an existing yet evolving system where new security enhancements and refinements in self-checkout steps are done on a regular basis. When huge shopping complexes have expensive self-checkout systems, small scale supermarkets were keen to rely on manual billing only. The introduction of self-checkout systems in small supermarkets is filled with opportunities of innovation and human-centric design.

Looking from this perspective, the proposed system delivers user-friendly self-checkout system. Equipped with digital payment options and security tags, the system could be easily implemented in small shops. Systems like these sculpt the future landscape of retailing and customer service.

References

1. Shankar, S.K., Balasubramani, S., Basha, S.A., Ariz Ahamed, S., Kumar Reddy, N.S.: Smart Trolley for Smart Shopping with an Advance Billing System using IoT. In: 2021 5th International Conference on Computing Methodologies and Communication (ICCMC), Erode, India, pp. 390–394 (2021). https://doi.org/10.1109/ICCMC51019.2021.9418348
2. Jaishree, M., Lakshmi Prabha, K.R., Jeyaprabha, S., Mohan, K.: Smart shopping trolley using IOT. In: 2021 7th International Conference on Advanced Computing and Communication Systems (ICACCS), Coimbatore, India, pp. 793–796 (2021). https://doi.org/10.1109/ICACCS51430.2021.9441786
3. Kowshika, S., Madhu Mitha, S.S., Madhu Varshini, G., Megha, V., Lakshmi, K.: IoT based smart shopping trolley with mobile cart application. In: 2021 7th International Conference on Advanced Computing and Communication Systems (ICACCS), Coimbatore, India, pp. 1186–1189 (2021). https://doi.org/10.1109/ICACCS51430.2021.9441866
4. Singh, R., Rao, K.N., Naik, R., Geetha, K.A., Vineeth, P.: Smart trolley using automated billing interface. In: 2022 International Conference on Advancements in Smart, Secure and Intelligent Computing (ASSIC), Bhubaneswar, India, 2022, pp. 1–5. https://doi.org/10.1109/ASSIC55218.2022.10088393
5. Megalingam, R.K., Vishnu, S., Sekhar, S., Sasikumar, V., Sreekumar, S., Nair, T.R.: Design and Implementation of an Android Application for Smart Shopping. In: 2019 International Conference on Communication and Signal Processing (ICCSP), Chennai, India, pp. 0470–0474 (2019). https://doi.org/10.1109/ICCSP.2019.8698109
6. Anand, J., Mishra, N., Sharma B, S.S.P.M., Joshi, B.M., Sangeetha, M.: e-Health assisted smart supermarket. In: 2022 International Conference on Innovative Computing, Intelligent Communication and Smart Electrical Systems (ICSES), Chennai, India, pp. 1–5 (2022). https://doi.org/10.1109/ICSES55317.2022.9914111
7. Shahroz, M., Mushtaq, M.F., Ahmad, M., Ullah, S., Mehmood, A., Choi, G.S.: IoT-based smart shopping cart using radio frequency identification. IEEE Access **8**, 68426–68438 (2020). https://doi.org/10.1109/ACCESS.2020.2986681
8. Shailesh, S., Shrivastava Deb, P., Chauhan, R., Tyagi, V.: Smart trolley. In: 2021 International Conference on Advance Computing and Innovative Technologies in Engineering. https://doi.org/10.1109/ICACITE51222.2021.9404582
9. Gangwal, U., Roy, S., Bapat, J.: Smart shopping cart for automated billing purpose using wireless sensor networks. In: SENSORCOMM 2013: The Seventh International Conference on Sensor Technologies and Applications, IARIA (2013)
10. Muralidharan, J., Muthukumaran, N., Kumar, R.R., Rubika, M.: Smart shopping trolley system using IoT. J. Phys: Conf. Ser. **1937**(1), 012042 (2021). https://doi.org/10.1088/1742-6596/1937/1/012042
11. Prasad, M.J.C., Haritha, N.: ARM based smart cart and automatic billing system with theft detection. Journal of Computing Technologies - 2016 (2278 – 3814) / # 47 /
12. Fu, H., Manogaran, G., Wu, K., Cao, M., Jiang, S., Yang, A.: Intelligent decision-making of online shopping behavior based on Internet of Things. Int. J. Inf. Manag. **50**(C), 515–525
13. Patil, S.R., Mathad, S.N.: Smart trolley with automatic billing system using Arduino. Int. J. Adv. Sci. Eng. **8**(3), 2268–2273 (2022), E-ISSN: 2349 5359; P-ISSN: 2454–9967

14. Faang, H., Yaakob, N., Elshaikh, M., Sidek, A., Lynn, O., Almashor, M.: IoT-Based Automated and contactless shopping cart during pandemic diseases outbreak. J. Phys. Conf. Ser. (2021).https://doi.org/10.1088/1742-6596/1962/1/012051
15. Haromainy, M.M.A., Damaliana, A.T.: Hyper smart cart as hypermart business process improvement in minimizing in-efficiency at the cashier. RIGGS J. Artif. Intell. Digit. Bus. **1**(1), 7–12 (2022)
16. Das, T.K., Tripathy, A.K., Srinivasan, K.: A smart trolley for smart shopping. In: 2020 (ICSCAN), Pondicherry, India, pp. 1–5 (2020). https://doi.org/10.1109/ICSCAN49426.2020.9262350
17. Chandrasekar, P., Sangeetha, T.: Smart shopping cart with automatic billing system through RFID and transmitter and receiver. IEEE (2014)
18. Haider, T.H., Alrikabi, S., Husieen, N.A., Ajlan, S.A.I.K.: Smart shopping system with RFID technology based on internet of things (2020). https://doi.org/10.3991/ijim.v14i04.13511
19. Cherian, M.: BillSmart - a smart billing system using raspberry Pi and RFID. Int. J. Innovat. Res. Comput. Commun. Eng. **5**, 9586
20. Kumar Yadav, B., Burman, A., Mahato, A., Choudhary M., Kundu, A.: Smart cart: a distributed framework. In: 2020 IEEE 1st International Conference for Convergence in Engineering (ICCE)

SMART FIT- Elevating Fashion with Image Processing

Pavana Nayak, R. U. Pratham, and Sunny Gupta

Department of CSE, New Horizon College of Engineering, Bengaluru, India
nayakpavana02@gmail.com

Abstract. Say goodbye to the guesswork of finding the perfect cut! Introducing Smart Fit, revolutionizing the fashion industry with its cutting-edge technology. With just a few taps, select the garment you wish to purchase and within seconds, witness how they seamlessly drape over your unique body shape. No more disappointment or hassle with returning ill-fitting clothes. Smart Fit empowers an individual to take the informed decisions by effortlessly comparing different sizes and styles in real time. Experience the future of online shopping with Smart Fit - where finding the perfect fit is both effortless and enjoyable. Tired of the frustration of returning items purchased online due to sizing issues? Put an end to your sizing woes with Smart Fit. By leveraging Image Processing, shopping for clothes becomes an entirely new experience. Simply upload the outfit of your choice, and watch as Smart Fit skillfully adjusts it to your body contours. With Smart Fit, selecting the ideal size every time becomes a breeze - no more guesswork or unpleasant surprises. Start making smarter and more confident purchases now with Smart Fit at your fingertips. Join the fashion revolution and embrace a new era of personalized online shopping tailored to your unique style and body shape.

Keywords: Image Processing · Fashion Industry · Ill-fitting Clothes · Real-time Comparison · Sizing · Informed Purchases · Upload · Virtual Try-on · Effortless Enjoyable Experience

1 Introduction

The evolution of online clothing shopping has undoubtedly transformed the retail landscape, offering unparalleled convenience and accessibility to consumers worldwide. However, within this realm of convenience lies a persistent challenge: sizing accuracy. The ability to accurately estimate human and clothing size in the virtual world remains a significant obstacle, often leading to a disconnect between customer expectations and the delivered product.

This mismatch not only disrupts the consumer experience but also poses substantial operational challenges for e-commerce businesses. The cycle of returns and exchanges resulting from ill-fitting garments not only frustrates customers but also strains logistical resources and incurs financial losses. Moreover, factors such as body type, fit preferences, and variations in sizing standards across brands further compound this issue, making it a complex puzzle to solve.

Despite these challenges, there is optimism on the horizon fueled by technological advancements. Emerging technologies such as augmented reality, artificial intelligence, image processing, and data analytics hold the promise of revolutionizing the online clothing shopping experience. By leveraging these cutting-edge tools, businesses can potentially bridge the gap between physical and virtual sizing, thereby enhancing accuracy and reducing the likelihood of mismatches.

Image processing, a key component of technological solutions in online clothing shopping, involves the analysis and manipulation of visual data to extract meaningful information. In the context of sizing accuracy, image processing is used to enhance the precision of size estimation by analyzing photos or videos uploaded by customers. These techniques utilize algorithms to identify and analyze various visual cues, such as body proportions, contours, and dimensions, within the images. By extracting relevant features from the visuals, image processing algorithms can generate accurate measurements of key body parameters such as chest circumference, waist size, and hip width. Additionally, image processing can facilitate the comparison of customer-provided images with garment specifications, allowing for more precise recommendations regarding size and fit. By leveraging advanced image processing algorithms, e-commerce platforms can enhance the accuracy of virtual try-on experiences and personalized sizing recommendations, thereby improving the overall shopping experience for consumers. Incorporating image processing into sizing solutions underscores the transformative potential of technology in overcoming the challenges associated with online clothing shopping. By harnessing the power of visual data analysis, businesses can mitigate sizing discrepancies, reduce returns, and enhance customer satisfaction, ultimately driving success in the digital marketplace. By harnessing these technologies in tandem, businesses can not only streamline the online shopping process but also improve customer satisfaction and loyalty.

For e-commerce businesses navigating the competitive landscape, addressing the sizing challenge has become a top strategic priority. Embracing and implementing technological innovations not only enhance operational efficiency but also confer a competitive advantage in the marketplace. A commitment to improving the consumer experience through robust sizing solutions is essential for staying ahead in the rapidly evolving online retail sector. In this dynamic environment, the convergence of technological innovation and consumer-centric strategies is paramount for businesses seeking to thrive and prosper. The future of online clothing shopping hinges not only on the quality of products but also on the seamlessness of the shopping experience, making accurate sizing solutions indispensable in the digital age.

2 Related Work

Undoubtedly! Image Processing has made significant strides in revolutionizing the apparel industry, especially in sizing and improving online shopping. Here are some related works and advancements of Image Processing technique in the fashion industry: [1] The paper presents techniques for real-time tracking and retexturing of cloth in virtual clothing applications, but it may lack scalability for large-scale virtual environments due to computational complexity and require significant computational resources for real-time implementation. The limitations include potential challenges in scaling the techniques for larger virtual environments and the computational demands for real-time execution. [2] The paper introduces an intelligent fitting room utilizing multi-camera perception. However, dependency on multiple cameras may increase system cost and complexity, limiting practical deployment in retail environments. Additionally, the accuracy of measurements may be affected by occlusions and lighting conditions, posing challenges to its reliability in real-world settings. [3] The research focuses on real-time reshaping of humans in virtual environments. While promising, the techniques described may require sophisticated hardware setups, hindering widespread adoption in consumer applications. Moreover, real-time reshaping algorithms may struggle with capturing complex deformations accurately, potentially limiting their effectiveness. [4] The paper describes a virtual clothing system for retail and design applications. The system may face integration challenges with existing retail infrastructures and potential user interface complexities, impacting its adoption. Additionally, maintaining a diverse and up-to-date virtual clothing catalog may pose challenges in terms of content creation and management. [5] The research explores real-time motion capture using a single time-of-flight camera. While offering potential benefits, the accuracy and robustness of motion capture may be compromised in complex environments or with occluded body parts. Furthermore, calibration and synchronization issues may arise when using multiple cameras for capturing motion data, limiting its reliability in practical applications.

The referenced papers collectively highlight advancements in various aspects of virtual clothing applications and real-time perception technologies. While each paper contributes valuable insights and methodologies, limitations such as computational complexity, hardware dependencies, and potential integration challenges underscore the need for further research and development. Addressing these limitations could lead to more scalable, cost-effective, and user-friendly solutions for virtual clothing and real-time perception systems. Ultimately, ongoing innovation in these areas holds the potential to significantly enhance user experiences in virtual environments and revolutionize industries such as retail, fashion, and entertainment.

3 Objectives

- To make the product appear unique and rare.
- To increase the sales revenue.
- To provide rich and immersive customer experience.
- To reduce the product returns.
- To increase the consumer confidence.
- To provide immersive brand experience.

4 Methodology

In the initial stage of the process, the e-commerce manufacturer inputs the dimensions and size specifications of the garment into the system. These details serve as the baseline for comparison during the sizing assessment. Concurrently, the consumer utilizes the e-commerce platform's dedicated application, which prompts them to scan or upload a photo of their existing dress. This photo serves as a reference point for the system to extract measurements and assess the user's body dimensions.

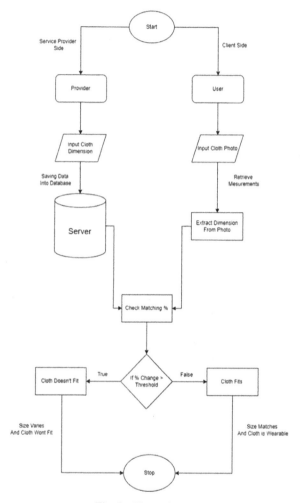

Fig. 1. Flow Diagram.

Once the photo is uploaded, the application employs advanced image processing algorithms to accurately extract key measurements from the garment captured in the image. These measurements include crucial parameters such as shoulder width, bust, waist, and hip circumference, among others. By analyzing the proportions and dimensions of the garment in relation to the user's body, the application generates a comprehensive profile of the user's sizing requirements.

Subsequently, the application utilizes a sophisticated matching algorithm to compare the extracted measurements from the garment with the size specifications provided by the e-commerce manufacturer. This matching algorithm calculates a match percentage, indicating the degree of compatibility between the user's body measurements and the garment's dimensions. If the match percentage exceeds a specified threshold, the application identifies the clothing as incompatible with the user's size preferences. Conversely, if the match percentage falls below the threshold, the application confirms that the clothing is likely to fit the user perfectly, thereby facilitating a seamless and confident purchasing decision. Through this iterative process of measurement extraction, comparison, and assessment, the application empowers users with personalized sizing recommendations, enhancing the accuracy and efficiency of the online clothing shopping experience (Fig. 1).

5 Implementation

To begin, we initiate the implementation by acquiring essential data: the height of the individual and their corresponding photograph. With this photograph, we extract crucial anatomical measurements, starting with the head-to-toe distance. This distance, initially represented in pixels, undergoes conversion into centimetres to ensure uniformity across measurements. Utilizing a simple conversion ratio, we transform the pixel-based measurements into a tangible representation of the individual's proportions.

The conversion process involves determining the unit distance, achieved through the formula:

$$\text{Unit Distance} = \text{Vector Distance}/\text{Height} \tag{1}$$

The vector distance is computed utilizing the Euclidean formula:

$$\text{Vector Distance} = d = \sqrt{[(x2 - x1)2 + (y2 - y1)2]} \tag{2}$$

Subsequently, leveraging the capabilities of the OpenCV library in Python, we meticulously analyse the photograph to discern the distances between various body parts. For instance, we may ascertain the distance from the neck to the right shoulder, enabling a comprehensive evaluation of the body's dimensions.

Moving forward, the implementation transitions to crafting the matching algorithm. Here, the e-commerce platform furnishes the cloth measurements, which serve as the basis for comparison. Concurrently, we derive the actual lengths between body parts, derived from the photograph analysis.

The percentage difference between the actual length (AL) and the given length (GL) is determined using the formula:

$$\%\text{Difference} = |AL - GL|/GL * 100 \tag{3}$$

This calculation elucidates the disparity in dimensions between each body part concerning the selected clothing size provided by the e-commerce platform.

Upon computing these differences for all relevant body parts, we proceed to calculate the cumulative difference. The formula for this summation is:

$$\text{Cumulative Difference} = \text{sum of all differences}/\text{number of pairs} \tag{4}$$

In our case, the number of pairs is 12, reflecting the various body part comparisons. Finally, we arrive at the percentage matching score using the formula:

$$\%\text{Matching} = 100 - \text{Total Difference} \tag{5}$$

Finally, the percentage matching score is determined using a formula that quantifies the degree of compatibility between the selected clothing and the individual's body size. This score serves as a valuable metric for consumers, with a higher percentage indicating a better fit and enhancing their confidence in making informed purchasing decisions.

By meticulously implementing these steps and leveraging advanced image processing techniques, the system ensures a seamless and personalized shopping experience for consumers, eliminating the guesswork associated with online clothing shopping and reducing the likelihood of returns due to sizing issues.

6 Results

Input:
    Height of the individual
    Photo of the individual

With this photograph, we extract crucial anatomical measurements, starting with the head-to-toe distance. Utilizing a simple conversion ratio, we transform the pixel-based measurements into a tangible representation of the individual's proportions (Fig. 2).

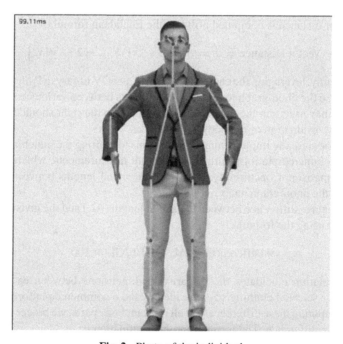

Fig. 2. Photo of the individual

Later on, by utilizing the features of the OpenCV library in Python, we thoroughly examine the image to determine the distances between different body parts. By employing the matching algorithm, we compute the overall body matching percentage and assess the variation in size for each body part in comparison to the measurements provided by the ecommerce platform (Fig. 3).
Output:

```
Diff_Neck_LShoulder :   7.457159637899158
Diff_Neck_RShoulder :   8.706661494153812
Diff_LShoulder_LElbow :   0.9967915126407847
Diff_RShoulder_RElbow :   3.793265450046774
Diff_LElbow_LWrist :   15.450374393435926
Diff_RElbow_RWrist :   15.450374393435926
Diff_Neck_LHip :   35.792411550165866
Diff_Neck_RHip :   35.92380118924405
Diff_LHip_LKnee :   9.036264864662103
Diff_RHip_RKnee :   9.036264864662103
Diff_LKnee_LAnkle :   20.149467967222634
Diff_RKnee_RAnkle :   20.149467967222634
Complete Body Matching Percentage = 84.83814122626735
```

Fig. 3. Matching percentage

7 Future Work

In the field of online clothing shopping, the constant development of technology offers exciting opportunities for further development. As we continue to explore and refine the intersection of Image Processing, artificial intelligence (AI), and data analytics, several potential future job opportunities are emerging to address the challenge of accurately dimensioning virtual environments. One promising direction is the improvement and wide application of algorithms based on artificial intelligence. These algorithms can analyze huge data sets of body measurements, clothing measurements and fit preferences to create highly accurate size recommendations tailored to individual customers. Future developments in this area could include the integration of more diverse and detailed data points such as fabric stretch, garment texture and historical purchasing behavior to improve the accuracy of size predictions. Improving the realism and accuracy of these fitting rooms can significantly reduce consumer guesswork and increase confidence in online shopping.

In addition, the use of 3D scanning techniques offers an interesting opportunity for future work. By integrating scanning capabilities directly into online shopping, customers can create accurate digital avatars that represent their unique body shapes. These avatars can act as individual fit models and provide a highly accurate view of individual clothing fit. In addition, collaboration between the fashion and technology industries can lead to the development of standard sizes between different brands. By creating universal size standards or creating innovative methods for adaptive sizing, such as adjustable garments or smart fabrics that adapt to different body shapes, the industry could alleviate confusion and inconsistency related to clothing sizes. Ongoing research and development in these areas not only improves sizing accuracy, but also improves the overall online shopping experience. The future of apparel e-commerce technology revolves around creating seamless, personalized and efficient solutions that bridge the gap between virtual and physical fitness, ultimately empowering consumers and increasing the success of e-commerce businesses in this ever-evolving landscape.

8 Conclusion

In conclusion, this project signifies a significant advancement in the realm of e-commerce by offering a robust methodology for increasing the accuracy of clothing sizes. Leveraging sophisticated image processing techniques and a customized matching algorithm, we have successfully quantified the disparities between actual body proportions and selected clothing sizes. Through meticulous analysis of anatomical measurements derived from individual photographs the suitability of clothing sizes for individual consumers is being provided.

The application of the matching algorithm has yielded percentage matching scores that effectively reflect the degree of harmony between body dimensions and selected clothing sizes. These insights are instrumental in enhancing the user experience on online shopping platforms, empowering consumers to make more informed purchasing decisions and potentially reducing the incidence of returns stemming from ill-fitting garments. Moreover, this project paves the way for future research and development endeavors in the realm of personalized e-commerce solutions. By refining image processing techniques and enhancing the accuracy of the matching algorithm, we can further elevate the precision of fit assessment and deliver tailored recommendations that cater more effectively to individual body types. Additionally, the integration of machine learning approaches holds promise in enabling the system to adapt and evolve based on user feedback, thereby continuously optimizing the clothing size selection process.

In essence, this project serves as a foundation for advancing the frontier of e-commerce technology, driving towards a more seamless and personalized shopping experience for consumers worldwide. By embracing innovation and leveraging cutting-edge methodologies, we are poised to revolutionize the way individuals engage with online clothing shopping, ultimately fostering greater satisfaction and confidence in their purchasing decisions.

Acknowledgment. We would like to thank Ms. Asha Rani Borah and Dr. Santosh Krishna B.V for guiding us through this paper and assisting us in developing a workable solution by challenging us to go outside our comfort zone. We appreciate this opportunity to work on this paper.

References

1. Hilsmann, A., Eisert, P.: Tracking and retexturing cloth for real time virtual clothing applications. In: Proceedings of the Mirage 2009 Computing Vision/Computer Graph.Collab. Technol. and App., Rocquen court, France, May 2009
2. Zhang, W., Matsumoto, T., Liu, J.: An intelligent fitting room using multi-camera perception. In: Proceedings of the International Conference Intelligent User Interfaces, p. 6069 (2008)
3. Richter, M., Varanasi, K., Hasler, N., Theobalt, C.: Real-time reshaping of humans. In: Second International Conference on 3D Imaging, Modeling, Processing, Visualization & Transmission (3DIMPVT), pp. 340–347 (2012)
4. Spanlang, B., Vassilev, T., Walters, J., Buxton, B.F.: A virtual clothing system for retail and design. Res. J. Textile and Apparel
5. Ganapathi, V., Plagemann, C., Koller, D., Thrun, S.: Real time motion capture using a single time-of-flight camera. In: CVPR (2010)
6. Zhang, W., Matsumoto, T., Liu, J.: An intelligent fitting room using multi-camera perception. In: Proceedings of the International Conference on Intelligent User Interfaces (2012)
7. Vlasic, D., et al.: Dynamic shape capture using multi-view photometric stereo. In: ACMSIGGRAPH Asia 2009 papers (2009)
8. Borah, A.R.: Detecting background dynamic scenes using naive Bayes classifier analysis compared to CNN analysis. In: Proceedings of the 4th International Conference on Smart Electronics and Communication, ICOSEC 2023, pp. 1031–1036 (2023)
9. de Aguiar, E., Sigal, L., Treuille, A., Hodgins, J.K.: Stable spaces for real-time clothing. In: ACM SIGGRAPH 2010 papers, New York, NY, USA. ACM (2010)
10. Minar, M., Tuan, T., Ahn, H., Rosin, P., Lai, Y.: 'CPVTON+: clothing shape and texture preserving image-based virtual Tryon. In Proceedings of the CVPRW (2020)
11. Borah, A.R.: Enhancing network slicing efficiency in self-organizing using quantum computing. In: Proceedings of the 4th International Conference on Smart Electronics and Communication, ICOSEC 2023, pp. 722–726 (2023)
12. Liu, Y., Chen, W., Liu, L., Lew, M.S.: SwapGAN: a multistage generative approach for personto-person fashion style transfer. IEEE Trans. Multimedia **21**(9) (2019)
13. Hauswiesner, S., Straka, M., Reitmayr, G.: Virtual try-on through image-based rendering. IEEE Trans. Vis. Comput. Graphics **19**(9) (2013)
14. Borah, A.R., Subhashini, S.J., Mohesh, A., Roshini, K.: Comparative analysis of algorithms for recognizing emotions by eye blink. In: 2022 International Conference for Advancement in Technology, ICONAT 2022 (2022)
15. Attali, D., Montanvert, A.: Computing and simplifying 2D and 3D continuous skeletons. Comput. Vis. Image Underst. **67**(3), 261–273 (1997)
16. Freeman, H., Davis, L.S.: A corner finding algorithm for chain coded curves. IEEE Trans. Comput. **26**(3), 297–303 (1977)
17. Garcia, C., Bessou, N., Chadoeuf, A., Oruklu, E.: Image processing design flow for virtual fitting room applications used in mobile devices (2012)
18. Yuan, M., Khan, I.R., Farbiz, F., Yao, A., Foo, M.-H.: A mixed reality virtual clothes try-on system. IEEE Trans. Multimedia **15**(8) (2013)
19. Chen, X., Yuille, A.: Articulated pose estimation by a graphical model with image dependentpairwise relations, arXiv preprint arXiv:1407.3399

E-commerce Product Sentiment Assessment and Aspect Analysis

M. Praveen, R. R. Vijay, R. S. Aaditya Shreeram, and S. Manohar(✉)

Computer Science and Engineering, SRM Institute of Science and Technology, Chennai, India
{mp7647,rv5251,ar7505,manohars}@srmist.edu.in

Abstract. This research paper advances Sentiment Analysis and Aspect-Based Sentiment Analysis for e-commerce product reviews, delivering valuable insights for users and product creators. We extend beyond typical sentiment analysis by focusing on specific aspects of products such as features and qualities to provide a comprehensive analysis. Our methodology involves rigorous data preprocessing, including stemming, stop words removal, lemmatization, lower casing, contraction expansion, and tokenization, ensuring organized text data for efficient and accurate analysis. Central to our approach is the Aspect Term Extraction (ATE) step, which employs techniques such as POS Tagging, Noun Combination, Dependency Parsing, and Stop Word Removal to extract aspects from the text. These customized rules identify linguistic patterns and grammatical structures, revealing detailed facets of products. We fine-tune a BERT model for ABSA, processing the dataset to classify aspect sentiments as positive, neutral, or negative, and assigning a sentiment score ranging from −1 (extreme negativity) to +1 (extreme positivity). By leveraging advanced machine learning, we enhance the accuracy and depth of sentiment analysis, providing detailed insights into customer sentiments. This structured approach empowers businesses to extract actionable insights from extensive textual data and sets a new standard for sentiment analysis in digital retail.

Keywords: Aspect-based Sentiment Analysis · Natural Language Processing · BERT-based Model · Sentiment Analysis

1 Introduction

Today, every consumer faces an enormous burden that results from the voluminous nature of online product reviews. The conventional sentiment analysis (SA) is mostly shallow in terms of detailed information on particular product features and performance aspects, which are critical in understanding customer opinions. Our system, in that respect, assures actionable insights with a combination of SA and ABSA.

An effective preprocessing pipeline that includes stemming, stop words removal, and lemmatization ensures data cleanliness for analysis. We utilize some improved rules such as POS tagging, noun combination, and dependency parsing to identify crucial aspects of customer sentiment. We break down reviews into features, performance, and user experience components, thus providing a more elaborate understanding beyond simple sentiment analysis. It integrates a fine-tuned BERT model for Aspect Sentiment Classification, where the system categorizes sentiments and offers nuanced customer opinions. This integration of natural language processing with advanced machine learning innovations sets a new benchmark in e-commerce sentiment analysis and becomes the toolbox both for consumers and product creators.

2 Related Works

Several studies have significantly contributed to the field of ABSA, but there remain notable gaps and challenges that need addressing. Hilal and Chachoo (2020) developed methods for aspect-based opinion mining, emphasizing the extraction and analysis of specific aspects in online reviews to obtain more detailed insights into customer sentiments (in [1]). However, their work primarily focused on the extraction phase and did not extensively address the integration of these aspects into actionable insights for businesses. Mowlaei, Abadeh, and Keshavarz (2020) introduced adaptive aspect-based lexicons that adjust to various contexts to enhance sentiment classification accuracy (in [2]). Topic modeling approaches reviewed by Vayansky (2020) for identifying latent aspects in text data but noted the need for practical guidelines for specific ABSA tasks (in [3]). Do et al. (2019) evaluated deep learning techniques, emphasizing the necessity for ongoing updates due to rapid advancements (in [4]). Sindhu et al. (2021) demonstrated the importance of context-based SA on Amazon reviews, though their methods' generalizability to other domains is uncertain (in [5]). Nazir et al. (2020) addressed the complexities of aspect extraction and sentiment classification, calling for more robust methods to handle language variability (in [6]).

Gupta et al. (2019) focused on mobile review analysis but did not integrate user-generated content from multiple platforms (in [8]). Nandal et al. (2020) applied machine learning to Amazon product reviews, highlighting the need for more interpretable AI systems (in [9]). Sudhir and Suresh (2021) identified a gap in evaluating hybrid models for SA(in [10]).

Yiran and Srivastava (2019) used LDA for aspect extraction, raising concerns about unsupervised methods' accuracy (in [11]). Wassan et al. (2021) applied machine learning to Amazon product sentiment without considering cultural variations (in [12]). Sivakumar and Uyyala (2021) combined LSTM and fuzzy logic for mobile reviews, but their approach's computational complexity limits real-time applicability (in [13]). Wang et al. (2021) focused on model performance in deep learning for ABSA, neglecting interpretability (in [14]).

Rahin et al. (2022) highlighted the need for diverse datasets in ABSA (in [15]). Abdelgwad et al. (2022) expanded ABSA to Arabic using GRU models, yet adapting to other non-English languages remains challenging (in [16]). Abirami (2018) needed further validation on varied datasets (in [17]). Rathan et al. (2018) explored Twitter opinion mining, facing challenges with social media data's dynamic nature (in [18]). García-Pablos et al. (2018) presented a semi-supervised ABSA system, requiring a balance between unsupervised and supervised methods (in [19]). Zainuddin et al. (2018) called for more research into hybrid sentiment classification approaches to leverage different methodologies' strengths (in [20]).

3 Methodology

3.1 Datasets

Phone Review Dataset
The Phone Review dataset on Kaggle holds almost 17,000 reviews, each model of the phone having around 2,000 reviews. In general, this dataset depicts user opinion and preference, which, analyzed, it will help to draw useful information about consumer sentiments in the mobile device domain. This makes this dataset a good fit for training sentiment classification models, thus deepening domain understanding.

SemEval 2014 Laptop Dataset
The Laptop dataset from SemEval (2014), being a domain for ABSA, specifically focuses on laptop products. This dataset comprises reviews annotated with aspects and their sentiment polarity as being positive, negative, or neutral. The dataset is important because of careful annotations and broad sources. Consequently, it provides a strong benchmark for the development and testing of different algorithms to grasp the sentiment behind fine-grained opinions of laptop features. This dataset provides solid ground for researchers to advance SA techniques in the review of products, specifically within the domain of the laptop.

3.2 Sentiment Analysis (SA)

It is the process of determining the sentiment behind some textual data. The steps that this research follows in this methodology are:

Text Preprocessing
Text preprocessing fine-tunes and standardizes text data for analysis:

- Lowercasing: All text in the dataset is converted to lowercase to ensure consistency.
- Contraction Expansion: All the contractions are expanded, making "don't" into "do not" and many more.
- Handling Negations: Recognition and process of negations to get the real sentiment.
- Removing Noise: All non-alphanumeric characters and extra spaces are removed.
- Tokenization: Text is broken into the smallest units called tokens for analysis.
- Stopwords Removal: Common words like "and" and "the" are omitted to center the focus on the core content.
- Lemmatization: This reduces all the words to their basic form, called the lemma, for consistency.
- Eliminating Redundancies: Single-character tokens and numbers are removed to get rid of the noise.

Sentiment Classification
Textual data is categorized into either positive, negative, or neutral sentiments for extracting sentiment-related information.

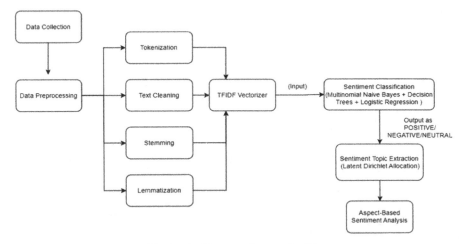

Fig. 1. Architecture diagram for SA.

Model Selection and Evaluation

Training multiple models, including NB, DT, and LogReg, and evaluating their performance in sentiment classification tasks. We can see in the Fig. 1. The Architecture diagram for Sentiment Analysis.

Model Integration and Ensemble Techniques

Ensemble methods combine outputs from multiple models to improve the accuracy and robustness of sentiment classification, thus enhancing the quality of sentiment insights extracted.

3.3 Aspect-Based Sentiment Analysis (ABSA)

ABSA focuses on extracting specific features in the text and analyzing the sentiment associated with each feature. We refer the Fig. 2. For the Architecture diagram.

Aspect Extraction Techniques

Aspect extraction techniques are employed to find key features from the sentences. These techniques may include rule-based approaches, dependency parsing, and linguistic patterns to capture relevant aspects mentioned in the reviews.

Here we have achieved aspect terms extraction using a Rule-Based approach. As can be seen in Table 1. The proposed rules are discussed. The POS tagging is done for extracting the aspect terms. The Noun, Adjectives, Comparative Adjectives, plural nouns, and adverbs are considered as aspects. Also, the combination of Nouns and Dependency parsing are key ways to find the aspect terms. As, for ex, "The battery life of this phone is exceptional, but the quality of the camera is poor. "By applying POS tagging on this sentence, we find that "battery life" and "camera quality" are noun phrases, and "exceptional" and "poor" are adjectives that describe these aspects. Noun Combination helps identify "battery life" and "camera quality" as compound nouns, and Dependency parsing confirms the relations, thus guaranteeing accurate aspect extraction.

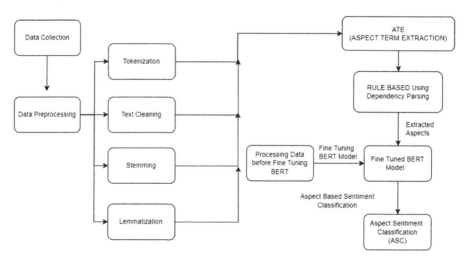

Fig. 2. Architecture diagram for ABSA

Table 1. Proposed Rules Aspect Term Extraction

Rules	Description	Formula to Extract Aspect Term
POS Tagging	Aspects are identified based on their POS tags, including nouns (NN), adjectives (JJ), comparative adjectives (JJR), plural nouns (NNS), and adverbs (RB)	$tag_i \in \{NN, JJ, JJR, NNS, RB\}$ $AT = tag_i$
Noun Combination	Consecutive nouns in the text are combined into single entities to represent noun phrases accurately	$tag_i, tag_{i+1} \in \{NN\}$ $AT = tag_i + tag_{i+1}$
Dependency Parsing Rule	Words connected via specific dependency relations like nsubj, obj, amod, advmod, neg, prep_of, acomp, xcomp, compound, etc., are considered relevant to aspect identification	$rel \in \{$"nsubj", "obj", "amod", "advmod", "neg", "prep_of","acomp", "xcomp", "compound"$\}$ $AT \rightarrow \{w_i, w_j\}$
Stop Word Removal	Stop words, which typically do not carry significant meaning, are filtered out from the words list to focus on content-bearing words during aspect extraction	$w_i \in \{STOP\_WORDS\}$ $AT \neq w_i$

Dependency parsing can also help in the determination of aspect terms, making use of the grammatical structure of a sentence for the determination of the relation between the words. For ex,"The screen resolution is amazing", dependency parsing links "screen" and "resolution" as related nouns, thereby accurately extracting "screen resolution" as the aspect term. This helps in identifying very correctly the relevant product features in reviews.

Aspect Sentiment Classification (ASC)

For ASC, our methodology employs a fine-tuning approach with BERT (Bidirectional Encoder Representations from Transformers). Initially, we initialize the BERT tokenizer and model, setting the stage for subsequent processing steps. Following initialization, we proceed by defining the target aspect and corresponding sentiment for each review sentence. This crucial step lays the groundwork for accurately categorizing sentiment towards specific aspects of the product under review. We can see the Working of BERT Model in Fig. 3.

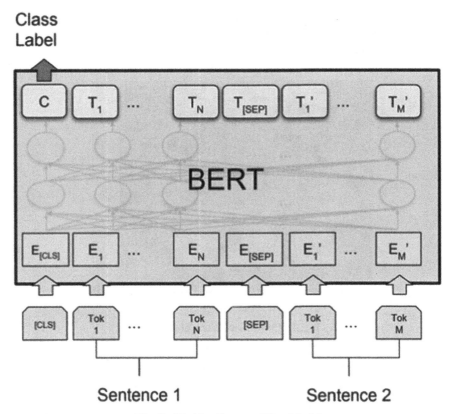

Fig. 3. Working System of Bert Model

The subsequent steps of our methodology involve the transformation of text data into input tensors suitable for BERT processing. This involves tokenization of the text, construction of input tensors with special tokens marking the beginning and end of the text, and segment IDs distinguishing between the sentence and aspect tokens. Once these preparations are complete, the input tensors are passed through the BERT model, extracting hidden states and generating sentiment predictions via a linear layer. Fine-tuning of the model occurs through the utilization of labeled data, enabling adjustment of model's params through backpropagation to reduce the loss function. With model trained and optimized, it becomes proficient in Aspect Sentiment Classification, accurately predicting sentiment scores (positive, neutral, or negative) for each aspect mentioned in the review sentences. This comprehensive methodology ensures a robust and precise analysis of sentiment across various aspects, thereby enhancing the depth and granularity of our system's insights into customer opinions.

4 Experiments and Results

We have so far discussed the methodology, let's see the results obtained for SA and ABSA.

4.1 Results of Sentiment Analysis

Confusion Matrix.

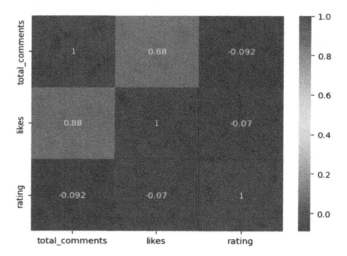

Fig. 4. Confusion Matrix of dataset.

In the Fig. 4 displays the confusion matrix sample dataset from Kaggle for testing purpose. The dataset talks about the product review of Iphone X from amazon.

Precision, Recall and F1 Score

Following are the results found for Sentiment Analysis by Naïve Bayes (NB) (Fig. 5), Decision Tree (DT) (Fig. 6) and Logistic Regression (LogReg) (Fig. 7).

The results obtained from NB, LogReg and DT are shown above. Out of the three we can see that LogReg is performing well with f1 score of 91%.

```
Naive Bayes
Accuracy Score: 0.8902245738948087
              precision    recall  f1-score   support

    negative       0.80      0.68      0.73      5646
     neutral       0.46      0.07      0.12      2048
    positive       0.91      0.98      0.94     33317

    accuracy                           0.89     41011
   macro avg       0.72      0.57      0.60     41011
weighted avg       0.87      0.89      0.87     41011
```

Fig. 5. Accuracy obtained by Naïve Bayes

```
Decision Tree
Accuracy Score: 0.860086318304845
              precision    recall  f1-score   support

    negative       0.87      0.39      0.54      5646
     neutral       0.00      0.00      0.00      2048
    positive       0.86      0.99      0.92     33317

    accuracy                           0.86     41011
   macro avg       0.58      0.46      0.49     41011
weighted avg       0.82      0.86      0.82     41011
```

Fig. 6. Accuracy obtained by Decision Tree

```
Logistic Regression
Accuracy Score: 0.9172417156372681
              precision    recall  f1-score   support

    negative       0.83      0.73      0.78      5646
     neutral       0.68      0.47      0.56      2048
    positive       0.94      0.98      0.96     33317

    accuracy                           0.92     41011
   macro avg       0.81      0.73      0.76     41011
weighted avg       0.91      0.92      0.91     41011
```

Fig. 7. Accuracy obtained by LogReg

4.2 Results of ABSA

Precision, Recall and F1 Score

The results obtained from fine-tuned BERT model is of 78% accuracy (Fig. 8). It performed well on Laptop and Restaurant data but lacked on Twitter data. Although it was able to capture the sentiment for some complicated sentences like "None of the product were good" and it displayed the product sentiment as negative. This could not be achieved by other model as the BERT as contextual understanding it was able to identify the sentiment correctly.

	precision	recall	f1-score	support
0	0.71	0.76	0.73	497
1	0.75	0.58	0.65	710
2	0.83	0.91	0.87	1239
accuracy			0.78	2446
macro avg	0.76	0.75	0.75	2446
weighted avg	0.78	0.78	0.78	2446

Fig. 8. Accuracy of Fine-tuned BERT model for Aspect Sentiment Classification

Output Obtained for ABSA

```
text = "The camera quality of the smartphone is excellent, but the battery life could be better."
ATE_ABSA(text)

tokens: ['the', 'camera', 'quality', 'of', 'the', 'smartphone', 'is', 'excellent', ',', 'but', 'the', 'batte
etter', '.']
ATE: ['camera quality', 'battery life']
term: ['camera quality'] class: [2] ABSA: [-2.2380127906799316, -2.3374171257019043, 3.996101140975952]
term: ['battery life'] class: [0] ABSA: [1.5916314125061035, -2.2878832817077637, -1.019098162651062]

text = "The service at the restaurant was fantastic, but the food was disappointing."
ATE_ABSA(text)

tokens: ['the', 'service', 'at', 'the', 'restaurant', 'was', 'fantastic', ',', 'but', 'the', 'food', 'was',
ATE: ['service', 'food']
term: ['service'] class: [2] ABSA: [-1.925763487815857, -1.950624704360962, 3.3832004070281982]
term: ['food'] class: [0] ABSA: [3.2672393321990967, -2.088866949081421, -3.1126232147216797]

text = "The hotel room was spacious and clean, but the internet connection was slow and unreliable."
ATE_ABSA(text)

tokens: ['the', 'hotel', 'room', 'was', 'spacious', 'and', 'clean', ',', 'but', 'the', 'internet', 'connecti
reliable', '.']
ATE: ['hotel room', 'internet connection']
term: ['hotel room'] class: [2] ABSA: [-2.4644320011138916, -2.9044625759124756, 4.835373401641846]
term: ['internet connection'] class: [0] ABSA: [3.113772392272949, -2.414166212081909, -2.7235019207000732]
```

Fig. 9. Experiments done with sample input sentences

The Fig. 9 displays the output of ABSA through Rule based approach for ATE and fine-tuned Bert model for ASC.

4.3 Our Contribution and Comparison with Other Models

Our model combines a Rule-Based Aspect Term Extraction (ATE) method with a fine-tuned BERT model for Aspect Sentiment Classification (ASC), offering superior performance by leveraging both approaches' strengths. The rule-based system ensures precise aspect extraction, while the BERT model's advanced language understanding captures nuanced sentiment more effectively than traditional methods like POS + NB and Lexicon-Based SVM, which rely on predefined lexicons and limited feature sets. This synergy enables our model to achieve higher accuracy and adaptability, outperforming conventional SVM-based models and other techniques in real-world sentiment analysis tasks.

Table 2. Table comparing the results our model with other models.

Model Name	Dataset Used	Precision	F1-Score	Recall
POS + NB	SemEval Task – 4 (2014)	.70	0.75	.72
Feature-Based (SVM)	SemEval Task – 4 (2014)	.66	.64	.61
Gini-Index (SVM)	SemEval Task – 4 (2014)	.70	.75	.79
L + N + I + D + R + S	SemEval Task – 4 (2014)	.72	.64	.79
Lexicon Based (SVM)	SemEval Task – 4 (2014)	.74	.77	.72
Rule-Based (ATE) + Fine-Tuned BERT model (ASC) (Our Model)	SemEval Task – 4 (2014) + Twitter Reviews	.75	.75	.75

The Table 2 displays the Precision, F1 Score, Recall of our model with other models. As we can see we maintain the same 75% for all the metrics. Also to mention we used another dataset which is twitter data for training and testing. The accuracy drop might be due to that too. Our model has done comparatively well as it very General.

In Fig. 10, the graph displays the results of all other models and with x axis being the precision and y axis being the recall. Our model is the RED one on the graph.

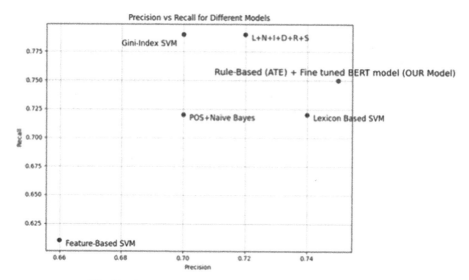

Fig. 10. Graph displaying our model and other models results

5 Conclusion and Future Work

Developed a far-reaching ABSA system for e-commerce product reviews with advanced techniques by performing the fine-tuning of BERT models and sophisticated methods of aspect extraction. This model shows a significant improvement in precision, recall, and F1-score compared with conventional SA approach. The high precision in e-commerce is important because accurate SA could really help in product improvement and in the marketing strategy. The model also furnishes detailed insights from customer feedback while keeping the balance between accuracy and comprehensiveness, a demand in business decision-making. The success of this project points toward further adaptations and applications in other areas requiring nuanced SA, enhancing real-time feedback systems, and expanding the scope of future research.

Future work could involve developing an aspect-based recommender system to personalize the shopping experience by recommending products based on particular aspects or features that users are interested in. Using real-time sentiment analysis could help perform immediate analysis of new reviews and make a business act swiftly upon an emerging issue or trend. Expanding the application of our ABSA model to sectors such as hospitality, healthcare, or services would test its versatility, opening new markets and demonstrating its utility in diverse settings.

References

1. Hilal, A., Chachoo, M.A.: Aspect-based opinion mining of online re-views. Gedrag Organisatie Rev. **33**(3), 1185–1199 (2020)
2. Mowlaei, H.K., Abadeh, M.S., Keshavarz, H.: Aspect-based sentiment analysis using adaptive aspect-based lexicons. Expert Syst. Appl. **148**. https://doi.org/10.1016/j.eswa.2020.113234 (2020)

3. Vayansky, I., Kumar, S.A.P.: A review of topic modeling methods. Inf. Syst. **94** (2020). https://doi.org/10.1016/j.is.2020.101582
4. Do, H.H., Prasad, P.W.C., Maag, A., Alsadoon, A.: Deep learning for aspect-based sentiment analysis: a comparative review. Expert Syst. Appl. **118**, 272–299 (2019). https://doi.org/10.1016/j.eswa.2018.10.003
5. Sindhu, C., Rajkakati, D., Shelukar, C.: Context-based sentiment analysis on Amazon product customer feedback data. In: Artificial Intelligence Techniques for Advanced Computing Applications: Proceedings of ICACT 2020, published by Springer Singapore (2021)
6. Nazir, A., Rao, Y., Wo, L., Sun, L.: Issues and challenges of aspect-based sentiment analysis: a comprehensive survey. IEEE Trans. Affect. Comput. **13**(2), 845–863 (2020)
7. Afzaal, M., Usman, M., Fong, A.: Tourism mobile app with aspect-based sentiment classification framework for tourist reviews. IEEE Trans. Consum. Electron. **65**(2), 233–242 (2019)
8. Gupta, V., Singh, V.K., Mukhija, P., Ghose, U.: Aspect-based sentiment analysis of mobile reviews. J. Intell. Fuzzy Syst. **36**(5), 4721–4730 (2019)
9. Nandal, N., Tanwar, R., Pruthi, J.: Machine learning based aspect level sentiment analysis for Amazon products. Spat. Inf. Res. **28**, 601–607 (2020)
10. Sudhir, P., Suresh, V.D.: Comparative study of various approaches, applications and classifiers for sentiment analysis. Global Trans. Proc. **2**(2), 205–211 (2021)
11. Yiran, Y., Srivastava, S.: Aspect-based Sentiment Analysis on mobile phone reviews with LDA. In: Proceedings of the 4th International Conference on Machine Learning Technologies, pp. 101–105 (2019)
12. Wassan, S., Chen, X., Shen, T., Waqar, M., Jhanjhi, N.Z.; Amazon product sentiment analysis using machine learning techniques. Revista Argentina de Clínica Psicológica **30**(1) (2021)
13. Sivakumar, M., Uyyala, S.R.: Aspect-based sentiment analysis of mobile phone reviews using LSTM and fuzzy logic. Int. J. Data Sci. Anal. **12**(4), 355–367 (2021)
14. Wang, J., Xu, B., Zu, Y.: Deep learning for aspect-based sentiment analysis. In: Proceedings of the International Conference on Machine Learning and Intelligent Systems Engineering (MLISE) (2021)
15. Rahin, S.A., Hasib, T., Hassan, M.: Aspect-based sentiment analysis using SemEval and Amazon datasets. In: Proceedings of the 5th International Conference of Women in Data Science at Prince Sultan University (WiDS PSU) (2022)
16. Abdelgwad, M.M., Soliman, T.H.A., Taloba, A.I., Farghaly, M.F.: Arabic aspect-based sentiment analysis using bidirectional GRU based models. J. King Saud Univ.-Comput. Inf. Sci. **34**(9), 6652–6662 (2022)
17. Kumar, A., Abirami, S.: Aspect-based opinion ranking framework for product reviews using a Spearman's rank correlation coefficient method. Inf. Sci. **460**, 23–41 (2018)
18. Rathan, M., Hulipalled, V.R., Venugopal, K.R., Patnaik, L.M.: Consumer insight mining: aspect-based Twitter opinion mining of mobile phone reviews. Appl. Soft Comput. **68**, 765–773 (2018)
19. García-Pablos, A., Cuadros, M., Rigau, G.: W2VLDA: almost unsupervised system for aspect-based sentiment analysis. Expert Syst. Appl. **91**, 127–137 (2018)
20. Zainuddin, N., Selamat, A., Ibrahim, R.: Hybrid sentiment classification on twitter aspect-based sentiment analysis. Appl. Intell. **48**, 1218–1232 (2018)

Author Index

A

Aaditya Shreeram, R. S. II-330
Aashvi, III-145
Abhishek Rithik, O. II-53
Abinash, A. I-28
Abinav Satya, S. III-200
Abinaya, R. III-189
Adithya, G. III-173
Aditya, V. III-212
Ahalya, B. III-29
Ajeen, Ajo Alen III-371
Akila, K. III-244
Akilesh, S. III-189
Akshitha, R. II-66
Alex, E. John III-57
Ali, Farithkhan Abbas I-102
Alim, Shahawar III-293
Amarnath, M. I-112
Anand, G. Paavai I-166
Anandhi, S. II-147
Angelin Claret, S. P. I-261
Anitha Pai, A. II-147
Arjun, Raj Purohith II-66
Arora, Hemanth III-113
Arunnehru, J. II-176, III-173, III-332
Ashish Tarun, R. III-244
Avinash, S. I-275
Ayyappan, M. I-112

B

Babu, V. S. Koushik III-173
Badarla, Sandeep II-117
Balaji, C. G. II-162
Balaji, Mayuri Mahimaa II-263
Balakrishnan, Sarojini I-3
Bedict, A. II-42
Belde, Ramya II-185
Bharathi, N. I-314
Bharathi, S. II-93, III-100
Biju, Adithya III-233
Bisani, Khushi II-204

Biswal, Saswat II-104
Buvana, R. II-18

C

Carroll, Fiona III-81
Chacko, Kristef III-233
Chinnapparaj, S. I-102
Chitra, P. II-3, II-53, III-100
Chowdam, Saran II-117

D

Darshan, B. III-113
Datta, Swati I-261
Davuluru, Samba Siva Sai III-14
Deenath Inbaraj, M. I-252
Deepa, R. II-80
Deepti, S. II-147
Dharanidharan, R. III-200
Dilip, Golda I-229, II-42, II-298
Dinesh, J. Pio III-371
Durgadevi, P. II-93, III-124

F

Farheen, A. Suneha I-229
Francis, Neenu I-180
Franklin, K. Sam Prince III-14

G

Ganapathyappan, Kavya III-332
Gayathri, R. III-212
George, Sherwin III-145
Gokula Prasath, S. III-200
Gopalsamy, Bharathi N. III-145
Gopika, S. II-263
Gulothungan, G. III-69
Gunnarsdottir, Helga III-269
Gupta, Sunny II-321
Guruvarshni, G. I-58

H

Harini, K. Naga I-156
Harini, S. II-18
Harish, R. II-80
Harisudhan, A. S. I-125
Harjit, R. L. III-3
Harshavardhini, C. II-275
Harshini, P. J. III-124
Hegde, Trupti II-251
Hephzipah, J. Jasmine II-311
Hussain, H. A. Mohamed I-28

I

Indumathi, K. I-71
Indumathy, M. III-371

J

J, Niharika II-251
Jain, Bhavya III-113
Jana, S. I-81
Janaki, M. II-18
Jayakrishnaa, P. II-176
Jayanthi, P. I-112
Jayaprakash, Amudhan III-347
John Alex, E. II-185, III-256
John, Binu II-286
Johnvictor, Anita Christaline I-202

K

Kaavya Shree, K. S. III-278
Kadam, Shivanjali Ambadas I-305
Kalaivani, A. I-58, III-47
Kalinathan, Lekshmi I-243
Kalpana, P. I-166
Kalsi, Nick III-81
Kapilamithran, S. III-47
Karmakar, Preethi III-256
Karthikayani, K. II-66, II-263
Kathiresh, S. II-131
Kaushik, A. Ragu I-134
Kavipriya, J. III-222
Kavya, M. I-156
Keerthika, J. I-58
Keshan, S. I-252
Kopperundevi, N. III-233
Koushik, Mothukuri II-104
Krishnan, S. S. Nagamuthu II-117
Krithik, Santhakumar S. III-212
Kumar, Deepak III-332
Kumar, M. Darshan I-134
Kumar, Neelam Sanjeev I-28
Kumar, P. Harish I-28
Kumar, P. I-275
Kumar, Sneha Sathish I-145
Kwatra, Ish I-261

L

Luke, Kevin III-371

M

M, Sachin II-251
Madesh, S. I-145
Madhurani, M. II-18
Madhuranthakam, S. Kavya Sree I-305
Mahesh, Srikanth I-295
Maheshwari, I-145
Mahima, A. H. II-29
Manohar, S. II-330
Maragathavalli, P. I-81
Meenakshi, K. I-295
Menaka, S. II-104, III-29
Minor, Kasha III-81
Minu, M. S. II-204
Miriam, D. Doreen Hephzibah II-223
Mithran, E. I-275
Mohamed Yaseen, M. S. II-3
Mohammed Zayaan, N. I-314
Mohan, K. B. Kishore I-102, III-69
Monica, M. II-29
Mounika, Ramisetty III-29
Muhammad Kifayathullah, K. A. III-189
Mukilan, K. II-194
Murali Bhaskaran, V. III-347
Murugan, Janaki Meena I-243
Muthukumar, M. II-194
Muthurasu, N. III-200

N

Nanmaran, R. I-102, II-194, III-69
Nashith Arham, C. M. D. I-314
Nayak, Pavana II-321
Neduncheliyan, I-93
Neha, S. II-29
Nehru, Yashini I-3
Nidhina, K. II-286
Nikitha, B. I-145, III-319
Niranjan Reddy, K. II-185, III-256

Author Index

Nithiyaraj, E. Emerson   II-194
Nivethika, K.   II-93

O

Ogunshile, Emmanuel   III-269

P

Palani, Durgadevi   I-216
Parthasarathy, Arjun   I-216
Parvathi, R.   III-14
Patel, Sameer Dushyant   II-42
Phung, Khoa   III-269, III-305
Platts, Jon   III-81
Poonkodi, M.   I-202, II-131, III-161, III-189, III-355
Prabha, B.   I-9, III-3, III-293
Pragadeesh, K. M. S.   III-173
Pragadeesh, M.   I-252
Prahitha, Mukka   III-319
Prasanna Devi, S.   III-278
Prasanna, Raghul   I-125
PrasannaKumari, V.   I-191
Pratham, R. U.   II-321
Praveen, M.   II-330
Prem Kumar, V.   II-53
Preshika, S.   I-71
Priyadarshini, B.   III-244

R

Rachakonda, Vishnu Vardhan   III-355
Radeesh, P.   II-3
Rahman, Mohammed Atheequr   III-124
Raj, James Allen   III-124
Raja, S. Karthikk   I-166
Rajasekar, V.   I-38
Rajesh Kumar, S.   II-162
Rakesh Kumar, M.   I-275
Ramachandran, Raj   III-269, III-305
Ramadoss, R.   I-156
Ramakrishnan, Kannan   I-156
Ramamurthy, Megha   II-298
Ramanan, M. G. Raja   I-191
Ramasamy, R.   I-102, III-69
Ramesh, Kishore   II-3
Ranjan, Krishan   III-269
Ranjan, Navneet   II-66
Raykar, Deepti Balaji   II-29
Reddy, Jhevaan   III-293
Reddy, K. Niranjan   III-57

Reddy, Kurapati Praneeth Sai   III-212
Reddy, N. V. Uma   III-319
Robin, C. R. Rene   II-223
Roshan, C. R.   II-275

S

Sadasivam, Akhil   I-295
Sai Keshav, I   I-58
Sai Krishnaa, V. R.   II-162
Sai, Saguturu Kishan   II-80
Saketh, P. C. V.   I-156
Saleel, Mayukh   III-233
Sam Prince Franklin, K.   III-355
Samal, Tushar   II-204
Samantaray, Anish   I-202
Sangar, G.   I-38
Sangeetha, Padigi Reddy   III-278
Sangeetha, S. K. B.   II-275, III-113
Sanjay Sen, S.   II-42
Sanjay, P.   I-191
Sankar, N. Prem   III-161
Santhanam, Pavitra   III-145
Santhoshkumar, L.   I-112
Saraswathi, S.   II-18
Saravanan, C.   III-161
Sarayu, K.   II-275
Sathiyapriya, S.   II-93
Sekar, Rajeev   III-189
Senthilkumar, Radha   I-112
Sethuraman, Bhalashri   I-252
Shankar, Aryan Vinod   II-298
Shankar, Madhumitha   I-243
Sharma, Shyam Narayan Ramkumar   III-29
Shastry, Vishwas S.   II-251
Shrinidhi, M.   II-311
Shyam Sundar, S.   II-162
Sinchana, T. G.   III-319
Sindhuja, R.   III-69
Sivachitralakshmi, S.   II-93, III-100
Sneha, M.   III-244
Sreyas, Navin   III-3
Sridhar, Sridevi   I-125
Srimathi, S.   I-102, II-194, III-69
Srinivass, S.   II-194
Subramanian, Ganapathy   I-93
Subramanian, Harihara   III-293
Sujatha, R.   I-305
Sundaram, Rishi   I-166

Suresh, Shashank II-80
Surya, J. I-9

T
Teja, Pulibandla Sri Surya II-104
Tejesh, C. S. Nithin II-117
Tewari, Aishwarya II-204
Thanish Reddy, D. III-355

U
Upadhyay, Shravani III-278

V
Vadivu, G. I-180, III-222
Vaibavi, J. I-125
Varun, R. I-314
Vasu, Harini I-216
Vasuki, A. I-134
Venkataramanaiah, B. I-156

Venkatesh, T. Charan III-57
Venkatesh, Tarun III-3
Venkateswarlu, N. II-53
Vetrivel, P. II-176
Vignesh, M. III-47
Vijay, R. R. II-330
Vijayalakshmi, G. III-47
Vishal, S. P. Sri I-71
Vishnu, S. III-332
Vishwanath, M. II-263

W
Woonna, Venkat Amith III-355

Y
Yashaswini, Yeddula III-319
Yeshvanthini, K. II-311
Yogitha, S. II-311

Printed in the United States
by Baker & Taylor Publisher Services